NMR TECHNIQUES IN CATALYSIS

CHEMICAL INDUSTRIES

A Series of Reference Books and Textbooks

Consulting Editor

HEINZ HEINEMANN
Berkeley, California

ADDITIONAL VOLUMES IN PREPARATION

NMR TECHNIQUES IN CATALYSIS

edited by

Alexis T. Bell
Alexander Pines

University of California and Lawrence Berkeley Laboratory
Berkeley, California

CRC Press
Taylor & Francis Group
Boca Raton London New York

CRC Press is an imprint of the
Taylor & Francis Group, an **informa** business

CRC Press
Taylor & Francis Group
6000 Broken Sound Parkway NW, Suite 300
Boca Raton, FL 33487-2742

First issued in paperback 2019

© 1994 by Taylor & Francis Group, LLC
CRC Press is an imprint of Taylor & Francis Group, an Informa business

ISBN-13: 978-0-8247-9173-5 (hbk)
ISBN-13: 978-0-367-40209-9 (pbk)

Library of Congress Cataloging-in-Publication Data

NMR techniques in catalysis / edited by Alexis T. Bell, Alexander Pines.
     p. cm. -- (Chemical industries; 55)
   Includes bibliographical references and index.
   ISBN 0-8247-9173-8 (acid-free)
   1. Catalysis.   2. Catalysts--Analysis.   3. Nuclear magnetic resonance spectroscopy. I. Bell, Alexis T. II. Pines, Alexander. III. Series: Chemical industries; v. 55.
   QD505.N67   1994
   621.39'5--dc20
                                         93-43400
                                              CIP

Visit the Taylor & Francis Web site at
http://www.taylorandfrancis.com

and the CRC Press Web site at
http://www.crcpress.com

Preface

Modern chemical processes and petroleum refining are critically dependent on the use of solid catalysts to achieve desirable selectivities and rates of product formation. In order to understand the function of such heterogeneous catalysts and to optimize their performance, it is essential to characterize the structure of catalytic materials and surfaces, to understand the interactions between molecules and catalysts, and to follow the dynamics of diffusion and chemical reactions of molecules during the course of catalytic processes.

In recent years, nuclear magnetic resonance (NMR) has emerged as a powerful, often unique, technique for the study of solid materials, and the area of heterogeneous catalysis is a natural beneficiary of this development. The evolution of modern high-resolution solid-state NMR results largely from novel theory and experiment, in particular through significant advances in resolution, sensitivity, and selectivity. Beyond the advantages of high magnetic fields and multidimensional spectroscopy, NMR in solids has been transformed by the advent of coherent averaging techniques, including multiple-pulse sequences, cross polarization, and sample spinning. Research in NMR spectroscopy continues to produce advances in resolution and sensitivity. Examples include the invention of variable-angle techniques to achieve high resolution and correlations, rotational echoes and multiple-pulse recoupling to measure interatomic distances, and the use of superconducting detectors and optical pumping to achieve high sensitivity. The recent combination of magnetic resonance with detection by scanning tunneling and atomic force microscopy provokes the contemplation of truly localized, single atom NMR, an appropriate goal in our imminent age of nano-scale and molecular catalyst design.

The "interface" between catalysis and NMR is currently an area of much activity, and it is the aim of the present volume to describe this interface. Topics have been selected to illustrate the scope of applications of NMR to the study of catalysts and catalytic processes. The contents of this volume provide illustrations of the principles of solid-state NMR, with applications to materials of relevance to catalysis, including zeolites and other molecular sieves, oxides, dispersed silica and alumina, supported metal systems, layered materials including clays, zirconium phosphates, metal sulfides and graphite, and to in-situ processes including diffusion and chemical reactions. The contributors to this volume are leaders in the field who have provided, following a brief introduction by the editors, seven chapters which we believe will form a useful resource for scientists and students interested in the principles and applications of NMR in heterogeneous catalysis.

Alexis T. Bell
Alexander Pines

Contents

Contributors

Alexis T. Bell Chemical Engineering Department, University of California and Chemical Sciences Division, Lawrence Berkeley Laboratory, Berkeley, California

Hellmut Eckert Department of Chemistry, University of California, Santa Barbara, California

Paul D. Ellis* Department of Chemistry, University of South Carolina, Columbia, South Carolina

C. A. Fyfe Department of Chemistry, University of British Columbia, Vancouver, British Columbia, Canada

Grant W. Haddix Analytical Chemistry, Shell Development Company, Houston, Texas

James F. Haw Department of Chemistry, Texas A&M University, College Station, Texas

---

Current affiliation: Molecular Science Research Center, Battelle Pacific Northwest Laboratory, Richland, Washington

Jörg Kärger   Department of Physics, Leipzig University, Leipzig, Germany

Jacek Klinowski   Department of Chemistry, University of Cambridge, Cambridge, England

G. T. Kokotailo   Department of Chemistry, University of British Columbia, Vancouver, British Columbia, Canada

Waclaw Kolodziejski   Department of Chemistry, University of Cambridge, Cambridge, England

Gary E. Maciel   Department of Chemistry, Colorado State University, Fort Collins, Colorado

K. T. Mueller*   Department of Chemistry, University of British Columbia, Vancouver, British Columbia, Canada

Mysore Narayana   Analytical Chemistry, Shell Development Company, Houston, Texas

Harry Pfeifer   Department of Physics, Leipzig University, Leipzig, Germany

Alexander Pines   Department of Chemistry, University of California and Materials Sciences Division, Lawrence Berkeley Laboratory, Berkeley, California

---

*Current affiliation: Department of Chemistry, The Pennsylvania State University, University Park, Pennsylvania

NMR TECHNIQUES IN CATALYSIS

Introduction

Alexis T. Bell and Alexander Pines

*University of California and Lawrence Berkeley Laboratory,
Berkeley, California*

I. CATALYSIS

Many of the chemical reactions used to produce the materials, fuels, chemicals, and pharmaceuticals needed by modern societies could not be carried out in a practical fashion without the intervention of catalysts [1,2]. The first catalysts used by humans were the enzymes present in yeast. However, it was not until the nineteenth century that studies by Berzelius, Faraday, and others revealed that many inorganic substances could also serve as catalysts. Today, it is widely recognized that the range of substances serving as catalysts is quite broad, encompassing electrons, protons, hydroxyl groups, gases, organometallic complexes, metals, oxides, chalcogenides, and enzymes. The common property shared by all catalysts is the ability to interact with one or more reactants in such a way as to accelerate the transformation of these species into products without consumption of the catalyst in the process. In reactions where one or more products might be formed from a given set of reactants, selectivity for the formation of a single desired product is often achievable by the action of a suitable catalyst.

The success of the fuel and chemical industries as we know them is a direct result of the discovery of catalysts for carrying out specific reactions or sets of reactions at high rates and with high selectivity for the desired products. It is, therefore, not surprising that over 90% of the chemical processes in use require one or more catalysts. The preponderance of solid, or heterogeneous, catalysts

1

(including, for example, zeolites, oxides, sulfides, and metals) is a result of the ease with which such materials can be separated from products and unconverted reactants. Consequently, it is to *heterogeneous catalysis* that the chapters of this volume are devoted.

II. STRUCTURE AND PROPERTIES OF HETEROGENEOUS CATALYSTS

Most heterogeneous catalysts exist in the form of microporous solids. The catalysts are usually produced in the shape of spheres, cylinders, or monoliths, such as those shown in Figure 1. The internal surface area is typically $10-10^3$ m^2/g. Catalysis occurs either on the surface of the microporous solid, as in the case of zeolites, or on the surface of microdomains of active material dispersed inside the microporous solid, as in supported metals, oxides, sulfides, etc. In either case, the high internal surface area of the microporous solid is used to obtain a high concentration per unit volume of catalytically active centers.

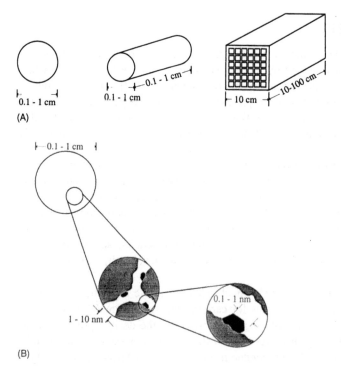

Figure 1

The reactants in a catalyzed reaction are present in either the gas or liquid phase, and in order to reach the catalytically active centers the reactants must diffuse through the pores of the catalyst, or catalyst support. Reactants in the immediate vicinity of the pore walls or the surface of the catalytically active components can interact with these surfaces by adsorption. This interaction ranges from weak (physisorption) to strong (chemisorption). Physisorption is governed primarily by dispersive forces and does not result in a chemical transformation of the adsorbate, whereas chemisorption is characterized by the formation of chemical bonds between the adsorbate and the adsorbent and can result in significant changes in the electronic properties of the adsorbate. The adsorbed reactants may undergo a variety of processes while on the surface of the catalyst. The simplest of these is surface diffusion in which the adsorbed material carries out a random walk on the catalyst surface. During diffusion, physisorbed material may become chemisorbed and vice versa. The adsorbed reactants may also leave the surface by desorption or, in the case of chemisorbed reactants, undergo chemical reaction. The reaction process can be unimolecular, as with the case of dissociation of a diatomic species, or bimolecular, as in the case of hydrogen abstraction or addition. The species formed via surface chemical reactions can serve as intermediates for other reactions, or they can desorb as final products. The products of catalyzed reactions can undergo all of the same processes experienced by reactants. Figure 2 illustrates two examples of such processes.

The observed activity of a catalyst is dictated by the intrinsic activity of the catalytically active component and by the effects of intraparticle mass transport.

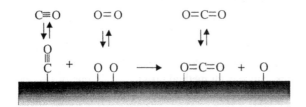

Figure 2

The intrinsic activity depends on the chemical and physical properties of the active component. For unsupported catalysts, the most important properties are the composition and structure of the catalyst surface and the presence, or absence, of special sites such as Brønsted or Lewis acid centers, anion or cation defects, and sites of high coordination. For supported catalysts, the size and morphology of the dispersed phase are of additional importance. If intraparticle transport of reactants occurs with a characteristic time that is short compared to that of the reaction, then the observed and intrinsic rates of reaction will be identical. When the characteristic time for intraparticle mass transport is less than that for reaction, the observed rate of reaction per unit mass of catalyst becomes less than the intrinsic value, and the reaction kinetics are dominated by the effects of intraparticle mass transport. The factors governing intraparticle transport are the diffusivities of the reactants and products and the characteristic distance for diffusion.

III. CHARACTERIZATION OF CATALYSTS AND CATALYTIC PROCESSES

Knowledge of catalyst composition and structure is crucial to an understanding of the factors that affect catalyst activity and selectivity. Such information makes it possible to determine which portions of a catalyst are active and how changes in catalyst synthesis and pretreatment affect the properties of the catalytically active sites. Catalyst characterization is also vital to understanding the changes that occur in the structure and composition of a catalyst following both use under reaction conditions and regeneration to reactivate the catalyst.

Some of the principal techniques used to characterize microporous catalysts are listed in Table 1. With the exception of electron microscopy and scanning tunneling microscopy, all the techniques integrate information over large portions of the sample volume or surface. As a result, nonuniformities in chemical and structural environments invariably contribute to a broadening of the observed spectroscopic signatures. Since each method provides only one or two elements of information, a combination of techniques can be used to achieve a comprehensive understanding of composition and structure.

The study of catalytic reaction mechanisms and kinetics, in particular the structure, dynamics, and energetics of reaction intermediates formed along the catalytic reaction path, can provide a basis for understanding the relationships between catalyst structure and catalyst activity and selectivity. Since it has been established that most catalysts undergo structural and chemical changes under reaction conditions, it is important that experimental investigations of reaction mechanisms be carried out in situ. This constraint requires analytical techniques that can operate while the catalyst is at high temperature and in the presence of reactants at moderate to high pressures. The principal techniques that meet these needs are infrared, Raman, ESR, and NMR spectroscopy.

Table 1 Experimental Techniques for Characterization of Catalysts and Adsorbed Species

Technique	Acronym	Type of information
Low-energy electron diffraction	LEED	Surface structure and registry
Auger electron spectroscopy	AES	Elemental analysis
X-ray photoelectron spectroscopy	XPS	Elemental analysis and valence state
Ion scattering spectroscopy	ISS	Elemental analysis
Ultraviolet photoelectron spectroscopy	UPS	Electronic structure, molecular orientation
Electron energy loss spectroscopy	EELS	Molecular structure
Infrared spectroscopy	IRS	Molecular structure
Laser Raman spectroscopy	LRS	Molecular structure
X-ray diffraction	XRD	Bulk crystal structure
Extended x-ray absorption fine structure	EXAFS	Bond distance, coordination number
Transmission electron microscopy	TEM	Crystal size, morphology, and structure
Scanning transmission electron microscopy	STEM	Microstructure and composition
Scanning tunneling microscopy	STM	Surface microstructure
Ultraviolet spectroscopy	—	Electronic states
Mössbauer spectroscopy	—	Ionic states
Electron paramagnetic resonance	EPR	Paramagnetic states
Nuclear magnetic resonance	NMR	Molecular structure and motion

IV. WHY NMR?

Among the techniques that can be used for the characterization of catalysts and adsorbed species, NMR is unique in its ability to provide information about both structure and dynamics [3,4]. Modern solid-state NMR makes it possible to detect signals from distinguishable sites in molecules and materials and to monitor the connectivities, correlations, and dynamics of these sites. Furthermore, NMR spectroscopy is essentially noninvasive and can be carried out in the presence of gases or liquids over a wide range of temperatures and pressures. While the principal use of NMR spectroscopy is to obtain information about the chemical environment of elements in catalysts or species adsorbed on catalysts, the technique can also be used to characterize atomic and molecular motions.

Figure 3 illustrates the types of information that may be derived from NMR signals:

1. *Site Identification*: Nuclear spins (of specific isotopes) at atomic sites in a molecule, in a crystalline lattice, in a glassy material, or on a surface can be associated with NMR spectral lines at characteristic frequencies that depend

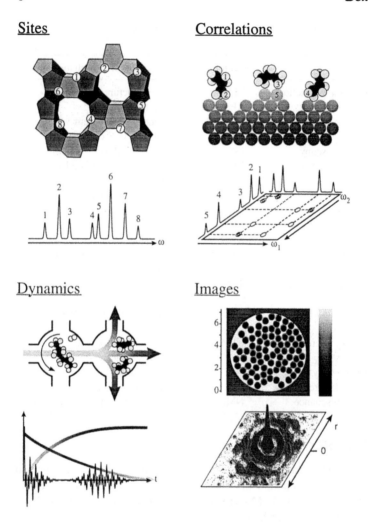

Figure 3

on the atomic environments of the spins. Among the interactions that determine the NMR frequencies are those between the nuclear spins and electrons in surrounding localized orbitals (chemical shifts) or metallic conduction bands (Knight shifts). In addition, the NMR frequencies are influenced by interactions between the nuclear spins and spins at other atomic sites (both electron-mediated scalar or J-couplings and direct dipole-dipole couplings) and those between the electric quadrupole moments of the nuclei and electric field gradients arising from the surrounding atomic charge distributions (quadrupole shifts and couplings). Quantitative measurements of the intensities of identified NMR lines can provide information about the occupancy or distribution of sites (e.g., how many carbon or silicon atoms are there of this or that type), an obviously important question in the study of structure and reactions. Chemical and Knight shifts, for example, can be used to distinguish different Si and Al sites in zeolites, to differentiate between surface and bulk sites in metals or metal oxides, and to characterize structural changes occurring in molecules upon interaction with catalysts.

2. *Intersite Correlations*: Beyond the question of site occupancies, it is possible to investigate interatomic correlations in chemical systems by observing spin-spin couplings and by exploiting the transfer of spin polarization and coherence between sites. Such correlations are often directly discernible in multidimensional spectra, in which signals acquired in a set number of time periods are Fourier-transformed to yield spectra in an equal number of frequency dimensions. Off-diagonal peaks between lines in a two-dimensional NMR spectrum, for example, imply some relationships between the sites associated with the lines in the two time periods. The relationship may be structural, implying bonding connectivity or spatial proximity, or it may be dynamic, resulting from molecular translation, reorientation, or chemical reaction. Correlations may also be determined between the isotropic shifts defining the high-resolution NMR spectra and the anisotropic shifts characterizing the spatially dependent couplings intrinsic to solid-state NMR. Indeed, the combination of information available from multidimensional NMR correlations and known molecular structures is often a valuable aid in line assignment and site identification. Examples of the utility of the information gained from correlations include the ordering of Si and Al atoms in zeolites, the bonding of atoms in reacting molecules, and the connectivities between atoms of adsorbates and catalyst surfaces.

3. *Dynamics and Reactions*: The effects of motion on NMR signals can be observed both directly in the spectra and indirectly through spin relaxation. The range of time scales accessible by NMR is enormous, from molecular reorientational correlation times of picoseconds to solid impurity diffusion times of hours. Over short time scales, it is the rapidly fluctuating local fields at the sites of nuclear spins, due primarily to molecular motions and paramagnetic centers, that are responsible for spin-lattice relaxation (the exchange of energy between spins and other molecular or condensed matter degrees of freedom) and spin-spin relaxation (the dephasing and broadening of NMR lines). Over intermediate to slow time scales, the effects of motion arising from the reorientation of molecules or from chemical exchange or reactions can be observed by multidimensional NMR spectra through the appearance and time dependence of off-diagonal peaks. It is also possible to

follow atomic and molecular translational diffusion both within a catalyst particle and between particles by observing the attenuation and phase shifts of spin echoes using pulsed magnetic field gradients.

4. *Imaging and Microscopy*: NMR in the presence of magnetic field gradients makes it possible to obtain images of spin density inside intact samples. In addition to the images themselves, imaging techniques allow the application of all the capabilities of NMR spectroscopy to specific localized regions of the sample. Thus localized NMR microscopy and spectroscopy make possible the determination of position-dependent composition and structure, and localized spin echo detection allows the recording of diffusion and flow profiles. Similarly, localized multidimensional NMR spectra permit the determination of position-dependent molecular dynamics and chemical reactions. Although it is the fluids contained in solid samples that are normally detected in NMR images, recent work has provided direct images and localized spectroscopy of solid materials as well. In some cases, rather than images, it is preferable to obtain directly a statistical measure of the distribution of density or motion, for example, the density correlation function. Imaging and statistical techniques currently achieve spatial resolution over the range of micrometers to meters, while spectroscopy provides information about distances and structures below about 100 Å.

V. OUTLINE

Rather than aim for a comprehensive review of NMR and catalysis, the present volume allows several leading practitioners to describe examples of materials and processes in heterogeneous catalysis under investigation by NMR techniques.

In Chapter 1, Fyfe, Mueller, and Kokotailo describe the applications of solid-state NMR to the study of zeolite molecular sieve catalysts and related systems. Zeolites provide an apt arena in which to demonstrate the capabilities of modern techniques such as sample spinning, cross-polarization, and multidimensional correlation spectroscopy. In Chapter 2, Kärger, and Pfeifer consider the question of molecular diffusion in catalyst systems. Applications of NMR techniques such as imaging, lineshape analysis, relaxation, pulsed field gradient echo spectroscopy, and NMR tracer exchange are described and compared with other, more traditional techniques such as radioactive tracing. In Chapter 3, Haw discusses the use of NMR to probe catalytic processes, showing how the combination of temperature control with novel NMR probes makes it possible to elucidate reaction mechanisms in situ.

Chapters 4 through 6 deal with the applications of NMR to three additional classes of materials relevant to catalysis. Eckert (Chapter 4) describes the study of oxide catalysts, showing how NMR is used to characterize structure both in the bulk material and on surfaces. Maciel and Ellis (Chapter 5) discuss dispersed silica and alumina catalysts, including supported metal systems, demonstrating, for example, how multiple-pulse and sample-spinning techniques can be used to characterize even the traditionally difficult hydrogen sites in such catalytic systems. Haddix and Narayana (Chapter 6) describe the applications of NMR to the study of layered materials, including clays and pillared clays, zirconium phosphates, metal sulfides, and graphite. In Chapter 7, Kolodziejski and Klinowski summarize the principles of some modern NMR techniques, including those already in use for the study of catalysis (and applied, indeed, in previous chapters). They conclude with a number of recent developments that hold promise for the future.

NMR appears to be alive and well in chemistry, never more so than in materials and catalysis. Along with the contributors to this volume, we look forward to the next generation of advances and surprises.

ACKNOWLEDGMENTS

We are grateful to Lyndon Emsley, Michael Munowitz, Jeffrey Reimer, and Jay Shore for helpful discussions and editorial advice, to Dione Carmichael for administrative assistance, and to Stefano Caldarelli, Peter Carlson, Russell Larsen, and Henry Long for help with graphics and illustration.

REFERENCES

1. C. N. Satterfield, *Heterogeneous Catalysis in Industrial Practice*, 2nd ed., McGraw-Hill, New York, 1991; B. C. Gates, *Catalytic Chemistry*, Wiley, New York, 1992.
2. A. T. Bell et al., *Catalysis Looks to the Future*, Panel on New Directions in Catalytic Science and Technology, National Academy Press, Washington, DC, 1992.
3. C. P. Slichter, *Principles of Magnetic Resonance*, 3rd ed., Springer-Verlag, New York, 1989. G. A. Webb, ed., *Nuclear Magnetic Resonance*, Specialist Periodical Report, Royal Society of Chemistry, London, 1992/1993.
4. A. Pines, *NMR in Physics, Chemistry and Biology: Illustrations of Bloch's Legacy*, Proceedings of the Bloch Symposium, (W. A. Little, ed.), World Scientific, Singapore, 1990; L. Emsley and A. Pines, *Lectures on Pulsed NMR*, Proceedings of the CXXIII School of Physics "Enrico Fermi," North-Holland, Amsterdam, 1993.

1

Solid-State NMR Studies of Zeolites and Related Systems

C. A. Fyfe, K. T. Mueller,* and G. T. Kokotailo

University of British Columbia, Vancouver, British Columbia, Canada

I. INTRODUCTION

In recent years, high-resolution solid-state nuclear magnetic resonance (NMR) spectroscopy has emerged as a powerful complementary method to diffraction techniques for the investigation of zeolite molecular sieve structures. The combination of NMR and diffraction methods is particularly appropriate as the former is most sensitive to local orderings and geometries while the latter probes long range orderings and periodicities. When taken together, these two techniques provide a more complete description of the overall structure. In this chapter we will present a description of the development of solid-state NMR as applied to zeolites and related systems, with particular emphasis on recent work and ongoing developments in the field. A comprehensive review of this topic up to 1986 has been presented by Engelhardt and Michel [1]. Although there has been some effort directed toward the study of adsorbed species and reactions of guest molecules within zeolite cavities, we will restrict our discussion to investigations of the molecular sieve frameworks themselves.

II. ZEOLITES AND RELATED MATERIALS

Zeolites are open-framework aluminosilicates, widely used in industrial applications as ion exchange resins, molecular sieves, sorbents, catalysts, and catalyst supports [2,3]. The first natural zeolite, stilbite, was discovered over 200 years

---

**Current affiliation*: The Pennsylvania State University, University Park, Pennsylvania.

ago, but very little interest in these new materials was shown until McBain [4] selectively adsorbed molecules in zeolites and coined the term "molecular sieves." Barrer also demonstrated the molecular-sieve character of zeolites [5,6] and derived the hydrogen forms of zeolites, which are strong acid catalysts [7]. Interest was generated in the synthesis of zeolites with Milton's synthesis of zeolite A [8], which is now used as a molecular sieve, sorbent, ion exchanger, and replacement for phosphates in detergents. Zeolite X, a large-pore zeolite, was converted into a petroleum-cracking catalyst by a group at the Mobil Corporation [9], and this development has had a tremendous economic impact. The synthesis of a high silica, medium-pore zeolite called ZSM-5 [10] was first reported in 1972, and the material is currently used in a number of catalytic processes.

Formally, zeolites can be represented as being derived from silica (SiO_2) by the replacement of SiO_4^{4-} tetrahedra with AlO_4^{5-} tetrahedra and described by the general oxide formula

$$M_{y/x}^{x+}[(AlO_2)_y(SiO_2)_{1-y}] \cdot nH_2O \tag{1}$$

where the part within brackets represents the lattice. Because of the difference in atomic charge between Al and Si, extra-lattice cations (M^{x+}) must be present to preserve electrical neutrality, a single positive charge being needed for each (AlO_2) unit. These cations are not part of the framework and may be easily exchanged. In addition, water of hydration is usually present but also is not part of the lattice structure. Importantly, the chemical composition of a group of zeolites may be formally the same (e.g., zeolites A and X), yet have unique and distinct crystal structures. Hence elemental analysis alone is insufficient for differentiation among zeolites.

The catalytic activity of zeolites is generated by converting them to an "acid form" by heating the "ammonium form" (where the cations are NH_4^+) to 450°C. This treatment causes decomposition of the ammonium ions and yields a material where the extra-lattice cations are, at least formally, H^+. The process is sketched out in Eq. (2). The acid form of the zeolite will now act as a very powerful acid catalyst in reactions of hydrocarbons such as isomerizations, alkylations, and hydrogen transfers:

$$
\begin{array}{ccc}
NH_4^+ & & 450°C \\
\text{"exchange"} & & \text{"calcine"} \\
\text{zeolite-Na}^+ \rightarrow & \text{zeolite-NH}_4^+ \rightarrow & \text{zeolite-H}^+ + NH_3 \uparrow \\
& \text{ammonium form} & \text{acid form}
\end{array}
\tag{2}
$$

The size- and shape-selective characteristics of zeolites come from their unique framework structures: in general they are formed from open arrangements of AlO_4^{5-} and SiO_4^{4-} tetrahedra linked by shared oxygen atoms (the Al and Si atoms are therefore often referred to as T-atoms). Strictly speaking, the designation zeolite refers to aluminosilicates only, but recent developments have expanded the compositions to include Be, B, Ge, Co, Fe, Zn, P, Ti, As, Mn, and Sn in addition to the previously known components Si, Al, and Ga. The term is now often used to describe these multicomponent compositions. If P, Al, and Si are incorporated in the structure with P and Al paired to give a charge balance, no extra-lattice cations are required to preserve electrical neutrality. Catalytic activity is subsequently lost when no cations are present since it is impossible to produce the acid form through cation exchange followed by calcination.

The lattice frameworks of several common zeolites can be assembled from a single building unit, as shown in Fig. 1. The truncated octahedron or sodalite cage subunit (Fig. 1a) has faces of four- and six-membered rings where the vertices are Si or Al T-atoms joined by linking oxygens (not shown). When two of these are joined via their four-membered rings with bridging oxygens, the structure shown in Fig. 1b is formed, and linking of two of these units together then gives the structure shown in Fig. 1c. This is the basic lattice structure of zeolite A, which is not known in nature but is widely available from synthesis and commonly known in research laboratories as "molecular sieve." There is a large central cavity accessible from three orthogonal straight channels. It is this pore and channel structure that allows the lattice to control the size of the molecules sorbed and the reaction products, thus establishing control of chemical

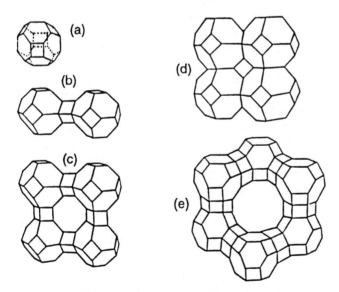

Figure 1 Lattice structures of typical zeolites formed from the sodalite cage building block (see text).

reactions involving molecules of 10 Å or smaller. The zeolite A based molecular sieves used in the laboratory are referred to by the size of molecules to which they are accessible. Depending on the size of the cations in the cavities, molecules of different sizes will be adsorbed. For example, if the cation is Na^+, then molecules up to 4 Å in diameter can enter. The ions K^+ and Ca^{2+} are present in 3 Å and 5 Å molecular sieves, respectively.

Figures 1d and e show two other zeolites made up from the same basic building block. When two units share a four-membered ring (Fig. 1d), the sodalite lattice is formed. In this case the structure is completely space filling (the central cavity being identical to the building unit), and this system shows no catalytic or molecular sieving properties. As opposed to zeolite A, sodalite occurs naturally. Figure 1e shows the lattice structure of faujasite, a very rare mineral species, whose synthetic counterpart is known as zeolite X or zeolite Y (depending on the Si/Al ratio). The synthetic versions are widely used in petroleum refining. There is again a large central cavity, but in this case the channels are no longer straight. There are four openings to each large cavity, giving a three-dimensional channel system.

A few practical examples provide some indication of the large number of uses for zeolites in research and industry, especially as catalysts and sorbents. The total annual zeolite consumption by industry is over 550,000 metric tons, of which more than 100,000 tons are used as catalysts, 375,000 tons are zeolites in detergent, and 40,000 tons are used as sorbents.

The main commercial applications of synthetic faujasite are for cracking catalysts to produce gasoline and fuel oil. Over 98% of the refineries in the world use some form of synthetic faujasite-cracking catalyst. Hydrocracking, another process using synthetic faujasite, produces kerosene, toluene, xylene, and jet fuels. ZMS-5 is used variously to produce low "pour point" lubes by dewaxing xylene and benzene from toluene, styrene from benzene alkylation, xylene from toluene disproportionation, and in the conversion of methanol to gasoline. It has also been used to control NO_x emissions with use of NH_3 as a reducing agent to convert NO_x to N_2 and H_2O. Zeolite A with Ca^{2+} cations (CaA) is used for air enrichment, removal of H_2S from "sour" gases, and the separation of iso- and *n*-paraffins. Zeolites have been used as well to remove radioactive Cs^+ and Sr^{3+} from nuclear-reactor-waste streams and ammonia from sewage and agricultural effluents.

Some zeolites have much more complex unit cells and lattice structures than zeolites A and X, each of which happens to have a single, unique lattice site. A commercially important example having a very complex unit cell is zeolite ZSM-5, one end member of a series of "pentasil" zeolites, the other being zeolite ZSM-11 [10,11]. The pentasil building unit consists entirely of five-membered rings (Fig. 2a). Joining these units in columns (Fig. 2b) and then

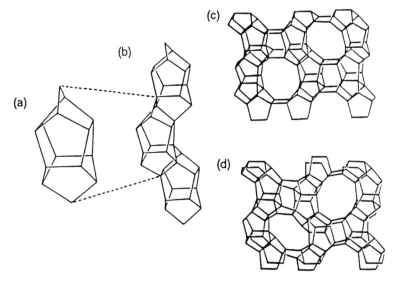

Figure 2 Schematic representation of (a) the pentasil unit, (b) a chain formed by interlocking pentasil units, and the structures of (c) zeolite ZSM-11 and (d) zeolite ZSM-5.

joining layers of columns, such that the neighboring layers are related by inversion symmetry, produces the zeolite ZSM-5 shown in Fig. 2d. Joining the layers so that they are related by a reflection plane yields the closely related ZSM-11 lattice structure (Fig. 2c). Since the two structures differ in only one projection and many of the repeat distances are the same, their powder x-ray diffraction patterns are quite similar. Zeolite ZSM-5 is a very important catalyst, being extremely size- and shape-selective towards organic molecules, and will be discussed in some detail in the present chapter.

An important related class of materials are the aluminophosphate molecular sieves, designated $AlPO_4s$ and originally reported by Wilson and coworkers [12]. These have exactly equal numbers of Al- and P-based T-sites and exact alternation of the Al and P atoms. Because of this structure, the frameworks are neutral, and they have no inherent catalytic capabilities. However, they do show molecular-sieving characteristics and can be made acidic by the introduction of a proportion of other elements into the framework. For example, Si substitution produces silicon aluminophosphates (SAPOs) with mild acidity. One recent $AlPO_4$ of special interest is VPI-5 [13] (Fig. 3). This sieve has a unidimemsional channel system with a particularly large diameter. It is made up of 18-membered

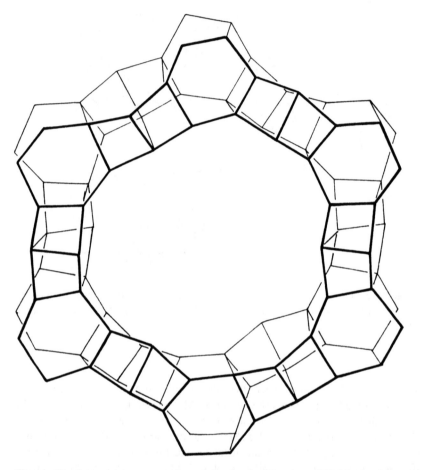

Figure 3   Schematic representation of the aluminophosphate VPI-5 lattice framework. (From Ref. 14.)

rings with internal dimensions of 12.5 Å and can accommodate organic molecules as large as porphyrins.

From the above discussion it can be seen that zeolites are unique catalytic materials combining a size and shape selectivity toward organic reactant and product species with high reactivity, as illustrated schematically in Fig. 4 [15], and in fact they show many of the characteristics of enzymes. It should be noted that all the chemical reactions in this figure are practical examples. Thus for a complete understanding of the catalytic properties of specific zeolites, a full and detailed description of the lattice frameworks that confer the molecular selec-

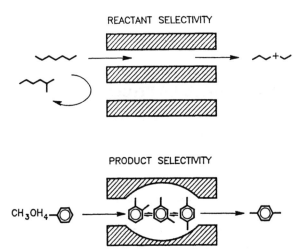

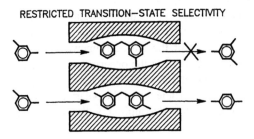

Figure 4 Different types of size and shape selectivity in reactions within zeolite cavities. (From Ref. 15.)

tivity is essential. There are particular difficulties, however, in determining zeolite structures: although zeolites are highly crystalline materials, the structural information accessible from diffraction measurements is severely limited. First, they are microcrystalline, yielding crystals that are too small (usually of the order of microns or less) for conventional single-crystal diffraction studies, and therefore most structure determinations must be attempted from much more limited powder-diffraction data [16]. Second, since Si and Al have relatively high atomic masses and differ by only one mass unit, their x-ray scattering factors are almost identical, and it is often not possible to clearly distinguish them. An additional complication is that the Si and Al atoms in zeolites are usually disordered over the T-sites, and so an average structure results from diffraction experiments. In general, diffraction measurements will be able to determine the overall lattice structure, but not define the placement of the individual Si and Al atoms over the T-sites within this structure [16].

III. NMR CHARACTERISTICS OF NUCLEI IN ZEOLITE AND ALUMINOPHOSPHATE LATTICES

All of the atoms making up the zeolite lattice [those enumerated in the bracketed part of Eq. (1)] have NMR-active isotopes and thus can be investigated by solid-state NMR. The important nuclei and their natural abundances are ^{29}Si (4.7%), ^{27}Al (100%), and ^{17}O (0.04%). Because of the low natural abundance of ^{17}O, isotopic enrichment is usually necessary. The 100% abundant ^{31}P nucleus is also present in the aluminophosphate systems discussed above. In addition, ^{1}H (100%) may be present in any of these framework structures in OH groups at defect sites. We will discuss the information regarding zeolite structure available through study of these nuclei and the interactions of their spins with local electric and magnetic fields.

Further insight can be gained from Eq. (1) regarding the appropriate experimental conditions as well as the information potentially available from solid-state NMR experiments. In general, most protons in the system are present only as water of hydration, which is not part of the lattice framework and is usually mobile. Thus the dipolar interactions between the observed nucleus and protons will be small or zero, and cross-polarization techniques for signal enhancement will not be widely applicable since the efficiency of the polarization transfer will be low. Conversely, high-power proton decoupling will not be required to remove the dipolar line broadening, and therefore the solid-state NMR experiment on zeolites usually reduces the simpler NMR techniques such as magic-angle spinning (MAS). Fortunately, MAS experiments can be easily carried out on a conventional high-resolution NMR spectrometer [17]. Further, the experiments are no more difficult at high magnetic fields and are often best carried out in as high a magnetic field strength as possible.

It must be stressed that there are fundamental differences in the natures of the different nuclei discussed above. ^{29}Si and ^{31}P are spin 1/2 nuclei, and MAS yields particularly simple spectra with complete averaging of the chemical shift tensor components. The average "isotropic" shift values are field independent and correspond to the solution chemical shifts. ^{27}Al and ^{17}O, however, are quadrupolar nuclei with nonintegral spins greater than 1, and their solid-state spectra are often much more complex. Since the quadrupolar interaction depends on local electric field gradients at the nucleus studied, solid-state NMR of quadrupolar nuclei can yield further information about the local environment. More complicated spinning techniques have been introduced to average the anisotropies in the spectra of quadrupolar systems, and these new methods have increased interest in the use of such nuclei as probes of local microstructure.

IV. INVESTIGATIONS OF LOW–Si/Al RATIO MATERIALS: SPIN-1/2 NUCLEI

A. ^{29}Si MAS NMR Studies of Zeolite Systems

As indicated above, ^{29}Si MAS spectra are particularly straightforward and, as we will demonstrate, may be related in a direct manner to the results of x-ray diffraction studies. In this section, we describe the development of these techniques.

Since the initial work of Lippmaa et al. [18], there have been many studies of the ^{29}Si MAS NMR spectra of low Si/Al ratio zeolites. The general features of these spectra are now well understood, and results from the early studies have been reviewed [1]. In general, the ^{29}Si MAS NMR spectra of simple zeolites contain a maximum of five reasonably well-resolved peaks, as illustrated in Fig. 5 for the zeolite analcite. It was demonstrated by Lippmaa et al. [18] that these five peaks correspond to the five possible distributions of Si and Al around a silicon nucleus at the center of an SiO$_4$ tetrahedron; namely, Si[4Al], Si[3Al,1Si], Si[2Al,2Si], Si[Al,3Si], and Si[4Si]. In addition, as shown in Fig. 6, the ranges of chemical shift over which the resonances occur are reasonably characteristic of the composition of the first coordination sphere, and the spectra may thus be used to probe the local Si/Al distributions in simple zeolite lattices. A particularly useful feature of these spectra is that the Si/Al ratio of the lattice can be calculated directly. This possibility has been investigated in detail for the previously described zeolite faujasite, which can be obtained by direct synthesis with the same basic structure but over a wide range of compositions (Si/Al = 1 –

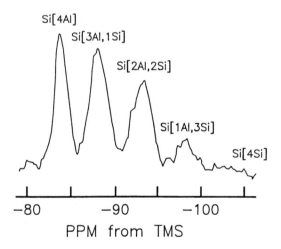

Figure 5 ^{29}Si MAS NMR spectrum of the zeolite anaclite (79.6 MHz).

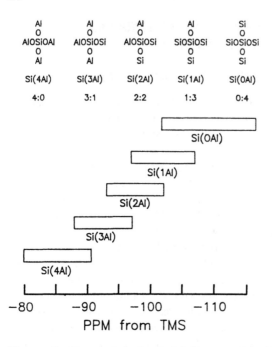

Figure 6 Characteristic chemical shift ranges of the five different local silicon environments.

3). Typical results are shown in Fig. 7 together with the Si/Al ratios calculated from the [29]Si spectra [19]. As the Si/Al ratio increases, there is a corresponding increase in the relative intensities of the high-field peaks, as would be expected from the peak assignments above. Under Loewenstein's rule [20], which postulates that Al-O-Al linkages are avoided if possible, the Si/Al ratio may be calculated from the five peak areas in the [29]Si spectrum according to

$$\frac{Si}{Al} = \frac{\sum_{n=0}^{4} I_{Si[nAl]}}{\sum_{n=0}^{4} 0.25 \, n \, I_{Si[nAl]}} \tag{3}$$

where I is the peak intensity and n indicates the number of coordinated Al atoms for a given peak.

Compared to a bulk chemical analysis, this method has the advantage that it detects the Al atoms indirectly from their effect on the Si atoms *in the framework*, and thus only detects framework Al atoms and the *Si/Al ratio for the framework*. The ratio from chemical analysis, by contrast, will include both

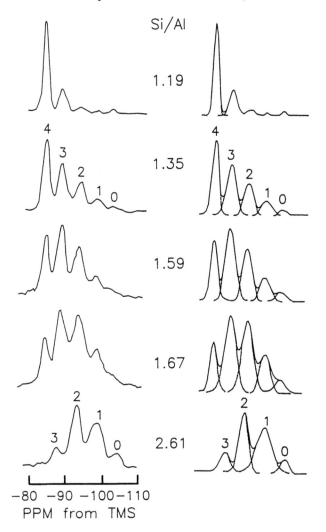

Figure 7   Observed and calculated ^{29}Si MAS NMR spectra of faujasite zeolites with the Si/Al ratios indicated. (From Ref. 19.)

framework Al and any Al occluded in the cavities or present as impurities and not an integral part of the system. The two analyses are thus complementary.

There is a natural tendency to attempt a deduction of specific Si, Al distributions within the lattice from the ^{29}Si NMR data, but the NMR results reflect the local environments of the Si atoms averaged throughout the lattice (and also over the many crystals in the sample) and as such do not necessarily imply any simple long-range ordering. However, they have been successfully used by Melchior [21] to probe differences in the formation mechanisms of zeolites A and Y that give quite different distributions of the local environments for identical Si/Al ratios of the final materials.

The simple MAS experiment yields spectra that are quantitatively reliable as long as care is taken to allow for complete spin-lattice relaxation between scans. In general, cross-polarization yields spectra of much poorer signal-to-noise ratio (S/N), but has an important application: the detection of Si $(OR)_3OH$ groupings that occur as defects (and which are of interest because of their possible involvement in catalysis) can be facilitated by the use of cross-polarization techniques that discriminate greatly in favor of silicon atoms close to the OH groups in the lattice [22]. This selectivity is illustrated in Fig. 8, which shows a marked increase in the relative intensity of the signals at ~ -104 ppm. However, care must be taken in the interpretation of such CP/MAS spectra, because they are nonquantitative and demonstrate only that some Si(OH) units are present, not that they are the only units contributing to the signal intensity at this chemical shift in the MAS spectrum. The corresponding spectra from 90° pulses indicate

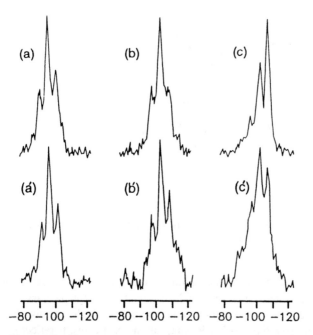

Figure 8 ^{29}Si MAS NMR spectra of dealuminated Y zeolites without (upper spectra) and with cross-polarization (lower spectra). a and a′ are the spectra from the starting NaY material. The other samples have been dealuminated to varying degrees.

that their *proportion* is small, but the CP technique is extremely useful in signaling their presence.

B. ^{31}P MAS NMR Studies of AlPO$_4$ Systems

^{31}P nuclei play a role in AlPO$_4$ systems similar to that of ^{29}Si nuclei in zeolites inasmuch as both atoms occupy tetrahedral-framework positions in the lattice. In contrast to most zeolites, however, the AlPO$_4$s have a 1:1 ratio of P and Al and an exact alternation of these elements. Thus, each ^{31}P atom is in a P(4Al) environment, and, in the simplest case, each Al is in a Al(4P) environment. The ^{31}P NMR spectra of most AlPO$_4$s therefore show single resonances [23] and do not provide the same kind of detailed structural information obtained from the ^{29}Si spectra of zeolites as described in Sec. IV.A.

By contrast, in AlPO$_4$ materials with more complex unit cells there can be several crystallographically inequivalent sites, and multiple ^{31}P resonances result [24]. Because the AlPO$_4$ systems are both highly crystalline and highly ordered, sharp resonances will be observed. Figure 9 shows examples of two such systems: VPI-5 and AlPO$_4$-8. A further complication in some of these systems (further discussed in Sec. V.A.) is that a portion of the aluminum atoms can become hydrated and change from four- to sixfold coordination. The resulting changes in the ^{31}P spectrum are shown in Fig. 10 for VPI-5. As further complex structures are synthesized, ^{31}P MAS NMR should continue to provide valuable information regarding T-site environments and framework structures, again complementing diffraction data.

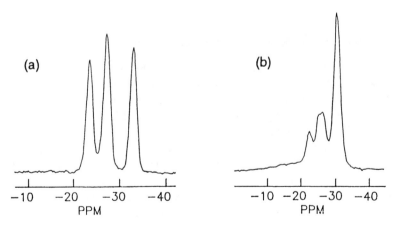

Figure 9 ^{31}P MAS NMR spectra of the aluminophosphates (a) VPI-5 and (b) AlPO$_4$-8. (From Ref. 24b.)

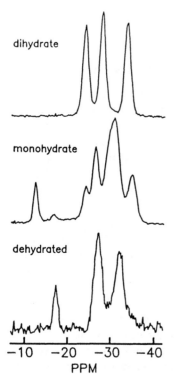

dihydrate

monohydrate

dehydrated

−10   −20   −30   −40
PPM

Figure 10 ^{31}P MAS NMR spectra of VPI-5 as a function of hydration. (From Ref. 24b.)

C. ¹H MAS NMR Studies of Acid Sites

One further, particularly informative experiment important to consideration of framework nuclei involves the detection and characterization of the proton sites within the zeolite structure, as developed by Freude et al. [25] (Fig. 11). Even when the protons are not major contributors to the overall lattice structure, they may be central to the catalytic reactions. Since they are relatively dilute in the lattice, a simple MAS experiment often yields spectra of sufficient resolution to identify the different functionalities. Spectra of this type will be critical in probing the catalytic natures of these systems and the optimization of techniques for their activation. Together with the use of DAS and DOR techniques (see Sec. V.C.), it should become possible to selectively obtain ^{17}O spectra of the acid sites themselves.

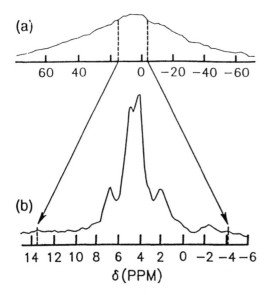

Figure 11 Line narrowing by the CRAMPS technique of the 270 MHz ^{1}H NMR spectrum of a dehydrated HY zeolite. (a) Static powder spectrum; (b) MAS spectrum at 2.5 kHz rotation frequency (sealed glass ampoule). (From Ref. 25c.)

V. INVESTIGATIONS OF LOW–Si/Al RATIO MATERIALS: QUADRUPOLAR NUCLEI

The other abundant nuclei in the zeolite framework, ^{27}Al and ^{17}O, show a more complex NMR behavior than ^{29}Si or ^{31}P, since they are quadrupolar nuclei with nonintegral spins (both with $I = 5/2$). The effect of the quadrupolar interaction on the Zeeman splitting is illustrated in Fig. 12 for a spin-5/2 nucleus such as ^{27}Al. As can be seen in the figure, the $+1/2 \leftrightarrow -1/2$ transition is unaffected by the quadrupolar interaction to first order (both levels being shifted by the same amount in the same direction), and it is this transition that is observed. The other allowed transitions are usually too broad and shifted too far from resonance to be observed directly. The lineshape of the $+1/2 \leftrightarrow -1/2$ transition is distorted and shifted owing to the second-order quadrupolar interaction [26], while the shift in the center of gravity is given by

$$\omega_{CG} - \omega_0 = \frac{1}{30} \frac{\omega_Q^2}{\omega_0} [I(I + 1) - \frac{3}{4}] (1 + \frac{\eta^2}{3}) \qquad (4)$$

Here $\omega_Q = 3e^2qQ/2I(2I - 1)h$, eq is the z-component of the electric field gradient tensor, eQ the quadrupole moment of the nucleus, η the asymmetry

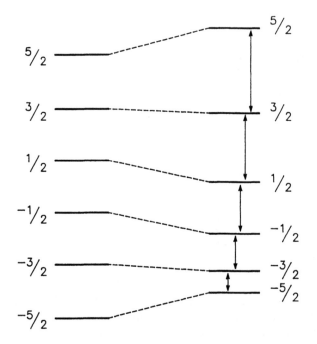

Figure 12 Energy level diagram for a spin-5/2 nucleus showing the effect of the first-order quadrupolar interaction on the Zeeman energy levels. The (m = 1/2 ↔ m = − 1/2) transition (shown in bold) is independent of the quadrupolar interaction to first order.

parameter, and ω_0 the Larmor frequency, which is directly proportional to the applied magnetic field. Therefore undesired shifts and lineshape distortions *in MAS spectra* will be minimized by working at high field. This effect is illustrated in Fig. 13 [16], which shows ^{27}Al MAS NMR spectra of a zeolite Y sample at 23.5 and 104.2 MHz (2.2 and 9.4 T magnetic fields, respectively). The high-field spectrum is much less distorted, and the apparent chemical shift is much closer to the correct isotropic-shift value. For quadrupolar nuclei, the nuclear spin is not quantized along the Zeeman field direction, and, accordingly, MAS greatly reduces but does not completely eliminate the interaction. If the second-order quadrupolar interaction completely dominates the spectrum (as for example in the case of ^{51}V in $VO_3{}^-$ and V_2O_5 [27]), spinning at an angle different from 54°44′ may be employed to minimize the quadrupolar contributions, although other interactions such as the shift anisotropy will increase. In each specific case a compromise must be sought, but, in any event, use of the highest possible field in MAS experiments will be advantageous, and these experiments may also be carried out using a conventional high-resolution spectrometer. More complete

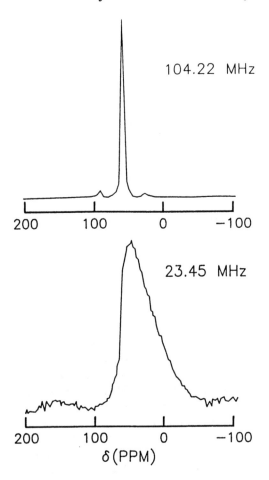

Figure 13 ^{27}Al MAS NMR spectra of zeolite Y obtained at 23.45 MHz and 104.22 MHz (proton frequencies 90 and 400 MHz, respectively). (From Ref. 17.)

averaging of both the shift anisotropies and the quadrupolar interaction by DAS and DOR experiments will be described below.

A. ^{27}Al Spectra

^{27}Al spectra of zeolites have been obtained at frequencies up to 154.5 MHz (600 MHz for protons) [28]. ^{27}Al MAS NMR spectra of the tetrahedral lattice aluminum in zeolites show only a single resonance, because every Al has the environment Al[4Si]; the Al chemical shifts of individual zeolites have, however, characteristic values at a particular field strength. Only in the case of zeolite omega at 154.5 MHz (^{1}H at 600 MHz) has MAS detected crystallographically inequivalent ^{27}Al sites [29]. This is because Loewenstein's rule forbids Al-O-Al linkages, and the primary information afforded by the ^{27}Al spectra relates to coordination. NMR discriminates between octahedral coordination, which gives a peak at about 0 ppm (with reference to $[Al(H_2O)6]^{3+}$.aq.), and tetrahedral coordination, which occurs in the range of 50–65 ppm. As before, it is not necessary to decouple the dipolar interaction with ^{1}H in these experiments because the protons present are generally mobile and do not contribute greatly to the final ^{27}Al linewidths. In order to obtain quantitative information, small ^{27}Al pulse angles ($\leq 15°$) should be used [30].

The sensitivity of the ^{27}Al spectra to coordination makes them ideal probes of reactions involving the zeolite lattices. For example, dealumination of the lattice can be achieved for some zeolites by treatment with silicon tetrachloride ($SiCl_4$). The overall reaction postulated to occur is

$$SiCl_4 + Na_y(SiO_2)_x(AlO_2)_y \rightarrow Na_{(y-1)}(SiO_2)_{(x+1)}(AlO_2)_{(y-1)} +$$
$$NaCl + AlCl_3 \tag{5}$$

$$NaCl + AlCl_3 \rightarrow Na^+AlCl_4^- \tag{6}$$

The ^{27}Al spectra shown in Fig. 14 confirm the basic mechanism of Eq. (5) but reveal it to be more complex [31]. The spectrum in Fig. 14b, taken after reaction with $SiCl_4$, shows that the tetrahedral (lattice) aluminum has greatly decreased, but there appears a very intense absorption at about $+100$ ppm, which can be identified as being due to $AlCl_4^-$ from the high-temperature reaction of NaCl and $AlCl_3$ indicated in Eq. (6). This peak disappears from the spectra after subsequent washing, but some of the aluminum remains behind as an octahedral species. Careful further washing and, if necessary, ion exchange are needed to remove more completely the nonlattice aluminum (Fig. 14d). The broad resonances under the spectra (c,d) indicate the presence of aluminum atoms in asymmetric sites. The presence of aluminum nuclei in very distorted environments is a difficult problem to tackle, since the quadrupolar broadening could easily become severe enough to preclude their observation. A number of authors have proposed treatments with acetylacetone (acac) to convert extra-lattice aluminum species to $Al(acac)_3$, which can easily be observed owing to its highly symmetric octahedral environment [32]. In other work, ^{27}Al nutation NMR experiments have been used [33,34], but neither of these approaches has yielded a completely satisfactory description of the range of aluminum environments in chemically modified zeolites.

For $AlPO_4$ molecular sieves, the ^{27}Al spectra provide a different type of information. In certain cases, some of the framework aluminum atoms can become hydrated, changing coordination from four to six and topology from

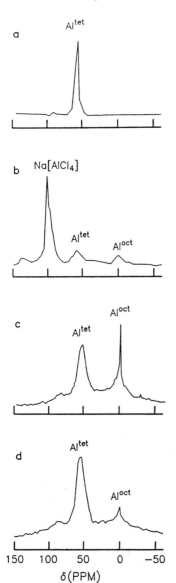

Figure 14 ^{27}Al MAS NMR spectra at 104.2 MHz obtained on zeolite Y samples at various stages of the SiCl$_4$ dealumination procedure. (a) Starting faujasite sample. (b) Intact sample after reaction with SiCl$_4$ before washing. (c) Sample as in (b) after washing with distilled water; (d) after several washings. (From Refs. 19 and 31.)

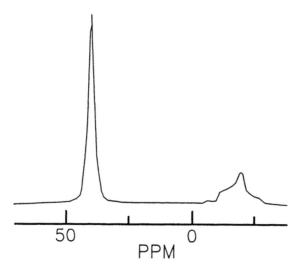

Figure 15 ^{27}Al MAS NMR spectra of VPI-5 showing resonances from tetrahedrally coordinated aluminum and octahedrally coordinated aluminum at 41 ppm and approximately -18 ppm, respectively. (From Ref. 24b.)

tetrahedral to octahedral. Such changes are clearly detectable in the ^{27}Al MAS spectra and are illustrated in Fig. 15 for the example of VPI-5.

B. ^{17}O Spectra

Particularly important from the point of view of the lattice structures are ^{17}O spectra, since the resonances will be affected by both chemical-shift anisotropy and the second-order quadrupolar interaction. As shown by Oldfield and co-workers [35], the important factors contributing to the spectra can be identified by investigation of selected reference samples. The ^{17}O spectrum of zeolite A, which has only Si-^{17}O-Al environments, shows a relatively small quadrupolar coupling, whereas a sample of siliceous zeolite Y, which contains only Si-^{17}O-Si environments, shows a pronounced quadrupolar interaction. The ^{17}O spectra of low Si/Al–ratio zeolites can be deconvoluted into the contributions of these two components (1 and 2 corresponding to Si-^{17}O-Si and Si-^{17}O-Al) as shown in Fig. 16 for zeolite Y with Si/Al = 2.74. It should be noted that it is the static spectra that are most informative since the quadrupolar effect is dominant. That is, "resolution" should be considered here as the relative separation of spectroscopic features, not simply the width of the resonances. The Si/Al ratios can be calculated from the relative Si-^{17}O-Al and Si-^{17}O-Si concentrations in a manner analogous to that described earlier for the ^{29}Si spectra and, again, will relate to the framework elements only.

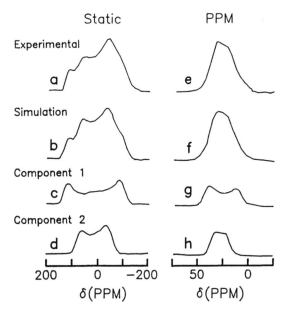

Figure 16 ^{17}O NMR spectra (67.8 MHz) and spectral simulations for NaY zeolite of Si/Al = 2.74. (a) Static spectrum; (b) simulation of (a) using parameters of (c) and (d); (c) component 1, Si-[17]O-Si; (d) component 2, Si-[17]O-Al; (e) MAS spectrum; (f) simulation of (e), using parameters of (g) and (h); (g) component 1, Si-[17]O-Si; (h) component 2, Si-[17]O-Al. (From Ref. 35.)

C. DAS and DOR Experiments

A most promising recent development relevant to both [27]Al and [17]O investigations has been the introduction of experiments in which the spinner axis undergoes a time-dependent trajectory with respect to the magnetic field axis. In this manner it is possible to average second-order as well as first-order interactions [36–38]. In double rotation (DOR) experiments, the axis of the rotor is moved continuously in a cone by placing the sample in a spinner within another spinner (Fig. 17A). By the correct choice of angles for the two spinner axes, second-order effects are averaged. Under dynamic-angle spinning (DAS), the sample is contained within a single spinner but the orientation axis of the spinner is switched between two discrete angles relative to the external magnetic field as in Fig. 17B. Sets of complementary angles are available, but different times must be spent by the sample at each of the orientations depending on the particular pair of angles chosen. For the complementary angles $\theta_1 = 37.38°$, $\theta_2 = 79.19°$, equal times are spent at both angles [39,40].

The power of this approach for removing second-order quadrupolar interactions compared to previous experiments is well illustrated in Fig. 18. The [17]O

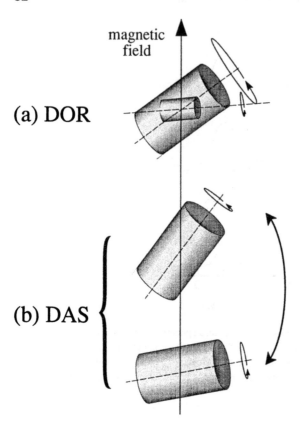

magnetic
field

(a) DOR

(b) DAS

Figure 17 Schematic representation of the two approaches for averaging second-order interaction: (a) double rotation and (b) dynamic-angle spinning.

DAS spectrum clearly shows the single oxygen site expected for this system [40]. More recently, DOR and DAS ^{17}O spectra have been reported for a variety of silicates [41]. In all cases, narrow lines were observed that could be related to the number of crystallographically inequivalent sites in much the same way possible for the ^{29}Si MAS spectra discussed earlier. Although ^{17}O results on zeolites have not been reported, studies on the highly siliceous systems described in Sec. VI should be straightforward.

Similarly, although there are no ^{27}Al data on zeolites, considerable line narrowing was observed for ^{27}Al spectra as a function of hydration in the closely related AlPO$_4$ system VPI-5 [42]. The spectra in Fig. 19 demonstrate this narrowing of the ^{27}Al resonances. Double rotation has also been used to study the molecular sieves AlPO$_4$-11 [43] and AlPO$_4$-25 [44], as well as the molecular sieve precursor AlPO$_4$-21 [44], which contains extremely distorted five-coordinated aluminum sites.

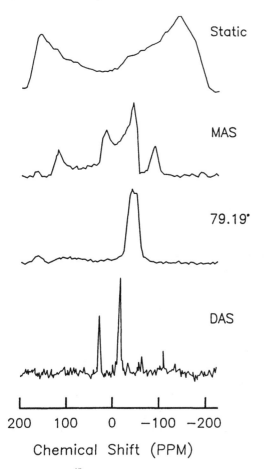

Figure 18 ^{17}O static, MAS, variable-angle spinning (at 79.19°), and one-dimensional DAS spectra for cristobalite (SiO_2). The single oxygen isotropic shift is at -16.6 ppm with respect to H_2O. (From Ref. 40.)

VI. INVESTIGATIONS OF HIGHLY SILICEOUS ZEOLITES

More recently, considerable information has been obtained from the ^{29}Si MAS NMR spectra of highly siliceous zeolites, where the Al content is so low that it does not affect the ^{29}Si spectrum [45]. Materials of this type may be synthesized directly and can also be obtained by removal of Al from low–Si/Al systems by chemical treatments. For example, $SiCl_4$ will replace the lattice aluminum atoms in some systems, or they may be removed by treatment with water vapor at elevated temperatures (hydrothermal dealumination). Where the Si/Al ratio is already high, these procedures may be used to remove any residual aluminum:

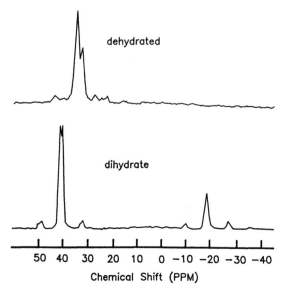

Figure 19 ^{27}Al MAS and DOR NMR spectra of VPI-5. (From Ref. 42.)

$$
\begin{array}{ccccc}
 & & \text{Si} & & \text{Si} \\
\text{Si} & & | & & | \\
| + & & \text{O} & & \text{O} \\
\text{O NH}_4 & & | & & | \\
| - & \text{H}_2\text{O} & \text{H} & \text{``Si(OH)}_4\text{''} & | \\
\text{Si-O-Al-O-Si} & \rightarrow & \text{Si-O-H} \quad \text{H-O-Si} & \rightarrow & \text{Si-O-Si-O-Si} \quad (7) \\
| & & \text{H} & & | \\
\text{O} & & | & & \text{O} \\
| & & \text{O} & & | + 4\text{H}_2\text{O} \\
\text{Si} & & | + \text{NH}_3 + \text{Al(OH)}_3 & & \text{Si} \\
 & & \text{Si} & &
\end{array}
$$

The hydrothermal dealumination procedure described by Eq. (7) is particularly versatile and can be used to produce a large variety of highly siliceous zeolites. Even zeolite A, which is known to be both thermally and chemically unstable, may be produced in a completely siliceous form by hydrothermal treatment of the high-silica analog ZK-4 [46]. Figure 20 shows the ^{29}Si MAS NMR spectra of the highly siliceous forms of some typical zeolites, together with those of the corresponding low–Si/Al ratio materials [47,48]. In all highly siliceous zeolites, the resonances observed are due to only Si[4Si] groupings and are extremely narrow. Because of their excellent resolution, these spectra can be exploited in a number of ways to obtain subtle information regarding zeolite structures not easily obtainable by other techniques.

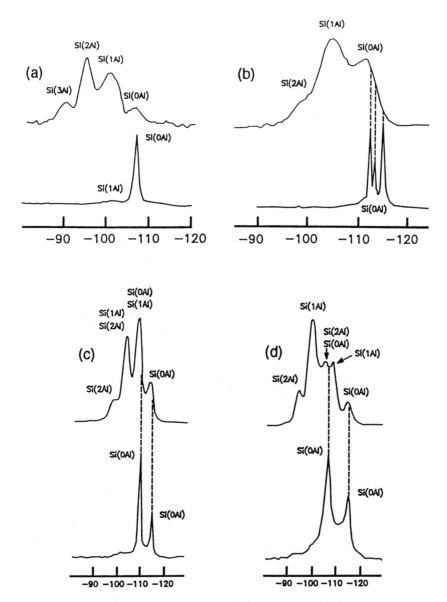

Figure 20 ^{29}Si MAS NMR spectra of the highly siliceous and corresponding low Si/Al ratio forms of (a) zeolite Y, (b) mordenite, (c) offretite, and (d) omega. (From Refs. 47 and 48.)

The line widths of the ^{29}Si resonances for low–Si/Al ratio materials are on the order of 5–8 ppm, which is surprisingly large considering their highly crystalline nature (^{13}C linewidths from highly crystalline organic solids are 1 ppm or less). From Fig. 20 it can be seen that the removal of Al from the zeolite framework reduces the linewidths to $\sim$1 ppm, indicating that the Al atoms in these systems are responsible for the line broadening. It is possible that the effect is attributable to residual dipolar couplings involving the quadrupolar ^{27}Al nuclei (interactions that are not completely removed by MAS), but this explanation is unlikely at the high fields used. Inspection of the spectra in Fig. 20 reveals that not only are the resonances narrowed, but they also shift to the high field extreme of the Si[4Si] peak in the corresponding low–Si/Al ratio spectra [48]. This upfield shift indicates that the residual broadening in the ^{29}Si MAS NMR spectra of low–Si/Al ratio zeolites is caused by a distribution of isotropic chemical shifts. The distribution results from the range of magnetic environments created by the many possible distributions of Si and Al in second- and further-nearest neighbor positions. Thus the broadening of the spectra reflects disordering in the lattice; it is not how much Al is present, but rather how it is distributed. This hypothesis is confirmed by the ^{29}Si MAS NMR spectra of the minerals albite and natrolite and of zeolite A, all of which are perfectly ordered and have narrow ^{29}Si resonances ($\sim$1 ppm), although the different silicons in these systems all have one or more aluminum atoms as first-nearest neighbors.

A. Crystallographically Inequivalent Sites from ^{29}Si Spectra

The ^{29}Si spectra of highly siliceous zeolites can be directly related to their framework structures. The numbers and relative intensities of the sharp resonances that are observed reflect directly the numbers and occupancies of the crystallographically inequivalent T-sites in the asymmetric unit of the unit cell.

In faujasite and zeolite A, there are single lattice sites, and the completely siliceous analogs both show single sharp signals. The mordenite structure has $16T_1$, $16T_2$, $8T_3$, and $8T_4$ sites, and there are three resonance lines in the NMR spectrum with relative intensities of $2:1:3$, which means that two of the resonances are accidentally degenerate. The spectrum of zeolite omega (mazzite structure with $24T_1$ and $12T_2$ sites) shows only two resonances with relative intensities $2:1$, which may be unambiguously assigned. In the same way, the two resonances in the offretite spectrum may be assigned to its $12T_1$ and $6T_2$ sites. For other highly siliceous zeolites, similar assignments can be made, giving a clear and direct correlation between the NMR and XRD experiments.

However, the spectra in Fig. 20 also indicate that interpretation of the ^{29}Si MAS NMR spectra of zeolites with more than one independent T-site in the unit cell may be much more complex than has been suggested up to this point, because the effect of site inequivalence as well as that of Al in the first coordination sphere must both be considered. (The systems originally investigated in the deduction of Fig. 6 were somewhat atypical in this regard since they contained only one T-site.) In the case of mordenite, the shift dispersion is limited and acts mainly as an additional line-broadening mechanism in the spectrum of the low–Si/Al ratio material. In the example of offretite, however, the shift difference due to T-site inequivalence (5.3 ppm) is almost exactly the same as the effect of Al in the first coordination sphere as discussed earlier, and there is exact overlap of the different peaks giving the assignments indicated in the figure. Particularly important is that the simple formula given in Eq. (3) for calculating the Si/Al ratio is invalid because it relates only to systems with single T-sites, and so more complex treatments must be undertaken. Note, though, that the offretite system is underdetermined, and hence the distributions of the silicon atoms over the two sites for each of the coordinations cannot be deduced from the ^{29}Si spectrum alone without some assumptions regarding the randomness of the distribution. From the case of zeolite omega, it is clear that such assumptions should not be made. In the ^{29}Si spectrum of zeolite omega, the shift difference between the T-sites is 8.5 ppm, which is larger than the first coordination sphere influence of Al [29]. Here, an unambiguous assignment of all the silicon atoms over the two T-sites for each of the local environments may be made using the Si/Al ratio found by chemical analysis as a correct estimate for the lattice, but without any assumptions at all regarding the nature of the Al distribution. Most importantly, the distribution of Al over the two T-sites is predicted to be 1.7:1, which can be directly confirmed from the ^{27}Al MAS NMR spectrum obtained at 14.1 T (600 MHz for protons), showing that the Si/Al distribution is nonrandom [29].

Good examples of the utility of the ^{29}Si spectra of highly siliceous systems, their direct relationship to diffraction structures, and sensitivity to local effects come from investigations of the ZSM-5 system whose structure has been described earlier. The ZSM-5 and ZSM-11 structures are very similar, and this is reflected in their XRD patterns. However, because the silicon chemical shifts are determined by local geometric effects, the ^{29}Si MAS NMR spectra of highly siliceous samples of these two materials are quite different and provide very sensitive fingerprints of the two unit cells [49]. In the case of ZSM-5, the relative contributions of the resolved low- and high-field resonances to the total spectral intensity in Fig. 21 [50] reveal that there is a total of 24 independent T-atoms in the unit cell of the (calcined) ZSM-5 lattice, thereby defining this form as having a monoclinic structure.

The ^{29}Si MAS NMR spectrum of dealuminated ZSM-5 is changed dramatically by the sorption of small amounts of *p*-xylene and other organic molecules, each of which gives a characteristic limiting spectrum with all the changes being completely reversible [51,52]. Molecules that are either very small or else too large to be accommodated in the lattice (e.g., *o*-xylene) cause no major changes in the spectra. The corresponding XRD patterns show small changes indicating a

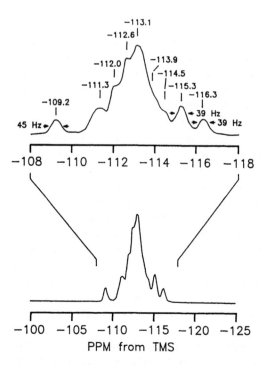

Figure 21 ^{29}Si MAS NMR spectrum (79.6 MHz) of highly siliceous zeolite ZSM-5. (From Ref. 50.)

transformation from monoclinic to orthorhombic symmetry with the loss of the characteristic "doublet" at $2\theta = 24.4°$. Figure 22 demonstrates that changes observed in the XRD lines at $2\theta = 23.3°$ and $2\theta = 23.8°$ clearly indicate, in agreement with the NMR data, that structural changes in the lattice characteristic of the sorbate are occurring, while the crystallinity is maintained.

The framework of highly siliceous ZSM-5 is also affected by temperature [53,54]. ^{29}Si MAS NMR spectra recorded between 300 and 377 K show gradual changes in the spectrum with movement of some resonances, which may be due to a general expansion of the lattice and then a discrete change between 355 and 370 K. The limiting high-temperature [29]Si spectrum indicates that the spectral changes are due to a phase transition from monoclinic (24 T-sites) to orthorhombic (12 T-sites) symmetry.

B. Ultra High-Resolution [29]Si Spectra of Highly Siliceous Zeolites

From the above discussion it can be seen that [29]Si spectra yield considerable detailed information on the structure of highly siliceous systems, and that the

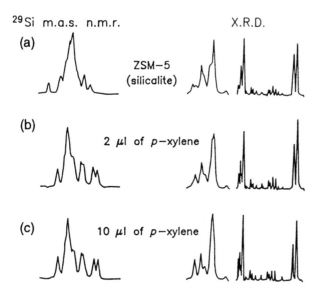

²⁹Si m.a.s. n.m.r. X.R.D.

(a) ZSM–5 (silicalite)

(b) 2 µl of p–xylene

(c) 10 µl of p–xylene

Figure 22 ²⁹Si MAS NMR spectra of highly siliceous ZSM-5 and this material treated with *p*-xylene together with the corresponding XRD patterns. (From Ref. 51.)

NMR results have a direct and complementary relationship to the diffraction data. The primary factor limiting application of the spectra is their resolution, which can be improved by very careful shimming of the magnetic field over the whole sample volume and by precise setting of the magic angle while optimizing sample spinning speeds. However, the linewidths will also be affected by the perfection of the local ordering. Thus, further improvements can be gained by the synthesis of extremely highly crystalline samples. The effect of careful optimization of all these different factors is shown in Fig. 23 for zeolite ZSM-5 [55,56]. The quality of the spectrum is now such that almost all of the 24 signals can be clearly observed, depending on the exact temperature of the measurement. The improved resolution of the spectrum makes it possible to investigate the effects of temperature and sorbed organics in considerable detail.

For comparison with the previous data, the spectra of ZSM-5 in the presence of sorbed organics and at 120°C are shown in Fig. 24. It is clear that the organics induce changes in the lattice that are quite characteristic of the organic sorbate present [56]. *p*-Xylene induces a change to a form with 12 inequivalent T-sites (Fig. 24a), whereas pyridine and acetyl acetone induce changes to new forms that still have 24 T-sites (Fig. 24c,d). Raising the temperature also causes a change to a 12 T-site structure (Fig. 24b), but the mechanism is different from the sorption of the organics.

The effect of *p*-xylene adsorption at ambient temperature on the ²⁹Si MAS NMR spectrum of highly siliceous ZSM-5 as a function of concentration is

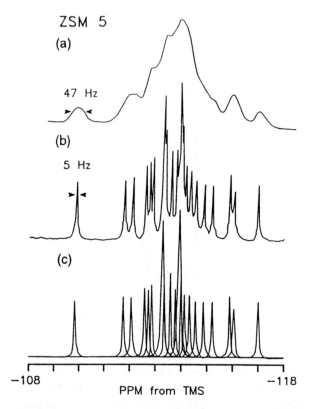

ZSM 5

(a)

47 Hz

(b)

5 Hz

(c)

−108   −118
PPM from TMS

Figure 23 ^{29}Si MAS NMR spectra of very highly crystalline ZSM-5 showing the effect of optimization of all experimental variables. (From Ref. 55.)

shown in Fig. 25. At 0.4 molecules/unit cell, the effect on the spectrum is minimal, with small shifts in individual peaks. At 1.6 molecules/unit cell, the spectrum is completely different and now shows only 12 resonances, making a complete transformation to the orthorhombic form in agreement with XRD data. The NMR spectra indicate that different proportions of both orthorhombic (12 T-atoms) and monoclinic (24 T-atoms) phases are present at intermediate loadings. The relative intensities of some of the better resolved resonances show the midpoint of the transition to be one molecule/unit cell, with complete transformation at somewhat less than two molecules/unit cell.

As with sorbed species, raising the temperature induces a phase transformation from monoclinic to orthorhombic. Detailed spectra at 10° intervals (Fig. 26a) show gradual shifts of individual resonances up to 353 K with a rapid change between 353 and 363 K (Fig. 26b). This is confirmed by synchrotron x-ray diffraction analysis of the phase transformation as a function of tempera-

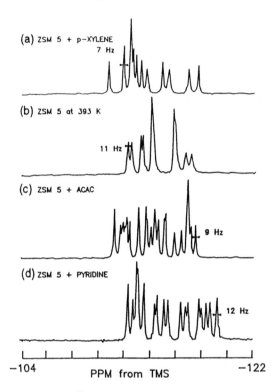

Figure 24 ^{29}Si MAS NMR spectra of ZSM-5. (a) 393 K; (b) 298 K plus *p*-xylene; (c) 298 K plus pyridine; (d) 298 K plus acetylacetone. (From Ref. 56.)

ture, which also shows that above the transition temperature only a single phase (orthorhombic) persists. The combined effect of temperature and organic sorbate is to lower the temperature of the phase transition and increase its width. In the temperature range of the phase transition, both phases now coexist and all the material is crystalline. Above the transition temperature, there are only minor changes reflecting lattice expansion due to temperature within the orthorhombic (12 T-atom) phase.

The NMR data may now be used to construct a three-dimensional phase diagram, that is, a plot of the amount of orthorhombic form present as a function of both sorbate and temperature (Fig. 27) [56]. This in turn determines the limiting conditions under which the different structures exist in phase-pure forms. Synchrotron x-ray data can now be collected on exactly these characterized structures.

Although the structural changes in ZSM-5 induced by temperature and sorbed organics are not shown by all zeolites, they are not restricted to ZSM-5 alone.

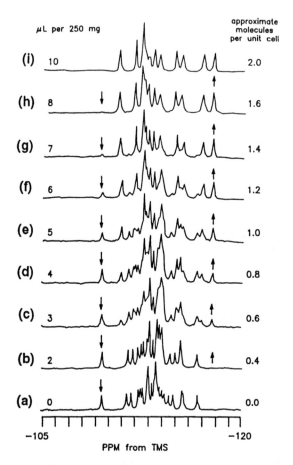

Figure 25 Effect of *p*-xylene on ZSM-5 at ambient temperature. (From Ref. 56.)

ZSM-11, the other end member of the pentasil family of zeolites, crystallizes in the tetragonal system with space group $\bar{I}4m2$ [57]. ^{29}Si MAS NMR spectra of a highly dealuminated and phase-pure ZSM-11 sample have been obtained as a function of temperature [58]. Again, elevated temperatures promote a transition to a phase of higher symmetry. At a temperature of 373 K (Fig. 28, top), the spectrum is resolved into six sharp Lorentzian peaks with relative intensities 1: (2 + 2):2:1:2:2. This distribution is consistent with the proposed tetragonal structure, with space group $\bar{I}4m2$, where the six resonances observed are due to the seven independent Si atoms in the ideal framework structure [16 $(T_2,T_3,T_4,T_5,T_7)8(T_1,T_6)$]. At room temperature, the symmetry is lower and the

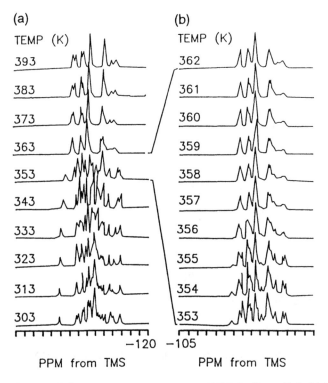

Figure 26 Effect of temperature on ZSM-5. (From Ref. 56.)

limiting structure is reached only at $-20°C$. X-ray data can be obtained on the limiting forms of the lattice as before. Rietveld refinement of the data from the high-temperature form yields exactly the previously proposed structure, but the room temperature data do not refine smoothly. This point will be addressed subsequently. As in the case of ZSM-5, the effect of sorbed molecules on the ^{29}Si MAS NMR spectra of ZSM-11 is thought to be a modification of the framework structure that alters the local T-atom environment. This is confirmed by small but significant changes observed in the XRD patterns that are reversible on desorption.

A particularly clear structural change is observed for zeolite ZSM-39, which is the silicate analog of the 17 Å cubic gas hydrate [59]. Its structure was proposed by Schlenker et al. [60] and refined from single crystal data in the space group Fd3 by Gies et al. [61]. However, x-ray powder patterns of the room temperature form show a few very weak extra reflections, indicating a space

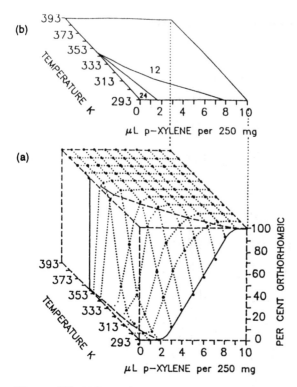

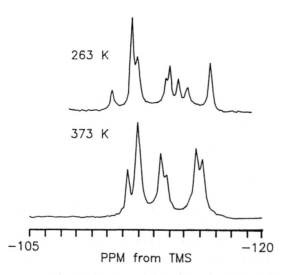

Figure 27 Effects of p-xylene and temperature on the crystal symmetry of ZSM-5. (From Ref. 56.)

Figure 28 ^{29}Si MAS NMR spectra of zeolite ZSM-11 at 263 and 373 K as indicated. (From Ref. 58b.)

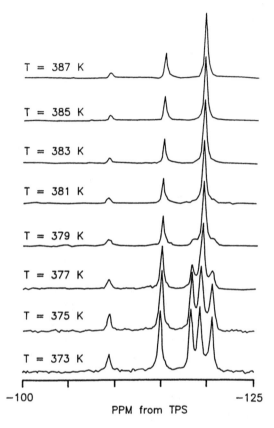

Figure 29 ^{29}Si MAS NMR spectra of zeolite ZSM-39 at the temperatures indicated. (From Ref. 62.)

group symmetry lower than face-centered cubic as noted by the authors [60]. There are three independent atoms in the ideal face-centered cubic structure of ZSM-39: 8(T_1), 32(T_2), and 96(T_3). In the ^{29}Si MAS NMR spectrum, the T_3 resonance is split into three resonances of equal intensity, making the relative spectral intensities 1:4:4:4:4. This splitting is most probably due to the removal of the threefold symmetry axis. The ^{29}Si NMR spectra obtained over the temperature range 373–387 K are shown in Fig. 29. All the resonances remain sharp as the temperature is raised, but there is a clear change over a 10° temperature range to a new spectrum showing three resonances with relative intensities 1:4:12. This pattern indicates face-centered cubic symmetry, at least locally on the NMR time scale. The changes are completely reversible and also apply to samples with less well-resolved spectra [62].

VII. THREE-DIMENSIONAL CONNECTIVITY EXPERIMENTS: ^{29}Si-O-^{29}Si LATTICE CONNECTIVITIES FROM TWO-DIMENSIONAL ^{29}Si MAS NMR EXPERIMENTS

As indicated above, ^{29}Si MAS NMR can provide a wealth of structural information on zeolites, limited in the case of highly siliceous samples only by the spectral resolution. In solution, the application of two-dimensional NMR techniques has provided considerable information on the two-dimensional connectivities between atoms within molecular structures [63]. For example, the HSC (heteronuclear shift correlation) sequence establishes heteronuclear connections such as ^{13}C/^{1}H, ^{29}Si/^{1}H, etc.; the COSY sequence defines homonuclear correlations, such as ^{1}H/^{1}H, ^{31}P/^{31}P, ^{29}Si/^{29}Si, etc.; the INADEQUATE sequence ^{13}C/ ^{13}C in natural abundance; while longer-range connectivities, such as ^{1}H/^{1}H/^{13}C, can be probed by the RCT (relayed coherence transfer) sequence.

A number of two-dimensional NMR experiments have been introduced in high-resolution solid-state NMR studies, designed, for example, to investigate chemical exchange processes [64], to retrieve chemical shift anisotropies [65] and dipolar couplings [66], and to probe spin-diffusion processes [67]. Opella has proposed an internuclear distance-determined, spin-diffusion mechanism in molecular crystals [68], and Benn and coworkers have demonstrated ^{13}C/ ^{13}C connectivities using the INADEQUATE sequence for the plastic crystal camphor and have used the COSY sequence for ^{29}Si/^{29}Si connectivities in the reference molecule Q_8M_8 [69].

At least in principle, two-dimensional NMR techniques can be used to establish connectivities in the solid state, and for crystalline three-dimensional framework structures these connectivities could be used to define the complete three-dimensional lattice itself. In this section we describe the use of two-dimensional ^{29}Si MAS NMR measurements involving scalar coupling interactions to establish three-dimensional ^{29}Si-O-^{29}Si lattice connectivities in zeolite framework structures. The first system investigated by these techniques was zeolite ZSM-39 [70,71]. To facilitate the measurements, a small quantity of this material was synthesized enriched to ~85% in ^{29}Si. From the known structure shown in Fig. 30 [60], the connectivities can be worked out. Neglecting the deviation from cubic symmetry that lifts the degeneracy of the T_3 site, the connectivities between the T-sites are: T_1 is connected to four T_2 sites; T_2 is connected to one T_1 and three T_3 sites; T_3 is connected to one T_2 and three T_3 sites. Within the unit cell there are (omitting self-connectivities) 32 T_1T_2 connectivities, 96 T_2T_3 connectivities, and no direct connectivities between T_1 and T_3. Figure 31 shows the results of a two-dimensional COSY experiment at 373 K. T_1T_2 and T_2T_3 cross-peaks are clearly observed in the spectrum, as would be expected from the known structure. The structure of the T_2T_3 cross-peak is real

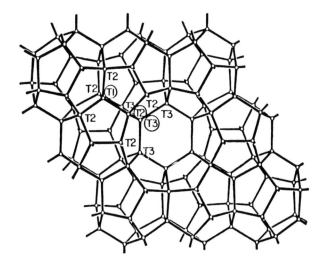

Figure 30 Schematic representation of the zeolite ZSM-39 lattice framework. The three crystallographically inequivalent tetrahedral lattice sites are indicated by T_1, T_2, and T_3 (inside circles), and in each case the identities of the four nearest neighbors are shown.

and is due to the partial resolution of the T_3 resonance. Identical connectivities were found by two-dimensional spin-diffusion experiments based on dipolar interactions, since these mirror one-bond J couplings because of their distance dependence. In a further study using ^{29}Si-enriched material, the three-dimensional bonding in zeolite DD3R was successfully investigated by both COSY and INADEQUATE experiments [72]. At high temperature, the experiments confirmed the symmetry of the structure as that proposed from a single crystal diffraction experiment [73], but the room-temperature structure was found to be of lower symmetry.

The use of ^{29}Si isotopic enrichment in the above investigations, however, is a very severe limitation, and subsequent experiments have been carried out using natural-abundance materials [74,75] on the typical zeolite systems ZSM-12 and ZSM-22. Figure 32 shows one projection of the lattice structure of zeolite ZSM-12. The structure is approximately that originally reported [76], although a recent refinement of synchrotron powder x-ray data by Gies and coworkers [77] establishes the unit cell symmetry as $C_{2/c}$ with a doubling of the c-cell dimension to 24.3275 Å. The number of crystallographically inequivalent silicons in the unit cell is unchanged (that is, there are seven T-sites with equal occupancies, as indicated in the figure). This is reflected in the ^{29}Si MAS NMR spectrum, which shows seven clearly separated resonances of equal intensities. Table 1 lists the predicted connectivities.

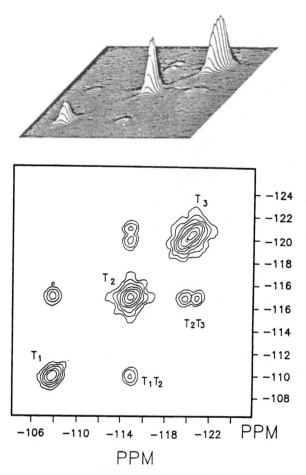

Figure 31 Contour and stacked plots of a two-dimensional COSY experiment on ZSM-39 at 373 K using a modified COSY pulse sequence with fixed evolution delays. The data were acquired using 128 experiments, 64 scans in each experiment, 5 kHz sweepwidth, 256 data points for acquisition, and a fixed delay of 5 ms. Sine bell apodization was used, and the data are presented without symmetrization or smoothing. The total experimental time was approximately 23 h. (From Ref. 71.)

From previous experiments with highly isotopically enriched materials, it was seen that the T_2^* value reflected by the linewidth was the determining factor in the experiments, limiting the maximum time available for the frequency-encoding and evolution steps quite independent of the magnitude of the J couplings anticipated. By careful shimming and setting of the magic angle, linewidths of approximately 8–10 Hz were obtained for the ZSM-12 sample. A

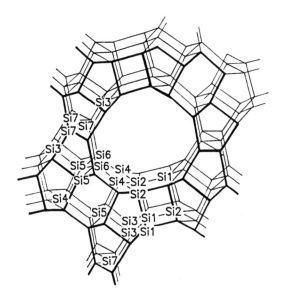

Figure 32 Schematic representation of the zeolite ZMS-12 lattice framework. The seven crystallographically inequivalent tetrahedral lattice sites are indicated by T_1, T_2 . . .T_7. (From Ref. 76.)

Table 1 T-Atom Sites, Their Occupancies, and Their Connectivities for the Asymmetric Unit in Zeolite ZSM-12

T-site	Occupancy	Connectivities
T_1	1	$2T_2 : 2T_3$
T_2	1	$2T_1 : 2T_4$
T_3	1	$2T_1 : 2T_5 : 1T_7$
T_4	1	$2T_2 : 1T_5 : 1T_6$
T_5	1	$1T_3 : 1T_4 : 2T_6$
T_6	1	$1T_4 : 2T_5 : 1T_7$
T_7	1	$1T_3 : 1T_6 : 2T_7$

two-dimensional COSY experiment [78] consisting of 64 individual experiments with a time increment of 834 μs was used together with a fixed delay of 5 ms for a total maximum t_1 period of 63 ms in the $[(90°)_{\phi 1}\text{-}t_1\text{-}(45°)_{\phi 2}\text{-acquire}]$-sequence. The results are shown in Fig. 32, with the data in the t_2 domain truncated to 256 data points before zero filling. Clear connectivities are observed in the unsymmetrized plot. T_5 may be unambiguously assigned as the resonance at $\delta = -112.3$ ppm, since it decreases significantly in intensity relative to the other

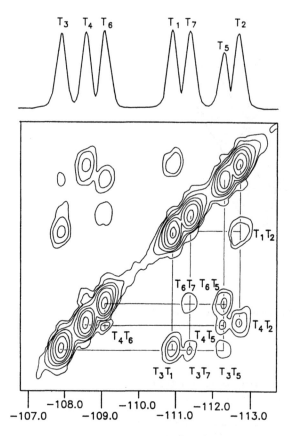

Figure 33 Contour plot of a COSY experiment on ZSM-12 with the projection in the F_2 dimension shown on top. The temperature was 300 K and 64 experiments were carried out with 160 scans in each experiment. A sweepwidth of 1200 Hz, a fixed delay of 5 ms, and 256 real data points were used for the data processing. (From Ref. 75.)

signals if long enough delay times are not used, thus indicating that it has a much longer T_1 value. A major contribution to the relaxation mechanism is ^{29}Si dipolar interactions with molecular oxygen, and T_5 is the only site that is not on the surface of a channel wall. From this assignment of T_5 one can proceed to assign the connectivities yielding the labeling of the cross-peaks given in Fig. 33.

When the number of points in t_2 is increased to 450, the two-dimensional plot shows the same connectivities but with a lower S/N (Fig. 34). Most importantly, doublet splittings are observed for the cross-peaks in F_2 (the limited resolution in F_1 precludes their observation in this dimension). Figure 35 shows plots of rows from this two-dimensional spectrum corresponding to the maxima in the diagonal peaks where the doublet structures are clearly observable. The observed split-

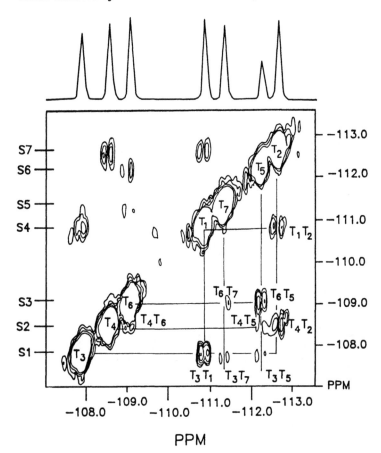

Figure 34   Contour plot of a COSY experiment on ZSM-12 with projection in F_2. The same conditions as those in Fig. 33 were used except that 80 scans were taken in each experiment. Four hundred and fifty real data points and power calculation were used in the data analysis. (From Ref. 75.)

tings are all between 9 and 16 Hz, but these are approximate values only. The data may be compared with the ^{29}Si-O-^{29}Si couplings observed in alkaline silicate solutions enriched in ^{29}Si [79]. Silicon atoms in four- and five-membered rings in these species show couplings of the order of 10 Hz, approximately those observed in the present work, and the present data represent the first direct observation of scalar ^{29}Si-O-^{29}Si couplings in the solid state.

Measurement of the scalar couplings is important in that it facilitates application of the two-dimensional INADEQUATE sequence. The conventional pulse sequence for such an experiment is $[(90°_x)\text{-}\tau\text{-}(180°)\text{-}\tau\text{-}(90°)\text{-}t_1\text{-}(135°)\text{-}t_2$, acquire] [80], where the value of τ is chosen to maximize the connectivities for a

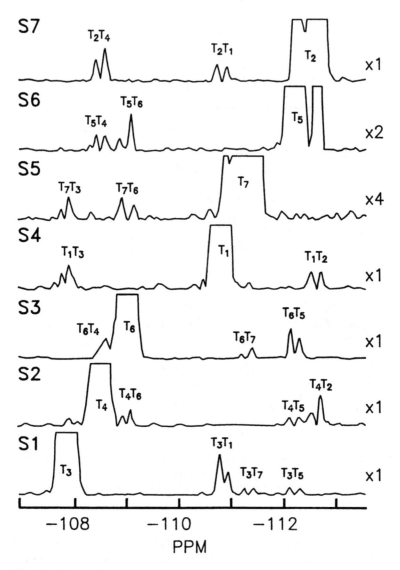

Figure 35 Row plots from Fig. 34. The numbers of the rows correspond to those indicated in Fig. 34.

particular J coupling value. Compared to the COSY experiment, the INADE-QUATE experiment has a number of advantages and disadvantages: in general, better S/N may be anticipated as there is a twofold decrease in the multiplicities of the cross-peaks, and main (noncoupled) center signals are subtracted, giving better dynamic range for the connectivities. Furthermore, it is much easier to

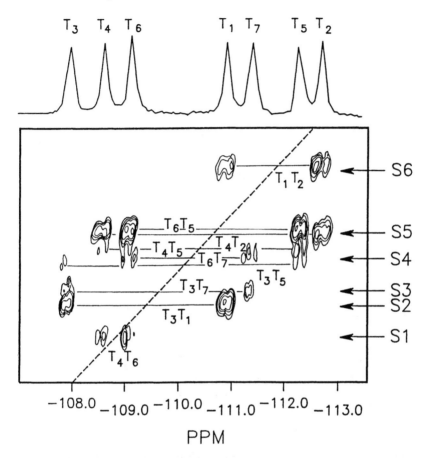

Figure 36 Contour plot of an INADEQUATE experiment on ZSM-12 at 300 K with the one-dimensional MAS NMR spectrum above. Fifty-two experiments were carried out, 64 scans per experiment, 800 Hz sweepwidth, fixed delay of 20 ms, and 450 data points collected during the acquisition. Sine bell apodization and power calculations were used in the data processing. (From Ref. 75.)

observe connectivities between closely spaced resonances near the large diagonal signals in the COSY spectrum. The main potential disadvantage is that one must have a reasonable estimate of the scalar J coupling to choose the appropriate value of τ. If this estimate can be made, then all of the individual frequency-encoding experiments contribute efficiently, resulting in good S/N. All connectivities found and assigned in the previous COSY experiments are confirmed in the INADEQUATE experiment (Figs. 36 and 37), including that between T_4 and T_6, which is clearly resolved here but was ambiguous in the COSY experiment owing to the proximity of the cross-peaks to the diagonal.

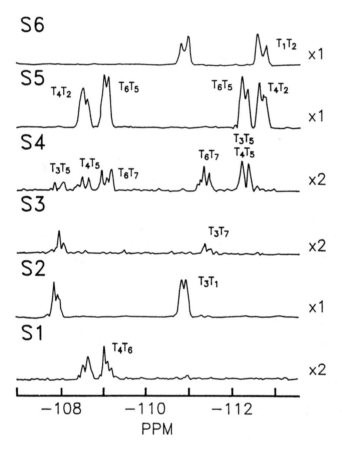

Figure 37 Row plots from Fig. 36. The numbers of the rows correspond to those indicated in Fig. 36.

It should be noted that the assignments previously discussed could have been made from the INADEQUATE experiment without having to know the identity of T_5. Observation of the T_4T_6 connection gives increased confidence in the assignment since every single connectivity is now accounted for. As previously, the intensities of the signals qualitatively reflect the numbers of connections involved. The doublet structures are clearly observable in this experiment, again confirming the previous results. The resolved splittings from single rows in the two-dimensional plot are in good agreement with the values previously found and do not change when the spinning rate is varied between 1.0 and 3.0 kHz. Similar to the earlier example with the COSY experiment, only the rows show doublet structure owing to the very limited resolution in F_1.

Similar agreement between the observed connectivities and those predicted from the structure are found in corresponding experiments on zeolite ZSM-22, giving further confidence in the reliability of these techniques. The demonstrated ability to perform them successfully on natural-abundance materials greatly extends their potential. In both cases, the two-dimensional INADEQUATE sequence gives superior performance and will usually be the experiment of choice.

Zeolite ZSM-5 has the most complex unit cell of any zeolite system, and its structure represents a very demanding test of the reliability of these techniques, there being either 12 or 24 T-sites depending on the phase. The different phases that have been investigated are summarized in Table 2 together with sources of the structural data. It should be noted that good single-crystal refinements have recently become available from the work of van Koningsveld and coworkers on the room-temperature, high-temperature, and high-loaded (eight molecules of p-xylene per u.c.) forms. The diffraction results can be used together with the NMR data [81,82,84].

Figure 38 shows the one-dimensional NMR spectra of the four phases along with detailed assignments of the resonances in terms of the corresponding unit cells shown in the figure [85]. Typical data from the orthorhombic form with p-xylene are presented in Fig. 39. There are 22 ^{29}Si-O-^{29}Si connectivities

Table 2 Description of the Samples of ZSM-5 Investigated by ^{29}Si Two-Dimensional NMR

Sample	Sample and conditions	Space group [ref]	T-sites in asymmetric unit	Name given in discussion
ZSM-5	Ambient temperature (300K)	Monoclinic form P2/n [81]	24	Monoclinic phase
	High temperature (430 K)	Orthorhombic form Pnma [82]	12	Orthorhombic phase (12 T-sites)
ZSM-5 with sorbed p-xylene	Low-loading with p-xylene (2 molecules/ UC, 300 K)	Orthorhombic form Pnma [83]	12	
	High-loaded (8 molecules/ UC, 293 K)	Orthorhombic form P2$_1$2$_1$2$_1$ [84]	24	Orthorhombic phase (24 T-sites)

Source: Ref. 85.

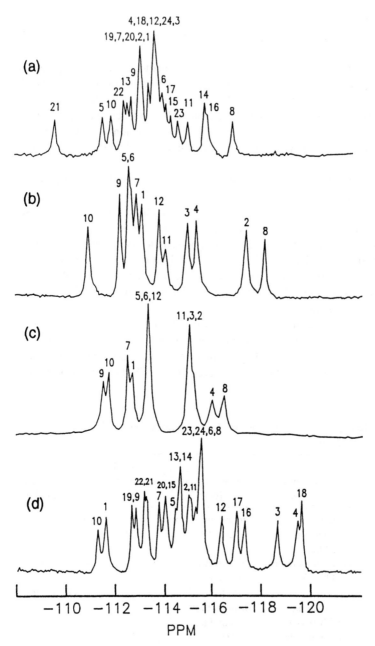

Figure 38 (a) ^{29}Si MAS NMR spectrum of ZSM-5 at 300 K. (b) ^{29}Si MAS NMR spectrum of low-loaded ZSM-5 (2 molecules of *p*-xylene per 96 T-atom unit cell) at 300 K. (c) ^{29}Si MAS NMR spectrum of ZSM-5 at 403 K. (d) ^{29}Si CP/MAS NMR spectrum of high-loaded ZSM-5 (8 molecules *p*-xylene per 96 T-atom unit cell) at 293 K. The assignments of the individual resonances of the four spectra are indicated. (From Ref. 85.)

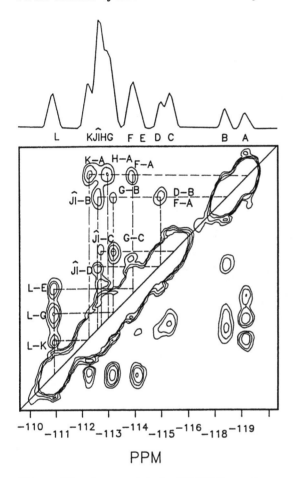

Figure 39 Contour plot of a COSY-45 experiment on ZSM-5 with 2 molecules of *p*-xylene per unit cell with the projection in the F_2 dimension shown above. The temperature was 300 K, and 64 experiments were carried out with 576 scans in each experiment. A sweepwidth of 1700 Hz, a fixed delay of 10 ms, and 220 real data points were used. Sine bell-squared apodization and power calculation were used for the data processing. (From Ref. 85a.)

expected in all for this phase, and a total of 12 are visible in the COSY experiment. The corresponding INADEQUATE spectrum (Fig. 40) is superior, with 21 of the 22 connectivities being clearly observed [85]. From the knowledge that only four silicons have self-connectivities and using general structural information from the x-ray data on all the phases, a complete self-consistent assignment is obtained.

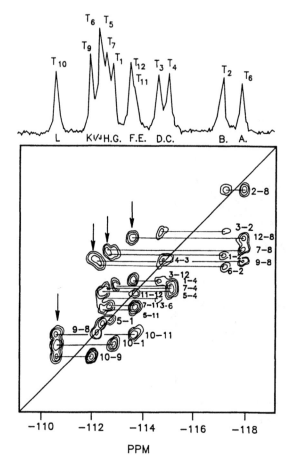

Figure 40 Contour plot of an INADEQUATE experiment of ZSM-5 with 2 molecules of *p*-xylene per unit cell carried out at 300 K with a one-dimensional MAS NMR spectrum shown above. Thirty-six experiments with 512 scans in each experiment were performed with a recycle time of 14 s, and the total time for the experiment was approximately 72 h. A sweepwidth of 800 Hz, a fixed delay of 15 ms, and 140 real data points were used. Sine bell and trapezoidal apodizations in the F_2 and F_1 dimensions, respectively, and a power calculation were used for data processing. (From Ref. 85.)

The reliability of these two-dimensional connectivity experiments is now well enough established that they can be used with confidence in the investigation of unknown structures in conjunction with diffraction studies. In a first step in this direction, a study of the three-dimensional bonding connectivities in ZSM-11 in both its high- and low-temperature forms has been reported [86]. As described earlier these are related to each other by a temperature-induced phase transition. The high-temperature form is known to have tetragonal symmetry with space group I4̄m2, while the details of the ZSM-11 structure at lower temperatures are not known to date.

Two-dimensional INADEQUATE experiments at high temperature (340 K) confirmed the postulated structure. All nine expected connectivities were observed (Fig. 41), permitting an unambiguous assignment of the spectrum. In the low-temperature (303 K) two-dimensional experiments, enough information was obtained so that, combined with the knowledge of the high-temperature structure, a structure could be suggested for the low-temperature form. The complete connectivity pattern of the 12 T-sites in the structure was established, and at least one space group, I4̄, is compatible with the NMR data. This result suggested a possible spacegroup for use in further diffraction investigations [86].

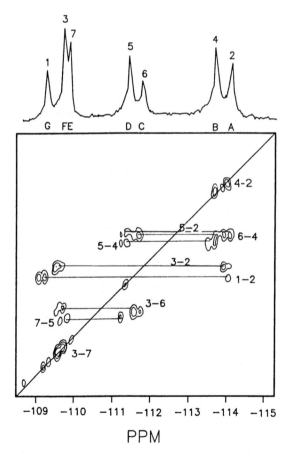

Figure 41 Two-dimensional INADEQUATE spectrum of ZSM-11 at 340 K. (From Ref. 86.)

VIII. COHERENCE TRANSFER FROM QUADRUPOLAR NUCLEI: ^{27}AL-O-^{31}P CONNECTIVITIES IN ALUMINOPHOSPHATES

Historically, NMR experiments such as cross-polarization [87], DEPT [88], and INEPT [89] have been used to transfer spin polarization from abundant spins (such as protons) to a dilute spin system. The few examples of coherence transfer experiments involving quadrupolar nuclei do make use of magnetization transfer from protons to nuclei such as ^{95}Mo [90], ^{23}Na [91], and ^{17}O [92], but, as noted in previous sections, zeolites and the related AlPO$_4$ structures contain many quadrupolar nuclei but few protons. Nevertheless, it is still possible to perform coherence transfer between other spin pairs in these systems, for example, between ^{29}Si ($I = 1/2$) and ^{27}Al ($I = 5/2$) in zeolites, or between ^{31}P ($I = 1/2$) and ^{27}Al ($I = 5/2$) in AlPO$_4$s.

The attainment of cross-polarization *from* quadrupolar nuclei is particularly important in materials research since quadrupolar nuclei usually have very short spin lattice (T$_1$) relaxation times, while spin-1/2 nuclei in inorganic systems may have T$_1$ values ranging from many seconds to even hours. Such long relaxation times greatly increase the time necessary for observation of the spin-1/2 species in many instances. By using cross-polarization from the fast-relaxing quadrupolar spins, the spectra of the spin-1/2 nuclei can be obtained in a relatively short time. Additional information regarding local structure and bonding in these systems may also be obtained through the distance dependence of the cross-polarization process.

An example of cross-polarization performed in both directions between spin-1/2 and quadrupolar nuclei is shown in Fig. 42 [93]. These spectra demonstrate the transfer of magnetization between the ^{27}Al and ^{31}P spins in the Al-O-P bonding units found in VPI-5. Use of magic-angle spinning averages the anisotropies of the chemical shifts for the ^{31}P nuclei to their isotropic values for the three crystallographically inequivalent ^{31}P sites in the unit cell. For the ^{27}Al nuclei in the sample, MAS partially averages the second-order quadrupolar interaction and two resonances are seen: one from the tetrahedrally coordinated aluminum sites (41 ppm) and a second from the octahedrally coordinated aluminum (approximately -18 ppm). The observed signal is attributable solely to cross-polarization and is not caused by direct irradiation, as proven by the appropriate cross-check experiments. In obtaining spectra (b) and (d) of Fig. 42, cross-polarization is avoided by removing the preparation and spin-locking pulses on the unobserved spin system. The expected null response is obtained.

A two-dimensional heteronuclear correlation experiment [94] using cross-polarization is performed by preparing the aluminum spins with a 90° pulse and then encoding the evolution frequencies of the aluminum spins in an initial time period. The aluminum polarization is subsequently transferred to the phosphorus spins with a spin lock, and a phosphorus free induction decay is accumulated

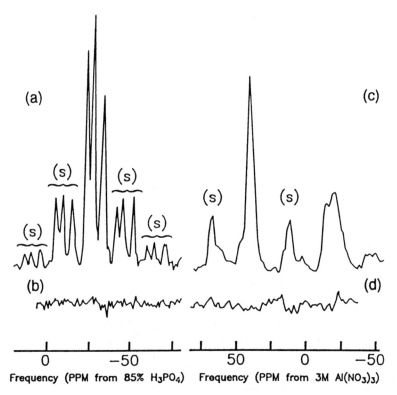

Frequency (PPM from 85% H_3PO_4)   Frequency (PPM from 3M $Al(NO_3)_3$)

Figure 42 $^{31}P \leftrightarrow {}^{27}Al$ cross-polarization MAS experiments in VPI-5. (a) $^{27}Al \rightarrow {}^{31}P$ CP experiment taking 16732 scans with a contact time of 0.8 ms, recycle delay of 0.5 s, and a 90° nutation time of 9.0 μs for the central transition of the ^{27}Al. Resonances from the three inequivalent phosphorus sites are observed with spinning sidebands (marked with an s) at multiples of the rotor frequency. (b) $^{27}Al \rightarrow {}^{31}P$ experiment as in (a) with no irradiation of the ^{27}Al nuclei. (c) $^{31}P \rightarrow {}^{27}Al$ CP experiment taking 3200 scans with a contact time of 1.6 msec, recycle delay of 30 s, and a 90° nutation time of 9.0 μs for the ^{31}P. (d) $^{31}P \rightarrow {}^{27}Al$ experiment as in (c) with no irradiation of the ^{31}P nuclei. (From Ref. 93.)

after each of a set of aluminum evolution times. Two-dimensional Fourier transformation of the resulting data array provides the correlation spectrum of Fig. 43. Peaks arise in this spectrum only when the resonances are from nuclei close enough to communicate through the dipolar interaction, which drops off rapidly as a function of internuclear distance. In the two-dimensional spectrum it is evident that each of the three ^{31}P resonances is connected to both the tetrahedral and octahedral ^{27}Al resonances, and this finding agrees with the proposed crystal topology of VPI-5 [13,14,24].

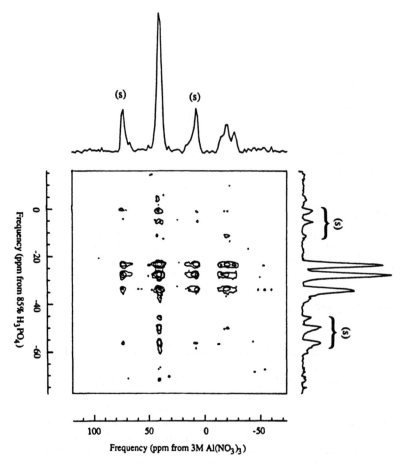

Figure 43 Two-dimensional cross-polarization experiment in VPI-5. The data were accumulated as described in the text with 640 scans from each of 256 cross-polarization experiments with an incremental increase in ^{27}Al evolution time of 12.5 μs. (From Ref. 93.)

From the success of the cross-polarization experiments, one may predict success for a variety of related experiments that depend on the dipolar interaction (such as those of Veeman [95], where spin echoes are observed under magic-angle spinning and double resonance conditions). Further experiments that allow determination of internuclear distances should also be possible, including rotational-echo double-resonance (REDOR) NMR [96], and transferred-echo double-resonance (TEDOR) NMR [97], the latter being related to INEPT experiments in liquids.

IX. CONCLUSIONS

In recent years there has been considerable progress in the development and application of solid-state NMR techniques to the investigation of zeolite structures, and a very direct relationship has been established between these investigations and x-ray diffraction studies. These efforts will surely continue in the future, with particular emphasis on homo- and heteronuclear connectivities, especially those involving quadrupolar nuclei, and there will be further efforts to apply them to systems of unknown or poorly described structure and to the identity of the active acid sites themselves.

REFERENCES

1. G. Engelhardt and D. Michel, *High-Resolution Solid-State NMR of Zeolites and Related Systems*, John Wiley and Sons, London, 1987.
2. W. M. Meier, in *Molecular Sieves*, Soc. Chem. Ind., London, 1968; D. W. Breck, *Zeolite Molecular Sieves*, Wiley Interscience, New York, 1974.
3. J. V. Smith, *Zeolite Chemistry and Catalysis* (J. A. Rabo, ed.), ACS Monograph 171, 1976; R. M. Barrer, *Zeolites and Clay Minerals as Sorbents and Molecular Sieves*, Academic Press, London, 1978.
4. J. W. McBain, *Sorption of Gases and Vapors by Solids*, Routlidge and Sons, London, 1932.
5. R. M. Barrer, *J. Soc. Chem. Ind. 64*: 130 (1945).
6. U.S. Patent 2,306,610.
7. R. M. Barrer, *Nature 164*: 112(1949).
8. U.S. Patent 2,882,243.
9. S. C. Eastwood, R. D. Drew, and F. D. Hartzell, *Oil and Gas Journal 60*: 152 (1962).
10. U.S. Patent 3,702,866.
11. G. T. Kokotailo, S. L. Lawton, D. H. Olson, and W. Meier, *Nature 272*: 437 (1978); G. T. Kokotailo and W. M. Meier, *Soc. Chem. Ind. London 10*: (1980).
12. E. M. Flanigen, B. M. Lok, R. L. Patton, and S. T. Wilson, *Pure Appl. Chem. 58*: 1351 (1986).
13. M. E. Davis, C. Saldarriaga, J. M. Garces, and C. Crowder, *Nature* (London) *331*: 698 (1988).
14. L. B. McCusker, Ch. Baerlocher, E. Jahn, and M. Bülow, *Zeolites 11*: 308 (1991).
15. C. M. Csicsery, *Zeolites 4*: 202 (1984).
16. W. M. Meier and D. H. Olson, *Atlas of Zeolite Structure Types*, Structure Commission of the International Zeolite Association, 1978; It has recently been demonstrated that the use of synchrotron x-ray sources permits single crystal studies to be carried out on much smaller samples. P. Eisenberger, J. B. Newsam, M. E. Leonowicz, and D. E. W. Vaughan, *Nature 309*: 45 (1984); The use of Rietveld refinement techniques allows better structural information to be obtained from powder diffraction data. W. I. F. David, W. T. A. Harrison, and M. W. Johnson, in *High Resolution Powder Diffraction*, Materials Science Forum, vol. 9 (C. R. A.

Catlow, ed.) Trans. Tech. Publication, Andermannsdorf, Switzerland, 1986, pp. 89–101.

17. C. A. Fyfe, G. C. Gobbi, J. S. Hartman, R. E. Lenkinski, J. H. O'Brien, E. R. Beange, and M. A. R. Smith, *J. Magn. Reson. 47*: 168 (1982).

18. E. Lippmaa, M. Magi, A. Samoson, A. R. Grimmer, and G. Engelhardt, *J. Am. Chem. Soc. 102*: 4889 (1980); E. Lippmaa, M. Magi, A. Samoson, M. Tarmak, and G. Engelhardt, *J. Am. Chem. Soc. 103*: 4992 (1981).

19. J. Klinowski, S. Ramdas, J. M. Thomas, C. A. Fyfe, and J. S. Hartmann, *J. Chem. Soc., Farad. Trans. II 78*: 1025 (1983).

20. W. Loewenstein, *Am. Mineralog. 39*: 92 (1954).

21. M. T. Melchior, *A.C.S. Symp. Ser.*, No. 218, 1983, p. 243.

22. G. Engelhardt, U. Lohse, A. Samoson, M. Magi, M. Tarmak, E. Lippmaa, *Zeolites 2*: 59 (1982).

23. C. S. Blackwell and R. L. Patton, *J. Phys. Chem. 88*: 6135 (1984).

24. a. M. E. Davis, C. Montes, P. E. Hathaway, J. P. Arhancet, D. L. Hasha, and J. M. Garces, *J. Am. Chem. Soc. 111*: 3919 (1989); b. P. J. Grobet, J. A. Martens, I. Balakrishnan, M. Mertens, and P. A. Jacobs, *Appl. Catal. 56*: L21 (1989).

25. a. D. Freude, M. Hunger, and H. Pfeifer, *Chem. Phys. Lett. 91*: 307 (1982); b. D. Freude, *Adv. Colloid Interface Sci. 23*: 21 (1985); c. H. Pfeifer, D. Freude, and M. Hunger, *Zeolites 5*: 273 (1985); d. M. Hunger, D. Freude, H. Pfeifer, and W. Schwieger, *Chem. Phys. Lett. 167*: 21 (1990).

26. R. V. Pound, *Phys. Rev. 79*: 685 (1950); M. H. Cohen and F. Reif, *Solid State Physics 5*: 321 (1957); G. M. Volkoff, *Can. J. Phys. 31*: 820 (1953).

27. M. D. Meadows, K. A. Smith, R. A. Kisney, T. M. Rothgels, R. P. Skarjunl, and E. Oldfield, *Proc. Natl. Acad. Sci. USA 79*: 1351 (1982).

28. C. A. Fyfe, G. C. Gobbi, W. J. Murphy, R. S. Ozubko, and D. A. Slack, *J. Am. Chem. Soc. 106*: 4435 (1984).

29. C. A. Fyfe, G. C. Gobbi, G. J. Kennedy, J. D. Graham, R. S. Ozubko, W. J. Murphy, A. Bothner-By, J. Dadok, and A. S. Chesnick, *Zeolites 5*: 179 (1985).

30, A. Samoson and E. Lippmaa, *Phys. Rev. B 28*: 6567 (1983).

31, J. Klinowski, J. M. Thomas, C. A. Fyfe, G. C. Gobbi, and J. S. Hartman, *Inorg. Chem. 72*: 63 (1983).

32. P. J. Grobet, H. Geerts, M. Tielen, J. A. Martens, and P. A. Jacobs, in *Zeolites as Catalysts, Sorbents and Detergent Builders*, Elsevier, Amsterdam, 1989, p. 721.

33. A. P. M. Kentgens, Ph.D. thesis, University of Nijmegen, Holland, 1987.

34. X. Hamadan and J. Klinowski, A.C.S. Symp. Ser., No. 398, 1989, pp. 448, 465.

35. H. K. C. Timken, G. L. Turner, J. P. Gilson, L. B. Welsh, and E. Oldfield, *J. Am. Chem. Soc. 108*: 7231 (1986).

36. A. Llor and J. Virlet, *Chem. Phys. Lett. 152*: 248 (1988).

37. G. C. Chingas, C. J. Lee, E. Lippmaa, K. T. Mueller, A. Pines, A. Samoson, B. Q. Sun, D. Suter, T. Terao, in *Proceedings, XXIV Congress Ampere, Poznan, 1988* (J. Stankowski, N. Pislewski, and S. Idziak, eds.), 1988, p. D62.

38. A. Samoson, E. Lippmaa, and A. Pines, *Mol. Phys. 65*: 1013 (1988).

39. B. F. Chmelka, K. T. Mueller, A. Pines, J. Stebbins, Y. Wu, and J. W. Zwanziger, *Nature* (London) *339*: 42 (1989).

40. K. T. Mueller, B. Q. Sun, G. C. Chingas, J. W. Zwanziger, T. Terao, and A. Pines, *J. Magn. Reson. 86*: 470 (1990).

41. K. T. Mueller, Y. Wu, B. F. Chmelka, J. Stebbins, and A. Pines, *J. Am. Chem. Soc. 113*: 32 (1991).
42. Y. Wu, B. F. Chmelka, A. Pines, M. E. Davis, P. J. Grobet, P. A. Jacobs, *Nature* (London) *346*: 550 (1990).
43. P. J. Barrie, M. E. Smith, and J. Klinowski, *Chem. Phys. Lett. 180*: 6 (1991).
44. R. Jelinek, B. F. Chmelka, Y. Wu, P. J. Grandinetti, A. Pines, P. J. Barrie, and J. Klinowski, *J. Am. Chem. Soc. 113*: 4097 (1991).
45. C. A. Fyfe, G. T. Kokotailo, G. J. Kennedy, G. C. Gobbi, C. T. DeSchutter, R. S. Ozubko, and W. J. Murphy, *Proc. Int. Symp. Zeolite 85* (B. Draj, D. Hocevar, and S. Pjenovich, eds.) Elsevier, Amsterdam, 1985, p. 219; G. T. Kokotailo, C. A. Fyfe, G. J. Kennedy, G. C. Gobbi, H. J. Strobl, C. T. Pasztor, G. E. Barlow, S. Bradley, W. J. Murphy, and R. S. Ozubko, *Pure Applied Chem. 58*: 1367 (1986).
46. C. A. Fyfe, G. J. Kennedy, G. T. Kokotailo, and C. T. DeSchutter, *J. Chem. Soc. Chem. Comm.*: 1093 (1984).
47. C. A. Fyfe, G. C. Gobbi, W. J. Murphy, R. S. Ozubko, and D. A. Slack, *Chem. Lett.*: 1547 (1983).
48. C. A. Fyfe, G. C. Gobbi, W. J. Murphy, R. S. Ozubko, and D. A. Slack, *J. Am. Chem. Soc. 106*: 4435 (1984).
49. C. A. Fyfe, G. T. Kokotailo, G. J. Kennedy, and C. T. DeSchutter, *J. Chem. Soc. Chem. Commun.*: 306 (1985).
50. C. A. Fyfe, G. C. Gobbi, J. Klinowski, J. M. Thomas, and S. Ramdas, *Nature* (London) *296*: 530 (1982).
51. C. A. Fyfe, G. J. Kennedy, C. T. DeSchutter, and G. T. Kokotailo, *J. Chem. Soc. Chem. Commun.*: 541 (1984).
52. G. W. West, *Aust. J. Chem. 37*: 455 (1984).
53. C. A. Fyfe, G. J. Kennedy, G. T. Kokotailo, J. R. Lyerla, and W. W. Fleming, *J. Chem. Soc. Chem. Commun.*: 740 (1985).
54. D. G. Hay, H. Jaeger, and G. W. West, *J. Phys. Chem. 89*: 1070 (1985).
55. C. A. Fyfe, J. H. O'Brien, and H. Strobl, *Nature* (London) *363*: 6110 (1987).
56. C. A. Fyfe, H. Strobl, G. T. Kokotailo, G. J. Kennedy, and G. E. Barlow, *J. Am. Chem. Soc. 110*: 3373 (1988).
57. U.S. Patent 3,709,979; G. T. Kokotailo, P. Chu, S. L. Lawton, and W. M. Meier, *Nature* (London) *275*: 119 (1978).
58. a. C. A. Fyfe, G. T. Kokotailo, G. J. Kennedy, and C. T. DeSchutter, *J. Chem. Soc. Chem. Commun.*: 306 (1985); b. C. A. Fyfe, H. Gies, G. T. Kokotailo, C. Pasztor, H. Strobl, and D. E. Cox, *J. Am. Chem. Soc. 111*: 2470 (1989).
59. U.S. Patent 4,287,166.
60. J. L. Schlenker, F. G. Dwyer, E. E. Jenkins, W. J. Rohrbaugh, G. T. Kokotailo, and W. M. Meier, *Nature* (London) *294*: 340 (1981).
61. H. Gies, F. Liebau, and H. Gerke, *Angew. Chem. 94*: 214 (1982); H. Gies, *Z. Kristallogr. 167*: 73 (1984).
62. H. Strobl, C. A. Fyfe, G. T. Kokotailo, and C. T. Pasztor, *J. Am. Chem. Soc. 109*: 7433 (1987).
63. R. Benn, and H. Günther, *Angew. Chem., Int. Ed. Engl. 22*: 250 (1983); A. Bax, *Two Dimensional Nuclear Magnetic Resonance in Liquids*, Delft University Press, Delft, The Netherlands, 1982; A. E. Derome, *Modern NMR Techniques for Chem-*

istry Research, Pergamon Press, Oxford, England, 1987; J. Sanders and B. Hunter, *Modern NMR Spectroscopy, A Guide for Chemists*, Oxford University Press, Oxford, England, 1987.

64. N. M. Szeverenyi, M. J. Sullivan, and G. E. Maciel, *J. Magn. Reson. 47*: 462 (1982).
65. A. Bax, N. M. Szeverenyi, and G. E. Maciel, *J. Magn. Reson. 51*: 400 (1983).
66. S. J. Opella and J. S. Waugh, *J. Chem. Phys. 66*: 4919 (1977). G. Bodenhausen, R. E. Stark, D. J. Ruben, and R. G. Griffin, *Chem. Phys. Lett. 67*: 424 (1979).
67. P. Caravatti, J. A. Deli, G. Bodenhausen, and R. R. Ernst, *J. Am. Chem. Soc. 104*: 5506 (1982).
68. M. H. Frey and S. J. Opella, *J. Am. Chem. Soc. 106*: 4942 (1984).
69. R. Benn, H. Grondey, C. Brevard, and A. Pagelot, *J. Chem. Soc. Chem. Commun.*: 102 (1988).
70. C. A. Fyfe, H. Gies, and Y. Feng, *J. Chem. Soc. Chem. Commun.*: 1240 (1989).
71. C. A. Fyfe, H. Gies, and Y. Feng, *J. Am. Chem. Soc. 111*: 7702 (1989).
72. C. A. Fyfe, H. Gies, Y. Feng, and H. Grondey, *Zeolites 10*: 278 (1990).
73. H. Gies, *Z. Kristallogr. 175*: 93 (1986).
74. C. A. Fyfe, H. Gies, Y. Feng, and G. T. Kokotailo, *Nature* (London) *341*: 223 (1989).
75. C. A. Fyfe, Y. Feng, H. Gies, H. Grondey, and G. T. Kokotailo, *J. Am. Chem. Soc. 112*: 3264 (1990).
76. R. B. LaPierre, *Zeolites 5*: 346 (1985).
77. C. A. Fyfe, H. Gies, G. T. Kokotailo, B. Marler, and D. E. Cox, submitted for publication.
78. L. Muller, A. Kumar, and R. R. Ernst, *J. Chem. Phys. 63*: 5490 (1975).
79. R. K. Harris, and C. T. G. Knight, *J. Chem. Soc. Farad. Trans. II*: 1539 (1983).
80. A. Bax, R. Freeman, T. A. Frenkiel, and M. H. Levitt, *J. Magn. Reson. 43*: 478 (1981); T. H. Mareci and R. Freeman, *J. Magn. Reson. 48*: 158 (1982).
81. H. van Koningsveld, J. C. Jansen, and H. van Bekkum, *Zeolites 10*: 235 (1990).
82. H. van Koningsveld et al., *Acta Cryst. B. 46*: 731 (1990).
83. H. Gies, B. Marler, et al., unpublished results.
84. H. van Koningsveld, F. Tuinstra, H. van Bekkum, and J. C. Jansen, *Acta Crystallogr. B45*: 423 (1989).
85. a. C. A. Fyfe, H. Grondey, Y. Feng, and G. T. Kokotailo, *J. Am. Chem. Soc. 112*: 8812 (1990); b. C. A. Fyfe, H. Grondey, Y. Feng, and G. T. Kokotailo, *Chem. Phys. Lett. 173*: 211 (1990); c. C. A. Fyfe, Y. Feng, H. Grondey, and G. T. Kokotailo, *J. Chem. Soc. Chem. Commun.*: 1224 (1990).
86. C. A. Fyfe, Y. Feng, H. Grondey, G. T. Kokotailo, and A. Mar, *J. Phys. Chem. 55*: 3747 (1991).
87. A. Pines, M. G. Gibby, and J. S. Waugh, *J. Chem. Phys. 59*: 569 (1973).
88. D. M. Doddrell, D. T. Pegg, and M. R. Bendall, *J. Magn. Reson. 48*: 323 (1982).
89. G. A. Morris and R. J. Freeman, *J. Am. Chem. Soc. 101*: 760 (1979).
90. J. C. Edwards and P. D. Ellis, *Magn. Reson. Chem. 28*: S59 (1990).
91. R. Harris and G. J. Nesbitt, *J. Magn. Reson, 78*: 245 (1988).
92. T. H. Walter, G. L. Turner, and E. Oldfield, *J. Magn. Reson. 76*: 106 (1988).
93. C. A. Fyfe, H. Grondey, K. T. Mueller, K. C. Wong-Moon, and T. Markus, *J. Am. Chem. Soc. 114*: 5876 (1992).

94. P. Caravatti, G. Bodenhausen, and R. R. Ernst, *Chem. Phys. Lett. 89*: 363 (1982).
95. E. R. H. van Eck, R. Janssen, W. E. J. R. Maas, and W. S. Veeman, *Chem. Phys. Lett. 174*: 428 (1990).
96. T. Gullion and J. Schaefer, *J. Magn. Reson. 81*: 196 (1989).
97. A. W. Hing, S. Vega, and J. Schaefer, *J. Magn. Reson. 96*: 205 (1992).

2

NMR Studies of Molecular Diffusion

Jörg Kärger and Harry Pfeifer

Leipzig University, Leipzig, Germany

I. INTRODUCTION

Any catalyzed reaction proceeds in parallel with a variety of molecular mass transfer phenomena, so that in many cases the observed activities and product distributions are linked to the diffusional behavior of the reactant, intermediate, and product molecules [1–3]. Knowledge of the intrinsic diffusivities is, therefore, particularly relevant for understanding the overall processes of molecular conversion, as well as for optimization of industrial plants.

Reactions catalyzed by zeolites and other solid catalysts proceed in highly heterogeneous systems. The determination of local diffusivities in such systems by conventional gravimetric or flow methods [3–5] is very complicated and incurs the risk of misinterpretation, since all these methods are sensitive to the response of the whole system rather than to the mobility of individual species within well-defined regions. With reference primarily to diffusion studies in zeolitic adsorbent-adsorbate systems, the present chapter will show how NMR spectroscopy is a most versatile tool for investigating molecular mass transfer phenomena in heterogeneous catalysis. Information may be provided with respect to both microscopic and macroscopic dimensions, involving the observation of molecular distributions as well as the diffusion paths of individual molecules.

II. RECORDING MOLECULAR DISTRIBUTIONS

A. NMR Tracer Exchange

A most straightforward way of applying NMR to study molecular distributions during dynamic processes such as adsorption, counterdiffusion, or chemical reaction is based simply on the NMR signal intensity being directly proportional to the number of resonating nuclei within the sample coil of the NMR probe head.

An example of how this measuring principle can be applied is in the determination of the exchange rate of molecules between an adsorbent and the surrounding atmosphere. An experimental arrangement for such measurements is presented in Fig. 1 [6]: the adsorbent-adsorbate system (in the given case about 200 mg of zeolite ZSM-5 loaded with C_6H_6 at pressure p) is contained in a small glass vial, which in turn is contained in a closed adsorption vessel filled with C_6D_6 at the same pressure p. In addition, the adsorption vessel contains a glass cylinder with a large amount (approximately 5 g) of granulated zeolite NaX acting as a buffer for the gas-phase composition and pressure. The exchange

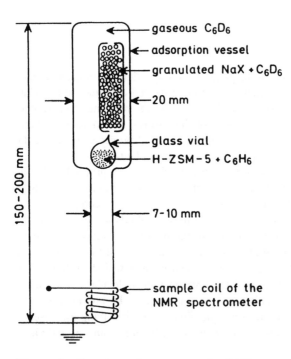

Figure 1 Experimental arrangement for NMR tracer exchange experiments. (From Ref. 6.)

experiment is started by crushing the glass vial, using the glass cylinder as a striker. The zeolite powder falls into the lower end of the vessel, which is part of the probe head of the NMR spectrometer. In this way, from the intensity of the ^{1}H NMR signal (recorded as a free induction decay following a $\pi/2$ pulse), one may directly determine the amount of the protonated species adsorbed, that is, the amount of tracer exchange, as a function of time. A typical pattern of the tracer exchange curves thus obtained is given in Fig. 2, with $\gamma(t)$ denoting the relative amount of C_6H_6 desorbed from the zeolite sample between the beginning of the exchange experiment and time t. Analysis of the exchange curves may be based on the method of statistical moments [7–10]:

$$\tau_{intra} \equiv M_1 = \int_0^\infty (1 - \gamma(t))dt \tag{1}$$

where τ_{intra} (coinciding with the first statistical moment) denotes the intra-crystalline mean lifetime. If molecular exchange is limited by intracrystalline diffusion, for spherical particles the intracrystalline mean lifetime is correlated with the coefficient of intracrystalline self-diffusion D by the formula [7–10]

$$\tau_{intra}^{Diff.} = \frac{<R^2>}{15D} \tag{2}$$

in which $<R^2>$ denotes the mean square crystallite radius. Representations showing the relation between the intracrystalline diffusivity and the mean life-time for various crystallite geometries may be found in Refs. 6–8. For cubes of edge length a, instead of Eq. (2) one has

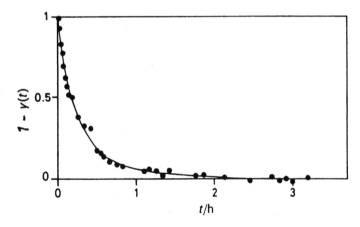

Figure 2 Typical time dependence observed in NMR tracer exchange experiments (benzene on ZSM-5 crystallites of $55 \times 55 \times 90$ μm^3, 4 molecules per unit cell, 386 K. $1 - \gamma(t)$ denotes the relative amount of C_6H_6 still adsorbed at time t. (From Ref. 6.)

$$\tau_{\text{intra}}^{\text{Diff.}} = \frac{a^2}{50D} \tag{3}$$

With a replaced by the radius $R_{\text{eq.}}$ of a sphere of identical volume, Eq. (3) becomes

$$\tau_{\text{intra}}^{\text{Diff.}} = \frac{R_{\text{eq.}}^2}{19.2D} \tag{4}$$

Since for a large variety of samples the shape of the zeolite crystallites tends to be even closer to that of a sphere, in general, Eq. (2) represents a reasonable estimate. It is used again in Sec. II.C., where another (microscopic) method to determine intracrystalline mean lifetimes is introduced.

Figure 3 compares the self-diffusivities determined on the basis of NMR tracer exchange measurements involving benzene in silicalite [6] with the results

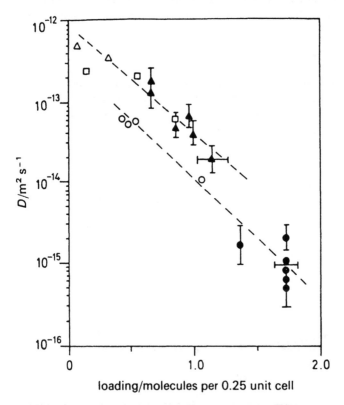

Figure 3 Self-diffusion coefficients of benzene in ZSM-5 determined by the NMR tracer exchange method at 293 K (●) and 386 K (▲), plus comparison with the results of uptake measurements at 303 K (○), 363 K (□), and 393 K (△), transformed into self-diffusivities via Eq. (5). (From Refs. 6 and 11.)

of conventional uptake measurements [11] for the identical specimen. Here the obtained transport diffusivities (D_t) have been transformed into self-diffusivities (D) by assuming the relation [4,5]

$$D_t = \frac{D \, dlnp(c)}{dlnc} \tag{5}$$

where $p(c)$ denotes the adsorbate pressure in equilibrium with concentration c (see Sec. III.B.3.). It turns out that for the given system, on the basis of Eq. (5), there is satisfactory agreement between the tracer exchange (i.e., equilibrium) and the uptake (nonequilibrium) measurements.

The great advantage of this type of measurement is that it may be carried out within a closed vessel. Mass transfer processes may be followed, therefore, over arbitrarily long periods, since any leakage of the system may be excluded. Thus, this method may be of great use for monitoring mass transfer in less mobile systems (in particular for single-file diffusion; see Sec. III.B.6.). Its application is clearly not confined to the observation of the exchange between different isotopes of the same chemical compound. It may be applied as well to study the exchange of different species, namely, counterdiffusion, in addition to chemical reactions, where a distinction between different species may be based on differences in either the nuclei or in the chemical shifts. In these applications the method is equivalent to a technique recently proposed to study transport diffusion and reaction by IR spectroscopy [12]. In addition to these possibilities, however, NMR spectroscopy allows measurement not only of the total concentration of a certain species within a bed of adsorbent particles, but also its spatial distribution, as described in the following section.

B. Spin Mapping by NMR (Dynamic Zeugmatography)

Under application of a space-dependent magnetic field

$$B = B_0 + gz \tag{6}$$

the distribution of NMR resonance frequencies

$$\omega = \gamma B = \gamma(B_0 + gz) \tag{7}$$

is determined by the spatial distribution of the nuclear spins with respect to the z coordinate (defined here as the direction of the applied magnetic field gradient). The NMR spectrum therefore represents the projection of the nuclear spin density on the z coordinate. Two- and three-dimensional images may be produced by combining projections taken for different field gradient orientations.

The resolution of an NMR image, Δz, is determined by the requirement that the gradient-imposed spread in NMR frequencies must dominate the spread in frequencies embodied in the normal line with $\Delta\omega$, according to the relationship

$$\gamma g \Delta z > \Delta\omega \tag{8}$$

In principle, arbitrarily high resolution should be attained by applying sufficiently large gradient intensities g, although the present state of the art of "NMR microscopy" only allows spatial resolution down to 1 μm [13,14]. One has to note, however, that an increase in the spatial resolution is possible only at the expense of measuring time. Hence for the observation of time-dependent concentration profiles, a compromise between time and space resolution is inevitable.

The method of NMR imaging was introduced by Lauterbur [15] in 1973 under the name zeugmatography. In subsequent years the spectacular progress of its application to anatomical imaging in medicine has made popular the name NMR tomography or NMR spin mapping [16,18]. The application of NMR imaging is by far not restricted to biomedicine, of course. An informative survey of the actual developments may be found in Refs. 13 and 14.

As an alternative to x-ray imaging [10], NMR spin mapping has been repeatedly applied to the determination of concentration profiles during the penetration of molecular species into both porous [19–21] and compact [22,23] media. It is just this method that is of particular relevance for determining molecular distributions within beds of catalysts or adsorbents. As an example of the procedure, Fig. 4 shows the distribution of butane molecules within a bed of crystallites of zeolite NaCaA as determined by ^{1}H NMR spin mapping during adsorption [19]. The zeolite crystallites were packed into an NMR sample tube to

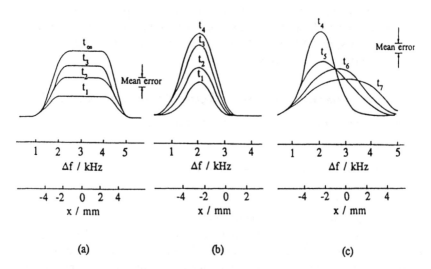

Figure 4 Concentration profiles of *n*-butane in a bed of 5A zeolite crystallites during adsorption from a limited volume. (a) Large crystallites (40 μm diameter) at times $t_1 = 3.5$ s, $t_2 = 10.5$ s, $t_3 = 35$ s, and at equilibrium (t_∞); (b,c) small crystallites (4 μm diameter) at times $t_1 = 4.5$ s, $t_2 = 9$ s, $t_3 = 50$ s, $t_4 = 3$ min, $t_5 = 20$ min, $t_6 = 50$ min, $t_7 = 85$ min. (From Ref. 19.)

a height of about 10 mm. After activation, the upper face was exposed to a butane atmosphere stemming from a reservoir of limited volume. Quite different uptake patterns are observed for the two samples considered: in the sample with the larger crystallites (Fig. 4a) the butane molecules are distributed over the whole bed nearly immediately after the onset of the adsorption process, whereas with the small crystallites the butane molecules at first are adsorbed solely within the upper layer, consuming virtually all the butane of the reservoir (Fig. 4b). Only in a second, much slower process is butane distributed over the whole bed (Fig. 4c). This difference may be easily attributed to the fact that butane molecules penetrating into the intercrystalline space are more readily swallowed by the small crystallites as a consequence of the larger external surface per bed volume and the enhanced tortuosity. Since the rate constants of the observed processes required measurements within seconds, the spatial resolution was confined to the mm region.

It is remarkable that even with these measurements in the early days of zeugmatography, notwithstanding the poor spatial resolution, an unambiguous discrimination between the two limiting cases of intracrystalline-controlled (Fig. 4a) and intercrystalline-controlled (Fig. 4b,c) transport was possible. From both a fundamental and practical point of view it would now be interesting to measure concentration profiles directly within the individual crystallites (see also Sec. III.B.3). In light of the most recent progress in both NMR microscopy [13,14] and zeolite synthesis of crystallites not far below the mm region [24–27], such research projects seem to be not unrealistic.

C. ^{129}Xe NMR Line Shift Measurements

In the last decade, xenon has proven to be an efficient sorbate for probing the pore structure and the internal surface of adsorbents by NMR spectroscopy [28–30]. The advantage of xenon in comparison with other adsorbates is brought about by the large chemical shifts of ^{129}Xe NMR as a consequence of the large electron shell and by the fact that xenon as a noble gas leaves the adsorbent structure essentially unaffected. In particular, in zeolite research ^{129}Xe NMR has been successfully applied to probe pore and channel dimensions [31], cation distributions [32], cation sites [33], and cation mobilities [34,35] as well as matter depositions and lattice defects [36–38].

Since the chemical shift of ^{129}Xe also depends on the nature and the concentration of molecules adsorbed in addition to xenon within the zeolite structure [30,39], ^{129}Xe NMR may be applied as well to map the sorbate concentration within adsorbent-adsorbate systems [40–43]. The principle of this method is illustrated in Fig. 5, which shows the ^{129}Xe NMR spectra and the corresponding distribution of the adsorbate under study (hexamethylbenzene) within a bed of zeolite NaY [42]. The spectrum shown in Fig. 5a was measured after a finite amount of liquid hexamethylbenzene was introduced into the sample, which

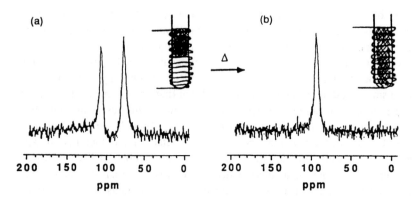

Figure 5 [129]Xe NMR spectra of xenon adsorbed on dehydrated zeolite NaY measured at room temperature and at an equilibrium pressure of 40 kPa, in which hexamethylbenzene, corresponding to an amount of 0.5 molecules per supercage, has been introduced from the upper face (a) after the sample has been heated for 2 h at 523 K and (b) after reheating for 4 h at 573 K. The shading in the diagrams of the samples reflects the concentration of the adsorbed hexamethylbenzene. (From Ref. 42.)

subsequently was heated up to 150°C and kept there for 2 hours. The existence of two distinct lines in the spectrum indicates that the adsorbate has been swallowed by the upper layer of the bed, while the lower part is still free of the adsorbate. Only after reheating the sample over 4 hours at a still higher temperature (200°C) is the adsorbate homogeneously distributed over the whole sample, leading to just one line at an intermediate position (Fig. 5b). The situation shown in Fig. 5a and b is equivalent to that of Fig. 4b and c, with the only difference being that at room temperature the gas phase pressure of hexamethylbenzene is so low that equilibration of the adsorbate over the sample would not be accomplished during a finite time interval.

In contrast to the methods of Secs. II.A. and II.B., which directly monitor the spatial dependence of concentration, [129]Xe NMR line shift measurements only provide information about the probability distribution of the sorbate concentrations within the sample. Thus the spectrum of Fig. 5a per se indicates merely that xenon is adsorbed within two different surroundings. In view of the situation, the only reasonable physical interpretation is represented by the indicated sorbate distribution. As a consequence of this difference, [129]Xe NMR is in fact able to monitor concentration differences over separations in the nm region. This possibility has been used in Ref. 44 for determining the distribution function of xenon encapsulated in the individual cavities of zeolite NaA.

Measurements of this type, however, are only possible if the mean xenon exchange time, τ_{12}, between the relevant regions is sufficiently large compared to the reciprocal of the difference of the corresponding Larmor frequencies ($\Delta\omega$):

$$\Delta\omega\tau_{12} > 2\pi \tag{9}$$

Estimating the mean exchange time between two positions at separation Δz by using Einstein's relation

$$\tau_{12} \approx \frac{(\Delta z)^2}{2D_{Xe}} \tag{10}$$

with D_{Xe} denoting the xenon diffusivity, we find from Eq. (9) that the requirement for space resolution is

$$\Delta z > (4\pi D_{Xe}/\Delta\omega)^{1/2} \tag{11}$$

For xenon atoms encapsulated in the individual cavities of zeolite NaA [44], τ_{intra} is equal to infinity and relation (9) is clearly fulfilled. In general, however, when used as a probe particle, xenon will not remain at a fixed position, and so relation (11) must be applied to estimate the range over which discrimination of concentration differences of the adsorbate molecules is possible.

In a first study using this principle for the measurement of *intra*crystalline concentration profiles, the adsorption/desorption process of benzene on zeolite ZSM-5 [45] has been considered. With $\Delta\omega \approx 5 \cdot 10^3$ s^{-1} [45] and $D_{Xe} \approx 10^{-11}$ m^{2}s^{-1} [46] (taking into account also that the xenon diffusivity is reduced by the presence of the benzene molecules [47]), relation (11) yields a lower limit of 0.2 μm for the spatial resolution. Thus, by applying crystallites with a mean radius of the order of 30 μm, in fact a determination of intracrystalline concentration profiles is possible. The experiments were carried out by simultaneously observing adsorption and desorption within one bed of crystallites. For this purpose the zeolite material was prepared in two fractions of identical quantity: one portion (fraction A) as a loose assemblage within the sealed NMR sample tube, loaded with benzene and xenon, and a second portion (fraction B) of unloaded zeolite contained in a glass vial, which was also located in the NMR sample tube together with a striker. The adsorption/desorption process was initiated by crushing the glass vial and vigorously shaking the sample in order to obtain an intimate mixture of the loaded and unloaded crystallites. It could be shown that both the mixing of the crystallites and the distribution of the xenon over the sample was accomplished within a time interval much less than the time constant of adsorption and desorption on the individual crystallites.

Figure 6 provides a comparison between measured spectra and theoretical spectra calculated under the assumption that the adsorption/desorption process is controlled by either intracrystalline diffusion (Fig. 6a) or external transport resistances such as surface barriers (Fig. 6b). For simplicity in the calculations, the crystallites have been assumed to be of nearly spherical shape with a concentration-independent transport diffusivity D_T or surface permeability α, respectively. Values of the intracrystalline mean lifetime are therefore given by

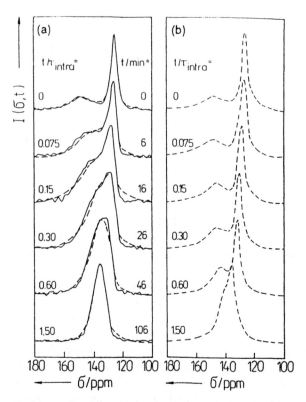

Figure 6 ¹²⁹Xe NMR spectra measured at room temperature during the adsorption/
desorption of benzene in a bed of ZSM-5 type crystallites with an initial concentration of 0
and 6 molecules per unit cell, respectively, and a mean xenon concentration of 16 atoms
per unit cell (a, full lines). Also, comparison with the simulated spectra (broken lines) for
the limiting cases of (a) diffusion-controlled and (b) barrier-controlled processes. (From
Ref. 45.)

Eq. (2) (with the self-diffusivity D replaced by the transport diffusivity D_T; see
Sec. III.B.3.) in the case of diffusion control, or by the relation [5]

$$\tau_{intra}^{Barr.} = \frac{R}{3\alpha} \qquad (12)$$

in the case of surface barriers. It turns out that the nature of the controlling
mechanism is distinctly reflected in the evolution of the spectra. Corresponding
to the fact that for barrier-controlled adsorption/desorption the intracrystalline
concentration within the sample will assume only two values, the distinction
between the two lines is preserved over nearly the whole process (Fig. 6b), while
for diffusion-controlled adsorption/desorption the wide range of intracrystalline

concentrations (brought about by the intracrystalline concentration gradients) leads to a rapid coalescence of the two lines.

The satisfactory agreement between the measured and calculated spectra in Fig. 6a indicates that the observed process of molecular exchange between the loaded and unloaded zeolite crystallites is controlled by intracrystalline diffusion. A comparison between the parameters t/τ_{intra} used in the calculations and the time t between the onset of the exchange process and the measurement of the spectra yields a transport diffusivity of the order of 10^{-14} m^{2}s^{-1}, which agrees satisfactorily with results of previous uptake measurements [10,49,50]. Moreover, analysis of the ^{129}Xe NMR spectra shows unambiguously that the adsorption/desorption process is controlled by intracrystalline diffusion, so that the values obtained are in fact real intracrystalline diffusivities. However, since the proposed method applies only under the assumptions that (1) the time constant of the observed process is larger than the time necessary for an intimate mixing of the adsorbing and desorbing crystallites and (2) the size of the crystallites is sufficiently large compared to the spatial resolution, its application is limited to systems with low molecular mobilities and large crystallites.

III. FOLLOWING THE DIFFUSION PATH OF INDIVIDUAL MOLECULES

A. Elementary Steps of Diffusion

The displacements measured by the PFG NMR method (see Secs. III.B. through III.E.) are much larger than the r.m.s. lengths of the elementary jumps of translational diffusion. While the probability distribution of molecular "large-scale" migration is given by a Gaussian function with a single parameter D (the coefficient of self-diffusion), the elementary steps of molecular motion generally must be characterized by more complicated relations since they also include molecular reorientations, which leave the molecule's center of gravity unchanged. It is not the aim of this review, however, to discuss NMR methods that yield information about the latter type of motion. We restrict ourselves instead to a determination of that quantity of the elementary steps that is directly connected with translational diffusion, namely, the mean residence time τ_t between two succeeding translational jumps of the molecule. Together with the coefficient of self-diffusion D measured by the PFG NMR method, it is then possible to determine the mean square jump length $<l^2>$ of the translational diffusion by inserting both quantities in the relation

$$<l^2> = 6D\tau_t \tag{13}$$

In studies of the adsorbed molecules, τ_t has been measured in most cases either by analysis of the proton magnetic relaxation or by analysis of the ^{2}H NMR line shape and relaxation behavior of deuterated molecules.

1. Proton Magnetic Relaxation Analysis

Since most molecules of interest in adsorption phenomena and in heterogeneous catalysis contain hydrogen atoms, and since the signal-to-noise ratio, which is of decisive importance in these studies [51], is maximum for the hydrogen nucleus, only proton magnetic relaxation shall be treated here. Relaxation of deuterium nuclei is included in the following section. In the case of adsorbed molecules, proton magnetic relaxation is mainly determined by magnetic dipole interactions with protons of the same molecule (intramolecular proton-proton coupling), with protons of other molecules (intermolecular proton-proton coupling), and with paramagnetic ions. These contributions can be separated by a relaxation analysis [52], an example of which is given in Ref. [53]. Except for adsorbents of extremely high purity (with paramagnetic impurities, mostly iron, less than 10 ppm) and in any case for commercial adsorbents and catalysts, longitudinal relaxation is controlled by magnetic dipole interaction with paramagnetic ions. This situation has advantages for a determination of τ_t, since here proton magnetic relaxation is affected by the translational motion of the molecules with respect to the paramagnetic ions, which are fixed in space. However, two conditions must be fulfilled: (1) the longitudinal electron spin relaxation time τ_1 must be much greater than τ_t, and (2) the paramagnetic ions must be located in sites where they do not affect the mobility of the adsorbed molecules (for example, surrounded by oxygen atoms of the adsorbent similar to the aluminum or silicon ions in the framework of zeolites). An example where the condition $\tau_1 \gg \tau_t$, which is often fulfilled, has been reversed to $\tau_1 \ll \tau_t$ by a reduction of the value of τ_1 can be found in Ref. [54]. The second condition is also valid relatively often, even for small polar molecules such as water [55].

In the following we shall discuss only the longitudinal (thermal) relaxation time T_1, there being an obvious extension to the other time constants of thermal relaxation like T_{1p}, T_{1d} and so forth (see Refs. 52 and 56). Neither shall we treat the transverse relaxation time T_2 or its reciprocal, which is the line width of the stationary NMR signal. The reason for this choice is that T_2 decreases with increasing value of τ_t so that even a few strongly adsorbed molecules (relaxation time T_{2s}), whose number N_s may be much less than that of the physically adsorbed molecules (N_p), will determine the mean relaxation time T_2 if only T_{2s} is smaller than $T_{2p}N_p/N_s$. In this case it follows that

$$T_2 = \frac{(N_s + N_p)}{N_s \cdot T_{2s}} \tag{14}$$

The quantity $1/T_{2s}$ is given by the product of τ_{ts}, the mean residence time of a strongly adsorbed molecule, with its mean square magnetic dipole energy E_{ms}. Accordingly, an observed change in T_2 may be due to changes in either the (small) concentration of the sites of high adsorption energy or in the arrangement of the hydrogen atoms around these sites (E_{ms}) or else in the microdynamical

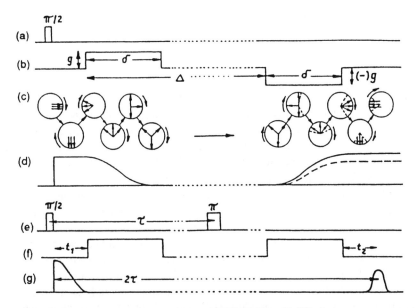

Figure 7 Schematic representation of PFG NMR self-diffusion measurements (a–d) and their application in the spin-echo technique (e–g). Broken lines in (c) and (d) indicate the behavior with molecular migration. (a,e) r.f. pulses, (b,f) field gradient pulses, (c) transverse magnetization of different regions, (d,g) total transverse magnetization, equal to the vector sum of (c). (From Ref. 73.)

If the z coordinates of the individual spins at the time of the second field gradient pulse differ from those during the first, then the dephasing of the first field gradient is not completely compensated. This leads to a decrease in the signal intensity, which is indicated in Fig. 7c and d by the broken lines. It is this signal attenuation that is analyzed in the NMR diffusion measurements.

For a quantitative treatment we have to consider that the spins of atoms that have been at positions z_0 and $z_0 + z$ during the first and second field gradient pulses suffer a net phase shift $\gamma\delta gz$, where the z coordinate has been assumed to be determined by the direction of the pulsed field gradients. Hence, summing over the contribution of all spins, one may write for the signal attenuation under the influence of the pulsed field gradients

$$\Psi(\delta g,\Delta) = \int \int \exp(i\gamma\delta gz)p(z_0)P(z_0 + z,z_0,\Delta)dz_0dz \tag{23}$$

in which $P(z_0 + z,z_0,\Delta)$ is the conditional probability density (the "propagator") that a spin will migrate from z_0 to $z_0 + z$ during the time interval Δ between the first and second field gradient pulse, and where $p(z_0)$ is the initial spin density function. Introducing the medium propagator

$$\bar{P}(z,\Delta) = \int p(z_0)P(z_0 + z,z_0,\Delta)dz_0, \tag{24}$$

we may write Eq. (23) as

$$\Psi(\delta g,\Delta) = \int \exp(i\gamma\delta zg)\overline{P}(z,\Delta)dz \tag{25}$$

with the Fourier transform

$$\overline{P}(z,t) = 1/2\pi \cdot \int \exp(-i\gamma\delta gz)\Psi(\delta g,\Delta)d(\gamma\delta g) \tag{26}$$

The propagator contains the maximum information attainable about the stochastic process of molecular migration. Equation (26) demonstrates that the mean of this function can be directly determined by a Fourier transform of the NMR spin echo attenuation with respect to the intensity δg of the applied pulsed field gradients [74,75]. As an example, Fig. 8 shows the propagator of ethane in zeolite NaCaA [74] for two different crystallite sizes at three different temperatures. For a given system and given temperature and observation time (i.e., the time interval Δ between the two field gradient pulses), the observed mean propagator may be determined exclusively by the rates of molecular redistribution within the individual crystallites (intracrystalline self-diffusion; see, for example, Fig. 8a at 153 K) or by migration through the bed of crystallites (long-range self-diffusion; Fig. 8b at 293 K). In the intermediate range between these two limiting cases, the propagator additionally depends on the rate of molecular exchange between the crystallites and the intercrystalline space. The specific information on mass transfer provided by PFG NMR in these three ranges of observation will be reviewed in Secs. III.B. through III.D.

An interesting feature of PFG NMR as represented by Eq. (25) is its analogy with diffraction [76–79]. Following this analogy, from the phase factor $\exp(i\gamma\delta gz)$ one may deduce as an equivalent of the wavevector in x-ray defraction the quantity [77]

$$q = \gamma\delta g \tag{27}$$

Under the assumption that molecular propagation is confined to microscopic subregions (e.g., pores) in the long-time limit the positions of the molecules at the application of the second field gradient pulse are uncorrelated with their starting points, being described by the same distribution function. Therefore it follows that

$$\lim_{\Delta \to \infty} P(z_0 + z, z_0, \Delta) = p(z_0 + z) \tag{28}$$

and the medium propagator [Eq. (24)] becomes

$$\lim_{\Delta \to \infty} P(z,\Delta) = \int p(z_0) \cdot p(z_0 + z)dz_0 \tag{29}$$

where the integration is to be carried out over the confining volume. Thus the mean propagator is exclusively determined by the shape and the spin distribution

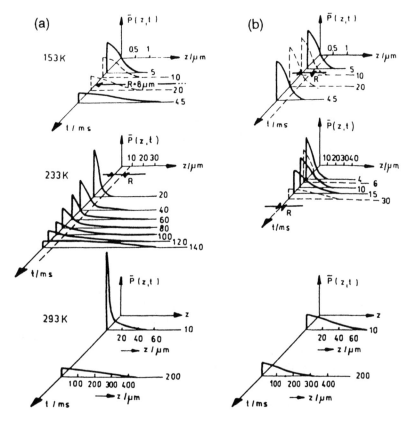

Figure 8 Representation of the mean propagator of methane in zeolite NaCaA = (a) 40 mgg⁻¹, R ≈ 8 μm; (b) 58 mgg⁻¹, R ≈ 0.4 μm. (From Ref. 74.)

function of the confining volume. A simple Fourier transform of the PFG NMR signal attenuation with respect to the "wavevector" $q = \gamma \delta g$ then yields information about the distribution of the spins within the confining regions (the spin density autocorrelation function [75]), equivalent to that of conventional NMR imaging (see Sec. II.B.). For comparable observation times, however, a higher resolution is attained since it is the whole sample rather than a single layer that is considered in this experiment [74–77]. Consequently the time-invarient mean propagator as observed, for example, for ethane in NaCaA (Fig. 8b, 153 K) under conditions of restricted diffusion (namely, for such low temperatures that the molecules are essentially unable to escape from the intracrystalline space into the surrounding atmosphere) is in fact nothing other than the autocorrelation function of the spin distribution within the zeolite crystallites as defined by Eq. (29). In fact, this type of "microimaging" has been repeatedly used in PFG

NMR studies with zeolites and has provided evidence that there are no substantial transport resistances in the interior of the zeolite crystallites [5,73].

It follows from Eq. (25) that any periodicity (l) in the propagator should give rise to a "coherence peak," that is, to a local maximum for $q = \gamma \delta g = 2\pi/l$ in the representation of the NMR spin echo intensity versus the intensity of the field gradient pulses ($\gamma \delta g$). Such behavior has indeed been observed in recent PFG NMR studies of water diffusion through the free space within an array of loosely packed monodisperse polysterene spheres [78].

In general, however, the mean propagator may be assumed to be given by a simple Gaussian function

$$\overline{P}(z,t) = (4\pi Dt)^{-1/2} \exp[-z^2/(4Dt)] \tag{30}$$

with an effective diffusivity D defined by the relation between the mean square displacement and the observation time t via Einstein's equation

$$<z^2(t)> = 2Dt \tag{31}$$

This is true in particular for diffusion in homogeneous (or quasihomogeneous) systems, comprising the limiting cases of intracrystalline (for $<r^2(t)>^{1/2} \ll R$) and long-range (for $<r^2(t)>^{1/2} \gg R$) diffusion in assemblages of zeolite crystallites. In these cases, Eq. (30) simply follows as the solution of Fick's second law for an infinite medium with an initial probability distribution given by Dirac's delta function. With Eq. (30) inserted into Eq. (25), the spin echo attenuation becomes

$$\Psi(\delta g, \Delta) = \exp(-\gamma^2 \delta^2 g^2 D \Delta) \tag{32}$$

The diffusivity D and [by using Eq. (31)] also the mean square displacement $<z^2(\Delta)>$ then may be determined straightforwardly from the slope of a logarithmic plot of the spin echo amplitude versus the squared gradient intensity.

Equation (23), and hence all the following equations, have been derived under the assumption that the molecular mean displacement during the field gradient pulses is negligible compared to the displacement during the observation time between the two field gradient pulses. Only under this condition is it justifiable to speak of the "position of a spin during a field gradient pulse." In the case of ordinary diffusion as characterized by the Gaussian propagator [Eq. (30)], the echo attenuation may be calculated without this assumption [70–73], leading to

$$\Psi = \exp[-\gamma^2 \delta^2 g^2 D(\Delta - \delta/3)] \tag{33}$$

Comparison with Eq. (32) reveals that the application of the "short-pulse approximation" as implied in this equation is justified under the condition $\delta/3 \ll \Delta$, which is in fact fulfilled in most PFG NMR measurements.

2. Limitations and Modifications of the PFG NMR Method

Because of their relatively large magnetogyric ratio and their nearly 100% natural abundance, protons are the best candidates for NMR self-diffusion studies. The most decisive parameter limiting the range of applicability of PFG NMR is the transverse nuclear magnetic relaxation time, T_2. Since in PFG NMR measurements one records the attenuation of the transverse nuclear magnetization brought about by diffusion in the time interval between the two field gradient pulses, this quantity cannot be made much larger than T_2.

If the longitudinal relaxation time, T_1, significantly exceeds T_2, then the range of observation times may be enhanced by the use of the stimulated echo ($\pi/2$-τ_1-$\pi/2$-τ_2-$\pi/2$-τ_1-stimulated echo) with field gradient pulses applied between the first and second $\pi/2$ pulse, and between the third $\pi/2$ pulse and the echo, respectively [70–73,80]. During this sequence, signal attenuation between the second and third rf pulse is determined by longitudinal rather than transverse relaxation. Since during the two other time intervals (τ_1) the decay of the NMR signal is determined by the transverse relaxation, however, T_2 must exceed a certain minimum value (0.2–0.5 ms). This condition follows because δ cannot be made arbitrarily small if the field gradients are to effect a measurable attenuation of the spin echo.

Typical values to be inserted into Eq. (32) are $g_{max} \approx 20$ Tm^{-1} for the maximum field gradient amplitude and $\delta \approx 5$ ms for the pulse width. Recognizing that an unambiguous determination of signal attenuation from the adsorbate bulk phase requires the exponent to be on the order of 1, we find with Eq. (31) that the minimum displacement necessary for observation by PFG NMR is on the order of 50 nm. Such small displacements have in fact been observed by polymers [81–83], where because of the high internal flexibility of the macromolecules a low translational mobility may occur simultaneously with sufficiently large transverse relaxation times (allowing, in particular, the application of gradient pulse widths of the above-indicated order of magnitude). In zeolites and other adsorbent-adsorbate systems, a drop in the mobilities is generally accompanied by a drop in the value of the transverse nuclear magnetic relaxation time (see Sec. III.A.1.). For those cases in particular where low mobility would necessitate large gradient intensities, the application of large pulse widths is prohibited by small values of T_2. A realistic lower limit for the molecular displacements in zeolites is therefore approximately 1 μm. For PFG NMR measurements with nuclei other than protons, according to Eqs. (31) and (32) this value must be multiplied by γ_p/γ, which, for example, in the case of ^{13}C or ^{129}Xe PFG NMR leads to an additional factor of about 4. However, this disadvantage may be partially compensated by the longer relaxation times for these nuclei [52]. For minimum displacements of 1 μm and an upper limit of about 100 ms for the observation time, the minimum self-diffusion coefficients for zeolites, still accessible by PFG NMR, are on the order of 10^{-12} $m^2 s^{-1}$. It should be emphasized, though, that such low values may only be attained under appropriate conditions, namely, for sufficiently long values of T_2 at low mobilities, as observed, for instance, for the *n*-alkanes in zeolite NaCaA in the temperature region between 100 and 300°C [84].

Attaining enhanced observation times and hence lower diffusivities is a major issue in the development of the PFG NMR measuring technique, and a number of modifications have been proposed for this purpose [85–92]. In general, however, these variants appear to be of substantial benefit only under rather special conditions, and, accordingly, most NMR diffusion measurements in adsorbent-adsorbate systems have been carried out either by the standard spin echo or by the stimulated echo sequence. It should be pointed out that attempts to enhance the observation time by using large numbers of accumulation incurs the risk that only a trifling amount of highly mobile molecules will be detected at the expense of everything else. The more mobile species, by virtue of their longer relaxation times, contribute more and more to the spin echo and do not reflect the diffusivity of the adsorbate bulk phase [93,94]. This source of error may be eliminated by starting the PFG NMR pulse sequence with the second half of the spin echo produced by a preparatory pulse sequence rather than with a $\pi/2$ pulse. If the measured diffusivities are in fact those of the adsorbate bulk phase, any change in the spacings within the preparatory pulse sequence must be without influence on the echo attenuation with increasing pulsed field gradient intensity [95,96].

An unambiguous measurement of intracrystalline self-diffusion requires that the molecular mean displacements during the observation time be sufficiently small in comparison to the crystallite dimensions. For attaining sufficiently short observation times one may apply a sequence of identical pairs of field gradients with alternating signs. In this case, the signal attenuation is the *n*th power of the attenuation given by the Eq. (32), with n denoting the number of applied pairs. Thus, echo attenuations due to very small molecular displacements, which with only two gradient pulses would be otherwise unobservable, may now become visible [97].

A most stringent confinement of the intensity of the field gradient pulses is demanded by the fact that any mismatch $d(g\delta) = g_2 \delta_2 - g_1 \delta_1$ between the two field gradient pulses leads to an additional damping Ψ_{add} of the spin echo amplitude, which is not due to diffusion, although it could be misinterpreted as such. According to the reasoning given in Sec. III.B.1., this effect may be intuitively understood as a consequence of an incomplete phase compensation of the precessing spins after the second field gradient pulse. A quantitative analysis yields for the additional attenuation of the spin echo [73,98,99]

$$\Psi_{add} = \frac{\sin[\gamma(l/2)d(\delta g)\sin\alpha]}{\gamma(l/2)d(\delta g)\sin\alpha} \tag{34}$$

if the constant field gradient is parallel to the sample axis and [99]

$$\Psi_{add} = \frac{2J_1[\gamma Rd(\delta g)\sin\alpha]}{\gamma Rd(\delta g)\sin\alpha} \tag{35}$$

if it is perpendicular. J_1 denotes the first-order Bessel function, l the sample length, R the sample radius, and α the angle between the directions of the pulsed field gradients and a constant field gradient. It follows from Eqs. (34) and (35) that this additional attenuation may be reduced by deliberately applying large constant field gradients parallel to the pulsed ones. Methods to overcome the experimental difficulties brought about by the influence of a mismatch of the field gradient pulses are described in Refs. 100 and 101.

Under the influence of a constant field gradient g_0 in addition to the pulsed ones, the spin echo attenuation in the PFG NMR experiments is given by the relation

$$\Psi = \exp\{-\gamma^2 D[2/3 \cdot \tau^3 g_0^2 + \delta^2(\Delta - \delta/3)g^2 - \delta(t_1^2 + t_2^2 + \delta(t_1 + t_2) + 2/3 \cdot \delta^2 - 2\tau^2)gg_0]\} \tag{36}$$

with t_1 and t_2 denoting, respectively, the time interval between the $\pi/2$ pulse and the onset of the first field gradient pulse and between the end of the second gradient pulse and the maximum of the spin echo. Equation (36) reduces to Eq. (33) under the condition $g\delta \gg g_0\tau$, and so for sufficiently intense field gradient pulses the constant field gradient is found to have no influence on the spin echo attenuation.

In nonuniform samples such as adsorbent-adsorbate systems, heterogeneity in the magnetic suceptibility produces internal field gradients which, according to Eq. (36), may interfere with the influence of the external field gradients [102, 103]. To circumvent such a complication, special pulse sequences have been proposed to suppress the disturbing influence of the internal field gradients [104–107]. For measurement of zeolitic intracrystalline diffusivities, it can be shown by both direct experimental evidence and quantitative consideration [108] that— at least for proton resonance frequencies below 100 MHz—the influence of internal field gradients may be neglected. As an example, Fig. 9 displays the results of ¹H PFG NMR self-diffusion measurements of *n*-hexane in zeolite NaX with varying crystallite sizes for two different resonance frequencies. The internal field gradients, being different for different crystallite diameters and field intensities, do not affect the data obtained.

3. Comparison with the Results of Other Techniques

PFG NMR is a method to study the rate of molecular redistribution under equilibrium conditions. For this purpose, prior to the measurements the samples are subjected to a reproducible pretreatment consisting of sample activation under vacuum (ca. 10^{-2} Pa) at elevated temperatures (ca. 400°C) and a subse-

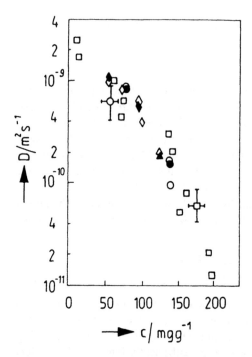

Figure 9 Concentration dependence of the self-diffusion coefficient of *n*-hexane in zeolite NaX with mean crystallite diameters of 55 μm (○), 20 μm(□), 15 μm (◇), and 4 μm (△) at 293 K. The proton resonance frequencies were 60 MHz (open symbols) and 16.6 MHz (full symbols), corresponding to external magnetic fields of 1.41 and 0.39 T, respectively. The error bars indicate the uncertainty in the diffusivities and concentrations. (From Ref. 108.)

quent introduction of the adsorbate, generally by freezing from a well-defined gas volume by means of liquid nitrogen. Afterwards the sample tubes are sealed. By repeating the measurements it may be easily checked whether equilibrium within the sample has in fact been attained. It is interesting to note that in a series of systems (e.g., water [52,109,110] and *n*-alkanes [61,108] in NaX) the data were found to be completely reproducible when measurements on the identical samples were repeated 10 years later.

In conventional sorption experiments it is the process of adsorption or desorption itself that is considered in determining the transport behavior of the adsorbate. Referring as they do to different physical situations (namely to equilibrium and nonequilibrium states), the coefficients of self-diffusion and transport cannot be expected to be identical. However, for various microdynamic models [4,5, 111–114] the relation between the two coefficients is found to be given by Eq. (5),

$$D_t = \frac{D \, dlnp(c)}{dlnc}$$

where c denotes the adsorbate concentration in equilibrium with the adsorbate pressure $p(c)$, and D_t is the transport diffusion coefficient. For sufficiently small amounts adsorbed, the sorbate concentration is proportional to the sorbate pressure (Henry region) so that according to Eq. (5) both coefficients should coincide. This result may be easily understood since, according to Fick's first law, the transport diffusion coefficient is the factor of proportionality between the gradient of intracrystalline concentration and the resulting adsorbate flux density. The coefficient of self-diffusion, on the other hand, is the factor of proportionality between the concentration gradient of labeled molecules and their flux density within an unlabeled surrounding; that is, under equilibrium conditions. If there is no mutual interaction between the adsorbate molecules, the flux density of the labeled molecules is clearly unaffected by the existence of the other (unlabeled) molecules, so that the coefficients of transport diffusion and self-diffusion must coincide. Deviations from this prediction are possible, however, if changes in the sorbate concentration are accompanied by structural changes in the adsorbent.

It was a remarkable result of the early PFG NMR measurements [52,115] that intracrystalline diffusivities were found to be up to five orders of magnitude larger than the values obtained from classical sorption rate measurements under apparently similar conditions. This observation led to a critical reexamination of the earlier sorption data, which revealed that, at least for some systems, the effect of external heat- and mass-transfer resistances in limiting the adsorption/ desorption rates was far greater than originally assumed [4,5,116–119]. This is critical in particular for highly mobile systems, since with increasing uptake rates the influence of processes different from intracrystalline diffusion is continuously increasing. Hence for the less mobile systems (see, for example, the data for benzene in ZSM-5 in Fig. 3), generally satisfactory agreement between uptake and self-diffusion data may be expected. However, self-diffusion measurements by the PFG NMR method are only possible for sufficiently mobile systems (see Sec. III.B.2.).

It is shown in Table 2 that, depending on the system under study, actual sorption measurements both agree and disagree with the PFG NMR data. An explanation of this situation is complicated by the fact that (as in the case of benzene in NaX) there is still some divergence in the results of different research groups obtained with apparently identical sorption methods and samples.

Table 2 includes as well the results of recent diffusion studies by molecular dynamics (MD) calculations and by quasielastic neutron scattering. Both methods consider the process of self-diffusion and reflect the transport properties over diffusion paths of typically a few nanometers. In view of the satisfactory agreement with the PFG NMR data, there should be no doubt that genuine

Table 2 Comparison of PFG NMR Diffusivities with the Results of Quasielastic Neutron Scattering and MD Simulations and of Macroscopic Nonequilibrium ("Sorption") Measurements[a]

Adsorbent	Adsorbate	T/K	PFG NMR	Neutron scattering	Molecular dynamics	Sorption techniques
NaCaA	methane	300	$4^{84,120}$	6^{121}		
	ethane	250	$0.6^{84,120}$			0.3^{122}
	propane	273	$0.003^{84,123}$			0.002^{123}
	butane	400	$0.008^{84,123}$			0.009^{124}
NaX	ethane	250	80^{61}	50^{125}		
	neopentane	187	0.2^{126}			0.3^{126}
	ethene	300	150^{127}	30^{128}		
	benzene	458	$8^{123,130}$	7^{129}		$3^{130};0.08^{131}$
HZSM-5	methane	250	70^{132}	50^{133}	50^{134}	1^{135}
	xenon	293	40^{46}		$20^{136};40^{137}$	0.8^{138}

[a]In the case of H-ZSM-5, the values represented are mean diffusivities $<D> = (D_x + D_y + D_z)/3$ (see Sec. III.B.5.). Data generally refer to the limit of small sorbate concentrations. The diffusivities are given in 10^{-10} m^{2}s^{-1}.

information about the rate of intracrystalline molecular transport is provided in this way. Moreover, it can be demonstrated that the NMR diffusivities are equally relevant for microscopic and macroscopic dimensions. This conclusion is reached by following molecular diffusion paths up to hundreds of microns [46,139] and by comparing the rate of intercrystalline molecular exchange (as determined by the method described in Sec. II.A.) with the intracrystalline diffusivities measured by PFG NMR within the identical sample [140,141]. We may exclude the possibility, therefore, that remaining differences between the PFG NMR data and the results of the sorption experiments (see Table 2) are due to differences in the scale of observation. A final clarification of the origin of the differences between the equilibrium and nonequilibrium data of molecular mass transfer in zeolites is one of the challenging problems in actual zeolite research.

4. Patterns of Concentration Dependence

The variety of diffusion mechanisms involved in intracrystalline molecular mass transfer is most vividly reflected in the different patterns of the concentration dependence of intracrystalline self-diffusion. A classification of the various concentration dependences so far observed by PFG NMR is presented in Fig. 10.

A monotonic decay with increasing concentration over the whole concentration range (type I behavior) is observed for the self-diffusion of alkanes in the wide and medium pore zeolites NaX (see, for instance, Fig. 9) and ZSM-5, respectively. For zeolite NaX, from analysis of the temperature dependence of the longitudinal proton magnetic relaxation times (see Sec. III.A.1.), the mean

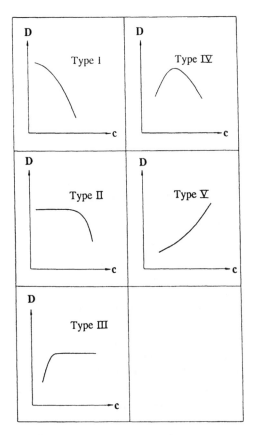

Figure 10   The different patterns of concentration dependence for intracrystalline self-diffusion. (From Refs. 139 and 145.)

lifetime τ_l between succeeding jumps was found to be independent of concentration. According to Eq. (13), the observed decrease in the diffusivities with increasing concentration must therefore be attributed to a decrease in the r.m.s. molecular jump lengths $<l^2>$. It is found that the decrease in the r.m.s. jump lengths thus determined is in satisfactory agreement with behavior to be expected from a generalization of the free volume model of diffusion [61]. By contrast, for zeolite ZSM-5 the mean lifetimes between succeeding jumps are found to decrease with increasing concentration, and follow nearly the same dependence as the diffusivities. With Eq. (13) one has to conclude therefore that the r.m.s. jump lengths will not change dramatically with varying concentration. The absolute values of the r.m.s. jump lengths are found to be of the order of the distance between adjacent channel intersections and are in satisfactory agreement with the results of neutron-scattering experiments [125,133].

Diffusivities for unsaturated hydrocarbons (alkenes [127] and aromatics [142]) in NaX are found to decrease only for the higher concentrations, remaining essentially constant up to medium pore filling factors (type II behavior). This pattern of concentration dependence may be correlated with the specific interaction between the cations and the double bonds of the molecules. Therefore, the constancy of the diffusivity up to medium concentrations may be caused by the fact that molecular jumps occur predominantly between well-defined adsorption sites, so that in this concentration range the r.m.s. jump lengths as well as the mean lifetimes are independent of concentration. It may also be possible that a mutual hindrance of the diffusants with increasing concentrations is compensated, more or less incidentally, by a decrease in the interaction energy.

For small polar molecules such as water and ammonia in NaX [110] (see also Sec. III.B.7. and Fig. 18), the existence of strong adsorption sites leads to a pronounced increase of molecular mobility with increasing concentration (type III behavior), corresponding to a progressive saturation of the adsorption sites. It is remarkable that the diffusivities are found to be essentially constant when starting from medium concentrations. For larger polar molecules (methanol [143,144] and acetonitrile [145]) at sufficiently large concentrations, however, again a decrease of the mobilities is observed (type IV behavior), in analogy to the patterns of types I and II.

A completely different concentration dependence is observed for the n-alkanes in zeolite NaCaA [84,120], where the diffusivity is found to increase monotonically with increasing concentration. A straightforward explanation of such a concentration dependence may be based on Eyring's theory of absolute reaction rates [5,122,146–148]. According to this theory, the rate of molecular jumps between adjacent cavities ($1/\tau_t$) is proportional to the ratio between the partition functions f^* and f_0 of the molecules in the activated state (i.e., during the passage through the window between adjacent cavities) and in the ground state (during the time of residence within one cavity). In turn, f_0 is proportional to the ratio between the molecular concentrations c and c_g in the adsorbed and gaseous phases, respectively. Hence, with l denoting the separation between the centers of adjacent cavities, one obtains from Eq. (13)

$$D \propto l^2/\tau_t \propto f^*/f_0 \propto c_g/c \tag{37}$$

With the exception of small concentrations the amount adsorbed increases less than linearly with increasing gas phase concentration [4], which according to Eq. (37) leads to the observed concentration dependence of type V.

5. Diffusion Anisotropy

Most zeolites are of noncubic structure, so that generally the intracrystalline diffusivities must be expected to be orientation dependent rather than isotropic. However, except for the pioneering diffusion studies of water in heulandite by Tiselius [149–151], the phenomenon of diffusion anisotropy has so far not been considered in detail. This is mainly because the microscopic size of the crystallites of synthetic zeolites complicates the measurement of orientation dependent diffusivities by macroscopic adsorption/desorption methods [152]. Since PFG NMR directly records the molecular mean square displacement in the direction of the applied field gradients (see Sec. III.B.1.), a measurement of diffusion anisotropy becomes possible if one can arrange the zeolite crystallites with a well-defined orientation relative to the magnetic field gradients. In the first experiments of this type [153], a specimen of large zeolite crystallites of ZSM-5 was introduced into an array of parallel capillaries such as used, for example, in artificial kidneys (Fig. 11). In this way, at 298 K and for a sorbate concentration of 12 molecules of methane per unit cell, the orientation-dependent diffusivities were determined to be $D_z = (1.6 \pm 0.4) \cdot 10^{-9} \text{ m}^2\text{s}^{-1}$ and $D_{xy} \equiv (D_x + D_y)/2 = (7.2 \pm 1.9) \cdot 10^{-9} \text{ m}^2\text{s}^{-1}$ in, respectively, the longitudinal extension of the crystallites and in the plane perpendicular to it. The orientation of the internal pore system with respect to the external appearance of a ZSM-5 type crystallite is represented in Fig. 12. In agreement with the expected behavior, the rate of molecular propagation in the plane formed by the two intersecting channels is found to be larger by a factor of about 5 than in the direction perpendicular to this plane, since there is no channel system in the latter direction. Molecular migra-

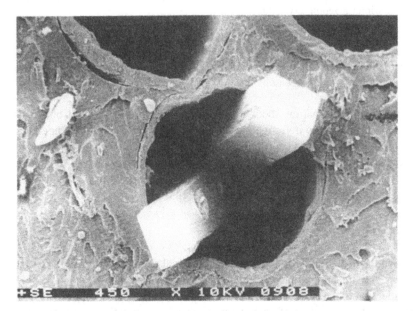

Figure 11 Scanning electron micrograph of the surface of a capillary matrix used for the PFG NMR measurements with aligned ZSM-5 crystals, showing two crystals at the orifice of a capillary. (From Ref. 153.)

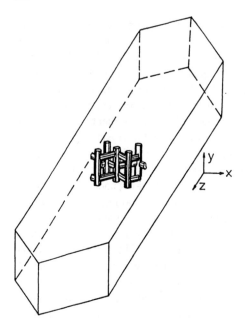

Figure 12   Schematic representation of the orientation of the internal channel system of ZSM-5 within a zeolite crystallite.

tion is only possible through alternating periods of propagation in either of the two channel types. Because an alignment of the crystallites is only possible with respect to their longitudinal extension, however, D_x and D_y cannot be determined separately in this way.

In principle, information about diffusion anisotropy also may be obtained from powder samples by investigating the shape of the NMR signal attenuation. In contrast to isotropic systems where the echo attenuation is found to be a simple exponential, now the echo attenuation results from a superposition of exponentials corresponding to the various orientations of the individual crystallites relative to the magnetic field gradient [72,73,154]:

$$\Psi(\vartheta,\varphi) = \exp\{-\gamma^2\delta^2g^2(D_x\cos^2\varphi\sin^2\vartheta + D_y\sin^2\varphi\sin^2\vartheta + D_z\cos^2\vartheta)\Delta\} \tag{38}$$

Here ϑ and φ denote the orientation of the principal axes of the diffusion tensor with respect to the applied field gradients. For quantitative analysis of Eq. (38) it is essential that in the case of ZSM-5 the three principal elements of the diffusion tensor are not independent of each other. Under the assumption that molecular propagation from one channel intersection to an adjacent one is independent of the diffusing molecule's history (in other words, independent of the original channel segment), it can be shown [155] that the diffusivities are correlated by the expression

$$\frac{c^2}{D_z} = \frac{a^2}{D_x} + \frac{b^2}{D_y} \tag{39}$$

where $a \approx b \approx 2$ nm and $c = 1.34$ nm denote the length of the unit cell in the x, y, and z directions, respectively. With this relation the shape of the signal attenuation as a function of the dimensionless quantity $\gamma^2 \delta^2 g^2 <D> \Delta$ turns out to depend on the single parameter D_y/D_x. $<D> = (D_x + D_y + D_z)/3$ stands for the mean diffusivity. Figure 13 gives a comparison between the theoretical dependence calculated for different ratios D_y/D_x and experimental data obtained by NMR signal attenuation at four different temperatures for methane in ZSM-5 at a sorbate concentration of three molecules per channel intersection [156]. For all temperatures the experimental data are found to be compatible with the theoretical dependence for ratios D_y/D_x on the order of 2 to 3. If we assume $D_y/D_x = 2.5$, then by combining this value with Eq. (39) and with the mean diffusivity as determined from the initial slope of the signal attenuation, the three principal values of the diffusion tensor may be determined. The Arrhenius

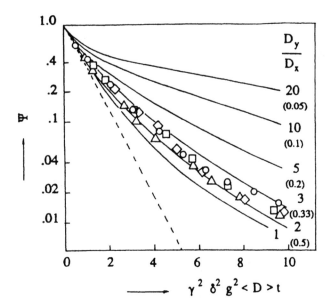

Figure 13 Theoretical dependence of the signal decay in PFG NMR experiments for different values of the ratio D_y/D_x, calculated from Eq. (38) by means of Eq. (39). Also, comparison with experimental data at 193 (O), 223 (□), 273 (◇), and 298 K (△) for methane in ZSM-5 and a concentration of 12 molecules per unit cell. (From Ref. 156.)

representation of these values is given in Fig. 14, and results of measurements with oriented samples are shown to be in satisfactory agreement with the data. For comparison, results of MD calculations carried out by three independent groups are also included. The averages of these latter values are in good agreement with the PFG NMR data. It turns out, however, that in Ref. 134 the anisotropy, as reflected in the ratio D_y/D_x, has been overestimated [159].

6. Single-File Diffusion

A large variety of zeolites (e.g., ZSM-12, -22, -23, -48, AlPO$_4$-5, -8, -11, and VPI-5) contain systems of parallel channels with diameters on the order of the molecular dimensions. Molecular propagation in this type of adsorbent represents a special case of diffusion anisotropy, since the main elements of the diffusion tensor referring to the plane perpendicular to the direction of the channel system are equal to zero. In the first PFG NMR diffusion measurements of methane in ZSM-12 and AlPO$_4$-5, only a lower limit on the order of 10^{-10} m^2s^{-1} could be determined for diffusivity in the channel direction [160]. This value is two orders of magnitude below the diffusivity of methane in the straight channels of zeolite ZSM-5 (see Fig. 14). Since the channel diameters of ZSM-12

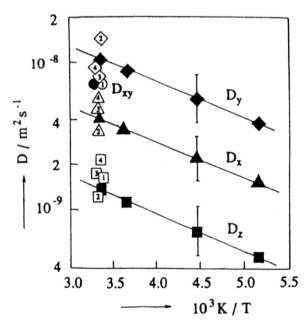

Figure 14 Principal values of the diffusion tensor for methane in ZSM-5 as determined on the basis of Fig. 13, plus comparison with results of measurements on oriented samples [(1), see Fig. 11] and MD simulations presented in Ref. 135 (2), Ref. 157 (3), and Ref. 158 (4) (open symbols with inserted numbers). (From Ref. 156.)

(0.55 nm × 0.59 nm) and AlPO$_4$-5 (0.73 nm) are on the order of or even larger than the diameters of the straight channels in ZSM-5 (0.53 nm × 0.56 nm), at first glance this result may be surprising. However, one must bear in mind that the stabilization effect [52] exerted by the zeolite framework on the molecules is much greater in the completely closed tubes of the undimensional zeolite structure than in zeolite ZSM-5, where molecular confinement is considerably reduced within the intersections of the two channel types.

An additional and dramatic decrease in the translational mobility of molecules in unidimensional channels is expected as soon as a mutual passage of the molecules within the channels is hindered or even completely excluded owing to steric reasons. The latter case, which reminds one of a string of pearls, implies a completely new pattern of molecular migration, which has been referred to as single-file behavior [112]. In contrast to ordinary diffusion where, according to Einstein's relation [Eq. (31)], the mean square displacement of the molecules is proportional to the observation time, in single-file systems the mean square displacement must be expected to increase less than linearly with t. This dependence arises because molecular displacements in succeeding time intervals are uncorrelated under ordinary diffusion, while in single-file diffusion a displaced molecule is more likely to return to its original position than to proceed further. The latter option would necessitate that the other molecules ahead also proceed in the same direction.

The limiting case of long observation times may be treated analytically by considering the "random walk" of the vacant sites rather than that of the individual molecules [161,162]. This model predicts that

$$<r^2(t)> \; = \; <l^2>(2/\pi)^{1/2}(t/\tau_t)^{1/2} \; \frac{(1 - \Theta)}{\Theta} \tag{40}$$

where $<l^2>$ and τ_t denote, respectively, the mean square step length and the mean time between succeeding step attempts (assumed to be successful only if the site toward which the step attempt is directed is vacant). Θ stands for the medium site occupancy. To get an idea of typical mean square displacements in single-file diffusion, one may estimate the quantity τ_t on the basis of Eq. (13) with the understanding that D would be the self-diffusivity of the system if the molecules were able to pass each other. Figure 15 provides a comparison of the observation times necessary to follow certain single-file mean diffusion paths $<r^2(t)>^{1/2}$ calculated in this way on the basis of the date for methane in the straight channels of ZSM-5 (with a diffusivity of 10^{-8} m^{2}s^{-1} and a r.m.s. step length of 0.5 nm, being on the order of the molecular diameter). Given that the lower limit of molecular displacements and the upper limit of observation times by PFG NMR range from 0.1 to 1 μm and from 10 to 100 ms, respectively (see Sec. III.B.2.), and also that methane is actually a very mobile molecule, Fig. 15 graphically demonstrates that tracing single-file diffusion by PFG NMR is a

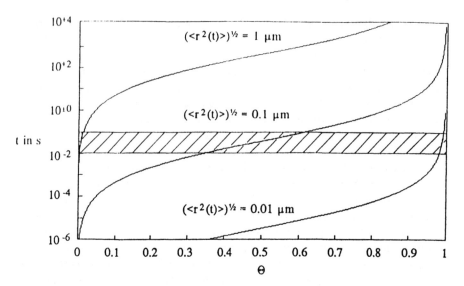

Figure 15 Observation times t necessary for following various single-file mean diffusion paths $\langle r^2(t)\rangle^{1/2}$ plotted against the site occupancy Θ, as calculated on the basis of Eqs. (40) and (13) with $D = 10^{-8}$ m²s⁻¹ and $\langle l^2\rangle^{1/2} = 0.5$ nm. The shaded area represents the region of maximum observation time accessible by PFG NMR. (From Ref. 162.)

challenging task for future research, one that will require the utmost sensitivity and resolution.

It is worthwhile to note that under conditions of single-file diffusion, the definition of the Thiele modulus

$$\Phi = L/2 \cdot (k/D)^{1/2} \tag{41}$$

for a reaction within a parallel sided slab of catalyst of thickness L of intrinsic reactivity k loses its sense. It is useful, therefore, to apply an alternative expression. Introducing the mean residence time within the slab ($\tau_{intra(Slab)} = L^2/(12D)$ [7,162]) into Eq. (41), one obtains

$$\Phi = (3k\tau_{intra})^{1/2} \tag{42}$$

With this more general equation, representation of the effectiveness factor in terms of the Thiele modulus also becomes possible for single-file diffusion. As an example, Fig. 16 shows the result of a computer simulation of diffusion and reaction within a single-file system consisting of $N = 100$ sites for different occupation numbers in comparison with the dependence

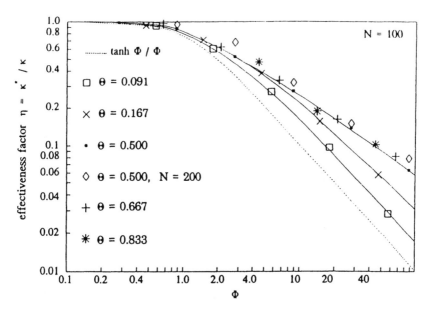

Figure 16 Effectiveness factor η of single-file reaction plotted against the generalized Thiele modulus as defined by Eq. (42). (From Ref. 162.)

$$\eta = \frac{\tanh\Phi}{\Phi} \tag{43}$$

for ordinary diffusion [162].

7. Multicomponent Diffusion

In technical applications zeolite molecular sieves and catalysts are generally used under conditions of multicomponent diffusion. Selective diffusion measurements of the individual components are therefore of immediate practical relevance. In the conventional adsorption/desorption method such measurements are complicated, however, by the requirement of maintaining well-defined initial and boundary conditions for any of the components involved. Being applied at equilibrium, such difficulties do not exist for PFG NMR. The traditional way to perform such experiments is to use deuterated compounds or compounds without hydrogen, thereby leaving only one proton-containing component, which then yields the ^{1}H NMR signal [163–165].

For example, Fig. 17 shows how the intracrystalline diffusivities of *n*-butane and butene-1 in zeolite NaX obtained in this way depend on the amount of co-adsorbed water molecules [110]. The decrease in the translational mobility by

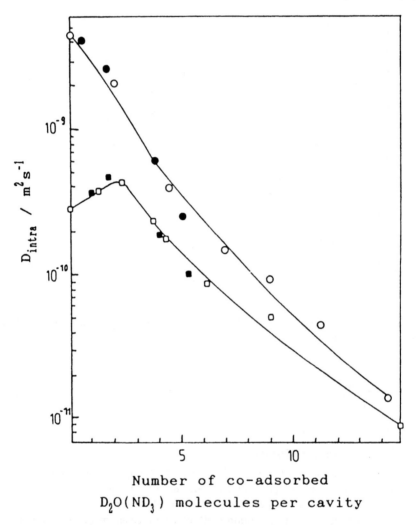

Figure 17 Values for the self-diffusion coefficient of *n*-butane ($\bigcirc$, $\bullet$) and butene-1 ($\square$, $\blacksquare$) in zeolite NaX in dependence on the amount of co-adsorbed D_2O (empty symbols) and ND_3 (full symbols) at 293 K. (From Ref. 110.)

nearly three orders of magnitude may be explained by the formation of water-cation complexes [166,167] in the windows between adjacent cavities. As is expected, saturated and unsaturated hydrocarbons are subjected to this hindrance in a similar way, and so for sufficiently high water concentrations the diffusivities of butane and butene approach each other. As a second consequence of the formation of the water-cation complexes, the specific interaction between the cations and the double bonds of the unsaturated hydrocarbons must be expected to be reduced. This fact is evidently reflected by an increase in the diffusivities of butene at small water concentrations. Since the formation of the cation-water complexes may be easily understood as a consequence of the interaction between the cation and the electric dipole moment of the water molecules, a similar effect should be observed for other polar molecules. Figure 17 demonstrates that this has indeed been found for co-adsorbed ammonia.

By applying deuterated hydrocarbons together with water or ammonia in the hydrogen-containing form, it is also possible to measure the mobility of water and ammonia. Figure 18 shows how the diffusivities of these molecules depend on their concentration, with and without co-adsorbed hydrocarbons. It turns out that the diffusivities of both ammonia and water follow a type III concentration dependence, being nearly unaffected by the existence of additional hydrocarbon molecules. In the range of higher water concentrations it is found in particular that the measured diffusivities are larger than those of the co-adsorbed hydrocarbons (see Fig. 17), though it is the existence of the water molecules that leads to

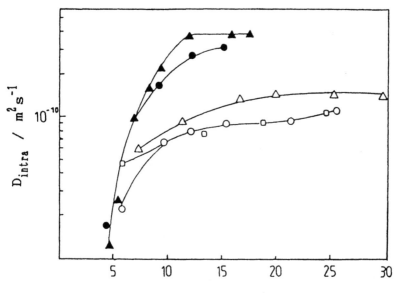

Figure 18 Concentration dependence of the self-diffusion coefficients of water (open symbols) and ammonia (filled symbols) without co-adsorbed hydrocarbons ($\triangle$, $\blacktriangle$) and with 0.8 molecules per cavity of co-adsorbed *n*-butane − d_{10} ($\bigcirc$, $\bullet$), and butene − d_8 ($\square$, 12% butene-1, 56% *trans*-butene-2, 32% *cis*-butene-2), respectively. (From Ref. 110.)

the low hydrocarbon diffusivities. This apparent contradiction may be explained, however, by realizing that obstruction of the windows might be brought about by only a small fraction of the water molecules, while the others are able to move much more freely than the larger hydrocarbon molecules.

Figure 19 compares the effect of co-adsorbed water molecules on the mobility of *n*-butane in zeolites NaX and NaCs(55%)X [110]. The diameter of the cesium ions (0.338 nm) is considerably larger than that of sodium (0.190 nm). What happens is that the cesium ions cause an effective obstruction of molecular propagation, leading to a decrease of the diffusivity by one order of magnitude compared to the pure sodium form. Therefore there is no dramatic influence on the butane diffusivity from the presence of a small amount of water molecules. By contrast, starting from that level of water concentration for which the butane diffusivity in NaX has dropped to the value in NaCsX, a further increase in the

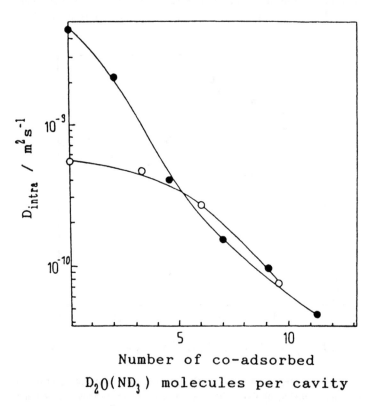

Figure 19 Comparison of the influence of co-adsorbed D_2O on the diffusivity of *n*-butane (0.8 molecules per cavity, 293 K) in zeolite NaX (●) and NaCsX (○). (From Ref. 110.)

amount of co-adsorbed water molecules effects a nearly parallel decrease of the butane diffusivity for both adsorbents.

This method of selective PFG NMR diffusion measurement clearly requires a tedious series of sample preparation. For study of a system containing n components at least n different NMR samples must be prepared, all with the same composition but each with a different compound in the hydrogen form. A more straightforward possibility for selective self-diffusion measurement is provided by Fourier transform NMR [168–171], whereby the total NMR signal is split into separate signals of the constituents if these exhibit different NMR spectra (chemical shifts). This procedure has been successfully applied to multicomponent liquids, where it was possible to measure the diffusivity of as much as eight different components [170]. In adsorbent-adsorbate systems, such experiments are complicated by a reduction of molecular mobility, which leads to broader NMR lines so that discrimination of the contributions from the various constituents becomes difficult. Initial ^{1}H PFG Fourier-transform NMR experiments of adsorbed molecules have been carried out with an ethane-ethene mixture adsorbed on zeolite NaX [172]. This system is especially convenient for such studies since the spectrum of either component consists of only one line and since the mobility of either component is still sufficiently high to ensure a line narrowing pronounced enough to allow separation of the two spectra. As an example, Fig. 20 shows ^{1}H NMR spectra obtained by Fourier transforming the second half of the NMR spin echo in PFG NMR experiments and illustrates the dependence on the width of the gradient pulses. Measurements here were made for a sorbate concentration of 1.5 molecules of ethane and 1 molecule of ethene per supercage at 293 K. Fitting Eq. (32) to the decay of either of these lines yields self-diffusion coefficients of $(4.6 \pm 0.9) \times 10^{-9}$ m^{2}s^{-1} for ethane and $(1.25 \pm 0.25) \times 10^{-9}$ m^{2}s^{-1} for ethene. In single-component PFG NMR measurements of ethane [62] and ethene [127] adsorbed in zeolite NaX with the same total sorbate concentration of about 2.5 molecules per supercage, diffusivities of 1.1×10^{-8} m^{2}s^{-1} and 1.25×10^{-9} m^{2}s^{-1} were obtained. Hence the mobility of ethene in the mixture is found to remain constant, while the diffusivity of ethane is reduced by a factor on the order of 2. The observation that the difference in the mobilities of the components within a mixture is smaller than that of the single components may be explained by the mutual interaction of the different types of molecules in the mixture and has likewise been observed with neat liquids [169,170].

C. Surface Barriers

1. The NMR Tracer Desorption Technique

In many cases the rate of molecular mass transfer through the bed of zeolite crystallites (see Sec. III.D.) is found to be so fast that the propagator determined in the PFG NMR experiment may be easily separated into its two constituents.

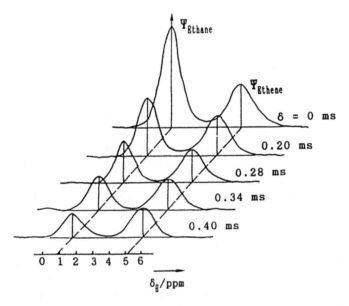

Figure 20 ¹H PFG NMR Fourier transform NMR spectra of an ethane-ethene mixture (1.5 molecules ethane and 1 molecule ethene per supercage) adsorbed in zeolite NaX at 293 K, with $g = 2.8 \, Tm^{-1}$ and $\Delta = 4$ ms. The chemical shift δ_H refers to TMS. (From Ref. 172.)

One part stems from molecules that have left their crystallites during the observation time (i.e., during the time interval Δ between the two field gradient pulses), and the other originates from those molecules which at the time of the second gradient pulse are in the same crystal as during the first gradient pulse. This is the situation depicted in the propagator representations given in Fig. 8a at 233 and 293 K. Since the activation energy of long-range diffusion, which generally coincides with the heat of adsorption (Sec. III.D.1.), is in general larger than that of intracrystalline diffusion, this condition may be fulfilled for most systems simply by enhancing the measuring temperature. The intensity of either of these distribution curves is directly proportional to the relative amount of molecules that have left their crystallites (broad line) and those that are still in the same crystallite (narrow line). Hence by varying the observation time one is able to determine the time dependence of molecular exchange between the intra-crystalline space and the surroundings. The information is completely equivalent to that obtained in conventional tracer exchange experiments (see Sec. II.A.), and therefore this method of analysing PFG NMR data has been called the NMR tracer desorption technique [48] or—in view of the very short observation times in the NMR experiments (ms to s)—fast tracer desorption [4]. In general, spin-echo attenuation in NMR tracer desorption experiments may be approximated by

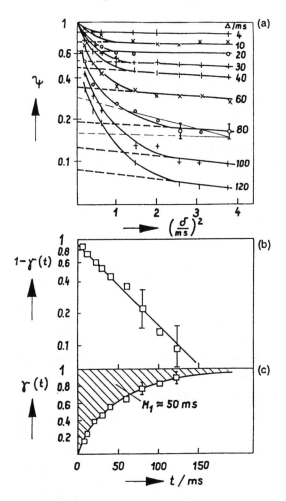

Figure 21 Analysis of PFG NMR data for NMR tracer desorption studies (butane/ NaX, 165 mgg^{-1}, $R = 25$ μm, 353 K). (From Ref. 48.)

the superposition of two exponentials in the form of Eq. (32), corresponding to two Gaussians for the two constituents of the propagator. The intensity of the two constituents is given simply by the two preexponential factors determined from the echo attenuation.

The usual way to analyze the PFG NMR data for NMR tracer desorption measurements is illustrated in Fig. 21 [48]: the intensity of the slowly decaying component (corresponding to the broader distribution within the propagator representation) coincides with the relative amount of molecules [1-γ(t)] which, for the given observation time Δ, have remained in their crystallites (Fig. 21a,b).

In general, the NMR tracer desorption data are represented in terms of the intracrystalline mean lifetime τ_{intra}, defined as the first statistical moment of the tracer desorption curve $\gamma(t)$ via Eq. (1) (Fig. 21c). With commercial samples, the range of measurement may be considerably enhanced by applying a large constant field gradient in addition to the pulsed field gradients. It has been shown [173,174] that in this case the NMR tracer desorption technique yields a time constant τ'_{intra}, which in the limit of small values coincides with the intracrystalline mean lifetimes (typically for $\tau'_{intra} \leq 5$ ms) and which increases with increasing values of τ_{intra} at a rate above that expected for a linear dependence. Hence differences in the kinetics of intercrystalline exchange in different samples may be determined more easily.

If molecular exchange is controlled by intracrystalline diffusion, then the intracrystalline mean lifetime is given by Eq. (2), where it is assumed that the crystallites may be approximated by spheres (Sec. II.A.). Clearly, $\tau_{intra}^{Diff.}$ coincides with the directly measured τ_{intra} if desorption is controlled by intracrystalline diffusion. If, however, the rate of molecular exchange is additionally reduced by transport resistances at the crystallite boundary (so-called surface barriers), τ_{iitra} may be much greater than $\tau_{intra}^{Diff.}$.

Equivalently, one may investigate the existence of a surface barrier by comparing the intracrystalline diffusivity as determined by PFG NMR with a quantity $D_{des.}$ derived from the NMR tracer desorption curve assuming intracrystalline diffusion control. In the absence of significant surface barriers, one should find $D_{intra} \approx D_{des.}$, whereas the existence of a barrier will give $D_{intra} > D_{des.}$.

2. Probing Surface Barriers with Different Molecules

Figure 22 gives a comparison of the results of NMR tracer desorption studies and self-diffusion measurements on short chain length paraffins in zeolite NaX [48]. For illustration, the complete tracer desorption curves are also given at selected temperatures. Covering the range from -140 to 200°C and chain lengths from one to six carbon atoms, the intracrystalline mean lifetimes are found to coincide with values of $\tau_{intra}^{Diff.}$ calculated via Eq. (2) from the NMR self-diffusion coefficients. This agreement indicates that molecular exchange is controlled by intracrystalline self-diffusion and that for the adsorbent-adsorbate systems considered there are no perceptible surface barriers.

Comparative investigations between the conventional adsorption/desorption method and PFG NMR have been carried out with aromatics in zeolite NaX. It was pointed out in Table 2 that there is still some divergence between the data obtained by both methods on intracrystalline diffusion. Table 3 compares the values for τ_{intra} and $\tau_{intra}^{Diff.}$ determined by the NMR methods [143,175,176]. ^{1}H PFG NMR measurements of these systems are complicated by the rather short transverse nuclear magnetic relaxation times, which range over milliseconds and lead to mean errors up to 50%. However, as with the n-paraffins in NaX, there is no indication of a significant enhancement of τ_{intra} in comparison with $\tau_{intra}^{Diff.}$ as

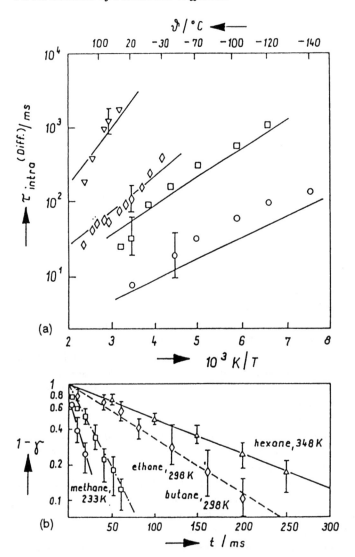

Figure 22 NMR tracer desorption measurements with methane ($\bigcirc$; 94 mgg^{-1}, $R = 25$ μm), ethane ($\square$; 113 mgg^{-1}, $R = 25$ μm), *n*-butane ($\diamondsuit$; 165 mgg^{-1}, $R = 25$ μm), and *n*-hexane ($\triangle$, $\triangledown$; 165 mgg^{-1}, $R = 10$ μm) in zeolite NaX. (a) Intracrystalline mean lifetimes as determined from the NMR tracer desorption curves on the basis of Eq. (1), plus comparison with the values of $\tau_{intra}^{Diff.}$ as calculated via Eq. (2) from the intracrystalline diffusivities determined by PFG NMR. (b) NMR tracer desorption curves for selected temperatures. (From Ref. 48.)

Table 3 Comparison of Values for the Intracrystalline Mean Lifetime τ_{intra} and the Quantity $\tau_{intra}^{Diff.}$, Calculated on the Basis of the Coefficients of Intracrystalline Self-Diffusion for Aromatic Compounds in Zeolite NaX

Mean crystallite radius (μm)	Sorbate	Sorbate concentration (molecules per cavity)	T (K)	τ_{intra} (ms)	D_{intra} (10^{-11} m^{2}s^{-1})	$\tau_{intra}^{Diff.}$ (ms)
2	benzene	4.7	393	12	15	18
2	benzene	4.3	423	20	2.0	13
18	toluene	2.1	463	37	40	54
18	meta-xylene	1.65	463	98	15	140

Source: Refs. 175 and 176.

determined on the basis of the intracrystalline diffusivities. The difference between the PFG NMR and adsorption/desorption data on intracrystalline diffusion therefore cannot be explained by the existence of surface barriers.

In contrast to the large pore zeolite NaX, where so far no indication for the formation of surface barriers has been observed, there are many PFG NMR studies showing unambiguous evidence for surface barriers in NaCaA and—to a lesser extent—in ZSM-5 (see Sec. III.C.3.–5.). One has to note, though, that the intensity of the transport resistance by the crystallite surface compared to that in the intracrystalline space may depend significantly on the method of sample preparation as well as on the nature of the probe molecule. Table 4 shows the results of combined PFG NMR and NMR tracer desorption studies of zeolite NaCaA as a powder and also in the granulated form with methane as an adsorbate [175]. The zeolite specimens have been activated under both "shallow bed" (SB: filling height $\leq$ 3 mm, heating rate 10 K/h, continuous evacuation, final pressure $\leq 10^{-2}$ Pa) and "deep bed" (DB: filling height $\geq$ 15 mm, heating rate 100 K/h, evacuation only at the final temperature, final pressure $\leq 10^{-2}$ Pa) conditions [177]. From the constancy of the values for $\tau_{intra}^{Diff.}$ it follows that neither the granulation procedure nor the mode of sample preparation significantly affects intracrystalline diffusion. However, only for the powder sample and only after moderate activation do the values of τ_{intra} and $\tau_{intra}^{Diff.}$ coincide. Thus both the deep bed activation and the process of granulation lead to the formation of a surface barrier. The common feature of the two processes is that the zeolite is subjected to hydrothermal conditions. This fact will be explored in more detail in Sec. III.C.3.

Table 5 shows the results of diffusion measurements with carbon monoxide, methane, and xenon in a comparative ^{13}C, ^{1}H, and ^{129}Xe PFG NMR study [178,179]. For all adsorbates the values of τ_{intra} and $\tau_{intra}^{Diff.}$ are found to be on

Table 4 Influence of Sample Activation and Granulation on Molecular Transport of Methane in NaCaA at 293K

Sample activation		$\tau_{intra}^{Diff.}$ (ms)	τ_{intra} (ms)	$D_{long\text{-}range}$ $(10^{-7}\ m^2 s^{-1})$
NaCaA (powder)	SB	0.3 ± 0.1	0.3 ± 0.1	18 ± 7
	DB	0.3 ± 0.1	2.5 ± 0.5	21 ± 8
NaCaA (granule)	SB	0.3 ± 0.1	1.8 ± 0.4	2.5 ± 1.0
	DB	0.3 ± 0.1	5.6 ± 1.1	2.8 ± 1.1

Table 5 Comparison of Values for the Intracrystalline Mean Lifetime τ_{intra} and for the Quantity $\tau_{intra}^{Diff.}$, Calculated on the Basis of the Intracrystalline Diffusivities via Eq. (2) for Different Molecules in Zeolite NaCaA (R ≈ 10 μm) at 293 K and a Sorbate Concentration of 3–4 Molecules (Atoms) per Large Cavity

	Adsorbate		
	Carbon monoxide, 0.38 nm[a]	Methane, 0.41 nm[a]	Xenon, 0.49 nm[a]
$D(m^2/s)$	$(2.0 \pm 0.5) \cdot 10^{-9}$	$(1.5 \pm 0.4) \cdot 10^{-9}$	$(0.3 \pm 0.1) \cdot 10^{-9}$
τ_{intra} (ms)	5 ± 2	4 ± 2	35 ± 20
$\tau_{intra}^{Diff.}$ (ms)	4 ± 2	5 ± 1	25 ± 8

[a]Gaskinetic diameter.
Source: Refs. 178 and 179.

the same order so that in no case a significant influence of transport resistances at the external crystallite surface is observed. It is difficult to say whether in the case of xenon the slightly larger value of $\tau_{intra}^{Diff.}$ in comparison with τ_{intra} may already be taken as an indication of a (though very small) surface barrier. Just as the intracrystalline diffusivity, so the surface permeability (and hence also its reciprocal, the surface transport resistance) depends on both the adsorbent and the adsorbate. It might be possible, therefore, that as a consequence of their larger diameter transport resistances in the surface layer of the crystallites are more readily probed by the xenon atoms. Only if the surface barriers are brought about by a total obstruction of a certain fraction of pores while the remaining pores remain unaffected, for a given adsorbent the ratio $\tau_{intra}/\tau_{intra}^{Diff.}$ must be expected to be a unique quantity for all adsorbates and temperatures. This possibility may be excluded, however, in view of the experimental finding that in systems with surface barriers, the activation energy of D_{intra} (coinciding with that of $\tau_{intra}^{Diff.}$) is smaller than that of $1/\tau_{intra}$ [180].

3. Zeolite Deterioration due to Hydrothermal Treatment

In their fundamental study of the influence of moisture on the transport properties of zeolites, Kondis and Dranoff [181] have noted a remarkable decrease in the adsorption/desorption rates of hydrocarbons in zeolite NaCaA after a hydrothermal treatment of the zeolites. Comparison of the values of τ_{intra} and $\tau_{intra}^{Diff.}$ in Table 4 indicates that this effect is attributable primarily to the formation of surface barriers rather than to a significant change of the intracrystalline mobility. Studies of the surface composition of the zeolite crystallites by x-ray photoelectron spectroscopy show that the formation and enhancement of the surface barrier is accompanied by a decrease of the cation content in the surface layer, thereby indicating a structural collapse of the surface layer of the zeolite crystallite [182].

Figure 23 shows results from a systematic study dealing with the influence of a hydrothermal pretreatment of granulated zeolite NaCaA on the three main transport parameters accessible by PFG NMR: the coefficients of intracrystalline and long-range diffusion, as well as the intracrystalline mean lifetimes [145,

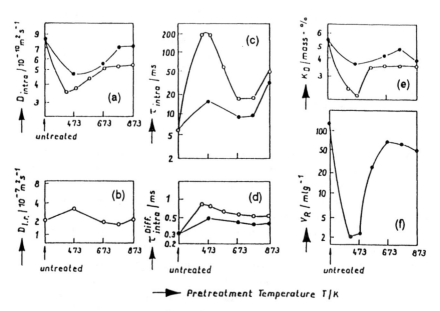

Figure 23 Coefficients of (a) intracrystalline and (b) long-range self-diffusion, and (c) intracrystalline mean lifetimes τ_{intra} and (d) $\tau_{intra}^{Diff.}$ for methane in granulated zeolite NaCaA at 293 K. Also, comparison with (e) the breakthrough capacities for a petroleum raffinate and (f) the specific retention volume for *n*-pentane all plotted against the temperature of hydrothermal pretreatment applied over a time interval of 7 h (●) and 14 h (○), respectively. (From Ref. 175.)

183,184]. Prior to the diffusion experiments, the zeolite had been subjected to an atmosphere of extreme humidity (p_{Water} = 90 kPa) at different temperatures (373–873 K) over a period of 7 and 14 hours, respectively. While $D_{long-range}$ is found to be essentially independent of the pretreatment conditions, changes in both τ_{intra} and D_{intra} are observed. However, a comparison with $\tau_{intra}^{Diff.}$ shows that it is exclusively the transport resistance of the surface barriers which controls intercrystalline molecular exchange. The most intense effect is evidently brought about at temperatures around 200°C. In subsequent studies using larger paraffins the decisive role of surface resistances on mass transfer could be confirmed [185]. In Fig. 23e and f these changes are seen to be accompanied by substantial deteriorations in the dynamic sorption properties of the zeolites. These latter properties are represented by the retention volume V_R for *n*-pentane at 523 K and the breakthrough capacity K_D of a hydroraffinate in the boiling range 461–594 K at 653 K [184].

The combined application of PFG NMR self-diffusion and tracer desorption experiments has thus proved to be an effective tool for studying the hydrothermal stability of A-type zeolites with respect to their transport properties [186]. It turns out that with commercial adsorbent samples there are considerable variations in hydrothermal stability between different batches of product and even between different pellets from the same batch. As an example, Fig. 24 shows the distribution curves [$N(\tau_{intra})$ versus τ_{intra}] measured with ethane as a probe molecule at 293 K for two different samples of commercial 5A zeolites. Evidently batch 1 is more resistant to hydrothermal deterioration, because the lengthening of τ_{intra} is less dramatic than with batch 2. Since the intracrystalline diffusivity was the same for all samples, the deterioration can be attributed to the formation of a surface barrier.

4. Zeolite Coking

One of the main factors determining the effectiveness of zeolites for catalysis and molecular sieving is their stability under working conditions. In many cases the deposition of carbonaceous compounds terminates a zeolite's lifetime, and therefore quantitative information about the localization of such coke deposits and the intensity of the mass transfer resistances they cause is of supreme importance for optimizing zeolite regeneration and replacement. Comparison of the quantities τ_{intra} and $\tau_{intra}^{Diff.}$ as determined by PFG NMR provides a straight-forward way to distinguish between coke depositions in the intracrystalline space and on the outer surface of the crystallites. Figure 25 shows, for example, the dependence of the relevant transport parameters on the time on steam for granulated zeolite NaCaA in a petroleum refinery [175,183]. In contrast to the intracrystalline diffusivity (D_{intra}, $\tau_{intra}^{Diff.}$), which essentially remains constant, there is a dramatic increase in the intracrystalline mean lifetime (τ_{intra}). One has to conclude, then, that deterioration of the dynamic properties of the zeolite, as reflected by the decreasing breakthrough capacities, arises from the formation of

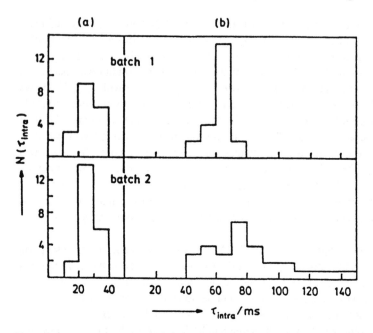

Figure 24 Distribution curves ($N(\tau_{intra})$ versus τ_{intra}) measured with ethane as the probe molecule at 293 K for two different samples of a commercial 5 A zeolite. (a) The initial material; (b) the same material after hydrothermal pretreatment. (From Ref. 186.)

surface barriers. This comes about from the preferential deposition of coke in a layer on or close to the external surface of the crystallites, rather than from a substantial reduction of molecular mobility in the interior of the crystallites.

Figure 26 compares values for the intracrystalline mean lifetime and $\tau_{intra}^{Diff.}$ for methane in ZSM-5 type crystallites after different coking times and of the values of $\tau_{intra}^{Diff.}$ [145,187]. Depending on the applied coking compound, completely different dependences are obtained. For *n*-hexane, τ_{intra} and $\tau_{intra}^{Diff.}$ coincide over a large range of coking times, whereas with mesitylene only τ_{intra} increases while the intracrystalline mobility represented by $\tau_{intra}^{Diff.}$ remains unaffected. Since the mesitylene molecules are too large to penetrate into the intracrystalline pore system, during mesitylene coking the carbonaceous compounds are thus found to be exclusively deposited on the external surface. For *n*-hexane, two stages of the coke deposition become visible. At shorter coking times *n*-hexane is mainly deposited in the intracrystalline space, thereby simultaneously effecting a retardation of intracrystalline diffusion and intercrystalline exchange. In a second stage, similar to the behavior observed with mesitylene, coke is predominantly deposited on the crystallite surface.

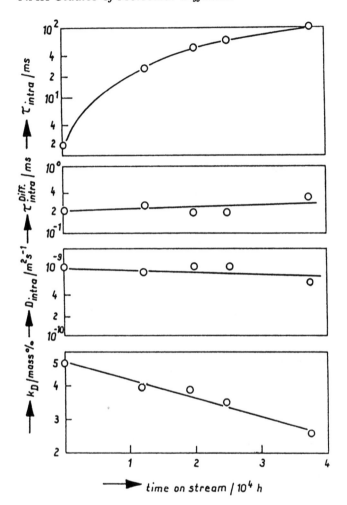

Figure 25 Parameters of molecular transport (methane, 293 K) in granulated zeolite NaCaA versus the time on stream in a petroleum refinery, plus comparison with the breakthrough capacity of the adsorber (k_D). (From Ref. 175.)

Figure 27 shows the ratio $\tau_{intra}^{Diff.}/\tau_{intra}$ for HZSM-5 crystals of different morphology as a function of the amount of coke deposited [187–189]: polycrystalline spherical particles and polyhedral crystallites. Once again, two stages of coke formation can be distinguished. The onset of the second stage is essentially the same for all polyhedral crystallites, while for the polycrystalline grains a distinct delay is observed. This experimental finding can be explained by

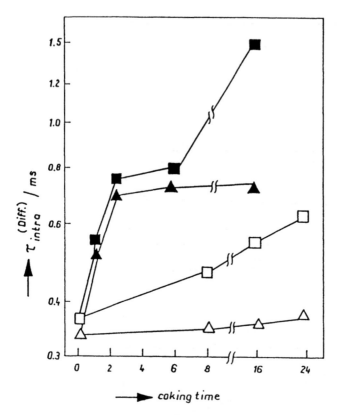

Figure 26 Values for the intracrystalline mean lifetime τ_{intra} ($\square$, $\blacksquare$) and the quantity $\tau_{intra}^{Diff.}$ ($\triangle$, $\blacktriangle$) versus time on stream for methane at 296 K and a sorbate concentration of 12 molecules per unit cell, in H-ZSM-5 coked by n-hexane (filled symbols) and mesitylene (open symbols). (From Ref. 145.)

the existence of the secondary pore system represented by the free space between the crystallites. Thus an additional amount of coke may be deposited on "neutral" spots outside the zeolite channel network, causing a delayed onset of the second period of coke formation.

5. Zeolite Modification by Phosphoric Acid

Modification of H-ZSM-5 zeolites by impregnation with H_3PO_4 has become a common technique to improve their activity and selectivity [190,191]. In parallel with these changes, however, the transport properties of the zeolite crystallites are also changed. As in the cases of hydrothermal treatment and coking discussed above, a combined application of PFG NMR to study both intracrystalline diffusion and intercrystalline molecular exchange may provide information about

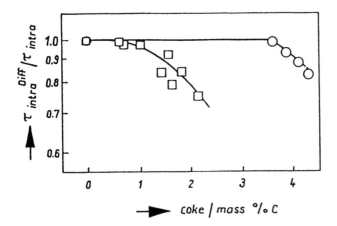

Figure 27 Ratio $\tau_{intra}^{Diff.}/\tau_{intra}$ for HZSM-5 samples of different morphology as a function of the amount of coke deposited: □ polyhedral crystals, ○ polycrystalline grains. (From Refs. 188 and 189.)

the structural changes during the process of zeolite modification. Figure 28 shows the effect of phosphorus incorporation on the intracrystalline diffusivity (D_{intra}) and on the rate of intercrystalline exchange (as expressed in terms of the apparent diffusivity, $D_{des.}$, determined from the NMR tracer desorption curves) [192]. The continuous decrease of D_{intra} with increasing phosphorus content indicates that phosphoric acid is progressively incorporated into the intracrystalline pore network. Since the apparent self-diffusivity decreases more steeply than D_{intra}, the process of impregnation must be assumed to lead to additional diffusion obstacles near the crystal surface, suggesting an enrichment of phosphorus species near the surface of the zeolite crystallites.

D. Long-Range Diffusion

1. PFG NMR Measurement of Long-Range Diffusion

Once the molecular root mean square displacements observed by PFG NMR are much larger than the mean crystallite diameters, a direct measurement of the rate of mass transfer through the assemblage of zeolite crystallites becomes possible. This rate is represented by the coefficient of long-range diffusion

$$D_{long-range} = \frac{<r^2(t)>_{inter}}{6t} \tag{44}$$

if the observation time t is sufficiently long so that $<r^2(t)>_{inter}^{1/2}$ is in fact much larger than the mean radius of the crystallites. As an illustration, Fig. 29 provides results from a PFG NMR study of the long-range self-diffusion of *n*-butane in an

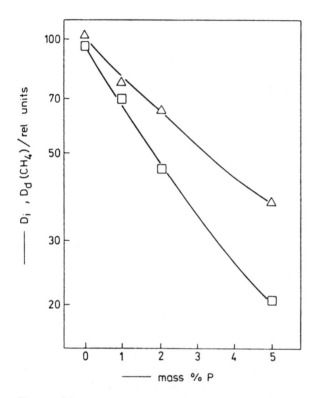

Figure 28   Values for the coefficients of intracrystalline self-diffusion ($\triangle$) and for the apparent diffusivity ($D_{des.}$), determined from NMR tracer desorption curves ($\square$) at 293 K for methane in H-ZSM-5 modified by impregnation with H_3PO_4. The dependence is on the phosphorus mass content. Self-diffusivity in the starting material amounts to 10^{-8} m^2s^{-1}. (From Ref. 192.)

assembly of crystallites of zeolite NaX [193]. The representation shows the quantity $\langle r^2(t)\rangle/6t$ as directly accessible in the experiments. For comparison, the value of

$$D_{restr.} = \frac{R^2}{5t} \tag{45}$$

where R denotes the mean radius of the crystallites, is given for all three observation times. The quantity $D_{restr.}$ may be understood as an apparent diffusivity, resulting from the PFG NMR measurements for sufficiently long observation times if molecular propagation is confined to the volume of a sphere of radius R [73,194]. It is shown in Fig. 29 that the measured quantities $\langle r^2(t)\rangle/(6t)$ coincide with $D_{long-range}$ as soon as $D_{restr.}$ is sufficiently small in

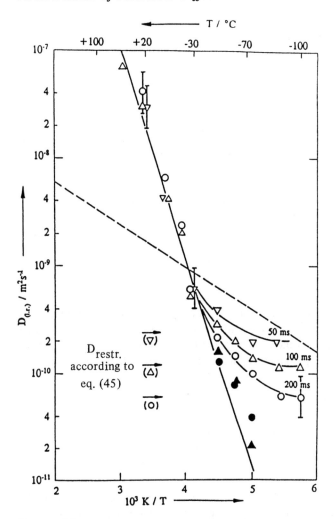

Figure 29 PFG NMR diffusivity data for *n*-butane in an assemblage of zeolite NaX with a mean crystallite diameter of about 16 μm and a sorbate concentration of 80 mgg^{-1}, represented by the quantity $D_{eff} = <r^2(t)>/(6t)$ (open symbols). Observation times are $t = 50$ ms ($\triangledown$), 100 ms ($\triangle$), and 200 ms ($\bigcirc$). In addition, comparison with values for $D_{restr.}$ determined according to Eq. (45). For $D_{eff} \gg D_{restr.}$, $D_{long\text{-}range}$ coincides with D_{eff}. For D_{eff} approaching $D_{restr.}$, $D_{long\text{-}range}$ has been estimated on the basis of Eq. (46) (filled symbols). The dashed line represents the values for the coefficient of intracrystalline self-diffusion. (From Ref. 193.)

comparison with these quantities. In the transition region between restricted and long-range diffusion, to first-order approximation the long-range diffusivity may be estimated by combining Eqs. (44) and (45)

$$D_{\text{long-range}} \approx \frac{<r^2(t)>}{6t} - \frac{R^2}{5t} \tag{46}$$

as indicated by the filled symbols in Fig. 29. In the temperature range where $D_{\text{restr.}}$ is larger than $D_{\text{long-range}}$ but smaller than D_{intra}, the pattern of molecular shifts observed in PFG NMR is completely determined by the shape of the crystallites. This is the situation that is reflected in the propagator representations of Fig. 8b at 153 K.

By both model considerations for mass transfer in composite systems [195–199] and simple random walk arguments, the contribution of intracrystalline mass transfer to long-range diffusion may be shown to be much less than that of mass transfer through the intercrystalline space [193]. The coefficient of long-range diffusion therefore may be represented in the form [52,145]

$$D_{\text{long-range}} = p_{\text{inter}} D_{\text{inter}} \tag{47}$$

with p_{inter} and D_{inter} denoting, respectively, the relative population of molecules in the intercrystalline space and their diffusivity. As a consequence of the large intracrystalline surface, in the case of gas phase adsorption one has in general $p_{\text{inter}} \ll 1$. In a first-order gas kinetic approach the diffusivity in the intercrystalline space is given by

$$D_{\text{inter}} = \frac{1/3 \cdot \lambda_{\text{eff}} v}{\tau_{\text{tort.}}} \tag{48}$$

with $v = (8 kT/\pi m)^{1/2}$ denoting the mean velocity of molecules of mass m, and with the effective mean free path λ_{eff} following by reciprocal addition as

$$\lambda_{\text{eff}}^{-1} = \lambda^{-1} + d^{-1} \tag{49}$$

from the mean free path in the gas phase (λ) and the mean diameter of the intercrystalline pore system (d). $\tau_{\text{tort.}}$ stands for the tortuosity factor (being typically of order 2 to 4) which takes into account that, in comparison to an unrestricted space, the diffusion path of the individual molecules is enhanced by the confinement brought about by the tortuosity of the intercrystalline pore system. It follows from Eqs. (47) through (49) that for sufficiently small gas phase concentrations (that is, for $\lambda \approx d$, the case of Knudsen diffusion) the temperature dependence of $D_{\text{long-range}}$ is determined by p_{inter}, so that in this case the activation energy of $D_{\text{long-range}}$ coincides with the heat of adsorption. In fact, the value of (44 ± 4) kJmol^{-1} for the activation energy of long-range self-diffusion of n-butane in NaX, as deduced from Fig. 29, is in satisfactory agreement with literature data for the heat of adsorption ($(42-44)$ kJmol^{-1} [200]).

2. Influence of the Pressure of Compaction

Figure 30 compares the long-range diffusivity of *n*-butane in a loose bed of NaX zeolite crystallites with that of the same sample after compaction under a pressure of 2.5 MPa. It is found that long-range diffusion in the compacted material is reduced by a factor on the order of 3, which may be attributed to a reduction of both p_{inter} (diminution of the intercrystalline void volume) and D_{inter} (decrease of the intercrystalline pore diameters and hence of the effective mean free path). This experimental result confirms that the contribution of intra-

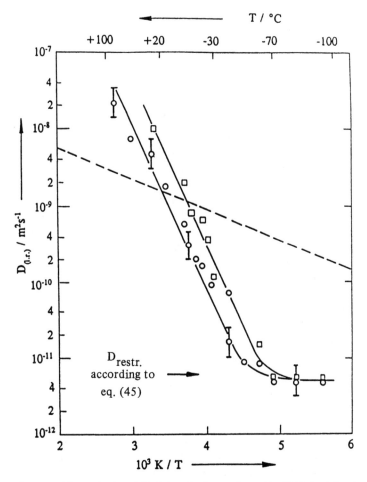

Figure 30 Effective diffusivity of *n*-butane in zeolite NaX, both in a loose assemblage (□) and after compaction under a pressure of 2.5 MPa (○). The mean crystallite diameter is 5 μm for a sorbate concentration of about 80 mgg⁻¹. (From Ref. 193.)

crystalline mass transfer to long-range diffusion is much smaller than that of mass transfer through the intercrystalline space, since otherwise an increase of the zeolite packing density must be expected to lead to an enhanced mobility. In both samples the transition of the PFG NMR data from the regime of long-range diffusion to restricted diffusion is observed to be determined by Eq. (45), indicating that during the process of compaction the crystallite size remains essentially unaffected.

A difference of even nearly one order of magnitude is observed on comparing the long-range diffusion in beds of zeolite crystallites with those in commercial granules (see Table 4). It is noteworthy that the long-range diffusivities appear to be unaffected by the hydrothermal treatment, whereas the rate of intercrystalline exchange is significantly reduced.

3. Influence of a Carrier Gas

In most technical applications, as well as in various experimental arrangements for diffusion studies, molecular diffusion proceeds under the influence of a carrier gas. Figure 31 shows the result of PFG NMR studies on how an inert argon atmosphere affects the coefficients of intracrystalline and long-range diffusion of cyclohexane in NaX [201]. Under the given conditions the intra-crystalline diffusivity is, in fact, found to be unaffected by the inert gas. This may be easily rationalized by realizing that argon is less favorably adsorbed than cyclohexane, so that the argon concentration in the intracrystalline space is negligibly small in comparison with that of cyclohexane. The substantial argon concentration in the gas phase, on the other hand, leads to a significant reduction in the long-range diffusivities, which is more pronounced the higher the pressure of the carrier gas atmosphere. In the sample without argon, in the initial temperature range (20–100°C) the activation energy of $D_{\text{long-range}}$ is determined to be (58 ± 6) kJmol^{-1}, which is in good agreement with the heat of adsorption for the same system (55 kJ/mol [202]). For higher temperatures, in consequence of the further increasing gas phase concentration, the effective mean free path is determined more and more by mutual encounters between the cyclohexane molecules rather than by encounters with the external crystallite surface (transition from Knudsen to gas phase diffusion) leading to a decrease of λ_{eff} [Eq. (49)] [and hence of D_{inter}, Eq. (48)] in parallel to the increase of p_{inter}. Therefore, $D_{\text{long-range}}$ increases less steeply as if D_{inter} remained constant.

The activation energy of (39 ± 5) kJmol^{-1} for long-range self-diffusion of cyclohexane under the influence of the carrier gas is smaller than that for the pure adsorbent-adsorbate system. This may be explained by the increasing argon concentration in the gas phase with increasing temperature, which leads to a decrease of λ_{eff} and hence of D_{inter} for cyclohexane. The total activation energy of $D_{\text{long-range}}$ must be given, therefore, by the difference between the heats of adsorption for cyclohexane and argon. With an adsorption heat of 12 kJmol for argon on NaX [207], one obtains a theoretical value of 43 kJmol^{-1}, which is in

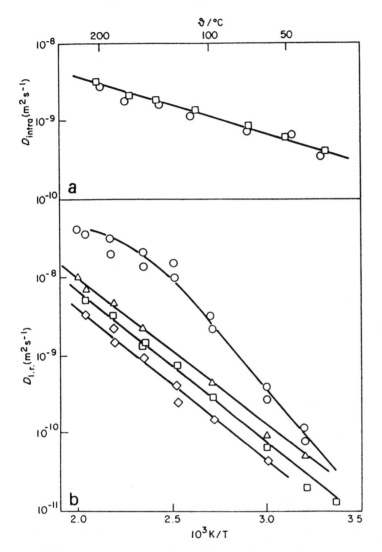

Figure 31 Values for the coefficients of (a) intracrystalline and (b) long-rang self-diffusion of cyclohexane in NaX at a sorbate concentration of 1.9 molecules per cavity in the pure adsorbent-adsorbate system ($\bigcirc$), and under the influence of an argon atmosphere of ≤ 0.06 MPa ($\triangle$), 0.13 MPa ($\square$), and 0.2 MPa ($\diamondsuit$). (From Refs. 145 and 201.)

satisfactory agreement with the experimental result. It is noteworthy that under the influence of the carrier gas, there is no deviation of the long-range diffusivities from the Arrhenius dependence, since over the whole temperature range considered the mean free path of the cyclohexane molecules is determined by encounters with the argon atoms in the intercrystalline space.

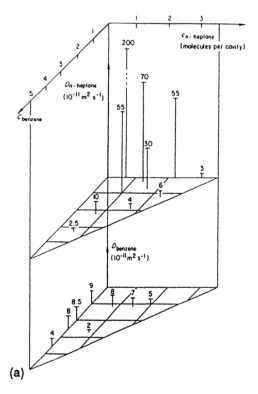

(a)

Figure 32 Values for the coefficients of (a) intracrystalline self-diffusion of the two components in *n*-heptane–benzene mixtures in NaX at 400 K and (b) long-range self-diffusion at 361 K, plus the separation factors calculated from them. (From Refs. 145, 163, and 165.)

4. Two-Component Diffusion Measurements

Figure 32 provides a comparison between the coefficients of intracrystalline and long-range self-diffusion of benzene and *n*-heptane at two-component adsorption in a bed of NaX zeolite crystallites [163–165]. In contrast to the regime of intracrystalline diffusion, where under conditions of two-component adsorption the diffusivities are found to decrease with increasing concentration (type I and type II patterns of single-component diffusion), here the long-range diffusivities tend to be enhanced with increasing concentration. This tendency comes about because with zeolites in general the amount adsorbed increases less than linearly with increasing gas phase concentration (Langmuir type isotherms), leading to an increase of p_{inter} with increasing concentration. Therefore as long as the gas phase concentration is small enough to ensure Knudsen diffusion, according to

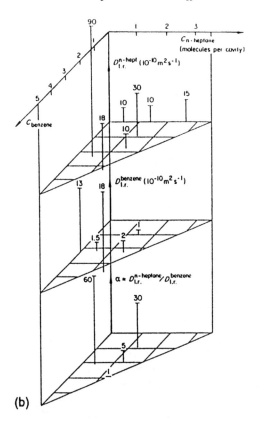

(b)

Eq. (47) the long-range diffusivity must be expected to increase with increasing concentration.

From the measurement of two-component long-range self-diffusivities, direct information about adsorption selectivity may be deduced. With $x_a^{(i)}$ and $x_g^{(i)}$ denoting the mole fractions of the ith mixture component in the adsorbed and gaseous phases, the sorption separation factor is defined by the relation [203]

$$\alpha = \frac{x_a^{(2)} x_g^{(1)}}{x_a^{(1)} x_g^{(2)}} \tag{50}$$

Since $p^{(i)}_{\text{inter}}$ is proportional to $x_g^{(i)}/x_a^{(i)}$, Eq. (50) may be transformed via Eq. (47) into

$$\alpha = \frac{D^{(1)}_{\text{long-range}} D^{(2)}_{\text{inter}}}{D^{(2)}_{\text{long-range}} D^{(1)}_{\text{inter}}} \tag{51}$$

For molecules of comparable mass and cross-sections, the values of D_{inter} are of the same order, so that the separation factor may be estimated from the ratio of the two long-range diffusivities. These values, likewise indicated in Fig. 32b, can be shown to be in satisfactory agreement with the separation factors determined in the traditional way from two-component desorption data [203].

5. Evidence on the Limiting Steps of Mass Transfer

Mass transfer from the void space between the individual pellets of a bed of granulated zeolites into the crystallites proceeds in a sequence of three transport processes: (1) mass transfer through the intercrystalline space, (2) penetration through the crystallite surface, and (3) intracrystalline diffusion [1–5]. Depending on the given system, each of these processes may be rate determining for the overall adsorption/desorption process. An analysis of the relative importance of each of these processes may therefore be of substantial relevance for ensuring optimum transport conditions for the application of zeolites in both fluid separation and catalysis.

Within the formalism of the statistical moments [4,5,7–9], the time constant of molecular exchange between the intracrystalline space and the atmosphere around the zeolite pellets [namely, the first statistical moment of the overall exchange curve, as defined by Eq. (1)] appears as the sum of the following contributions: (1) from intracrystalline diffusion [$\tau_{cryst.} \equiv \tau_{intra}^{Diff.} = R_c^2/(15D_{intra})$, Eq. (2)], (2) from the penetration through the crystallite surface [$\tau_{s.-barr.} = R_c/(3\alpha)$, with α denoting the surface permeability with respect to the intracrystalline concentration, Eq. (12)], and (3) from long-range diffusion [$\tau_{pellet} = R_p^2/(15D_{long-range})$, Eq. (2)], where both the crystallites and pellets have been assumed to be of nearly spherical shape with radii R_c and R_p, respectively.

It is noteworthy that, in principle, each of these time constants may be determined by the PFG NMR method: $\tau_{cryst.}$, which is identical with $\tau_{intra}^{Diff.}$ and τ_{pellet} directly from the measurement of the coefficients of intracrystalline and of long-range diffusion, and $\tau_{s.-barr.}$ by combining the result of the NMR tracer desorption measurements (τ_{intra}; see Sec. III.C.) with $\tau_{intra}^{Diff.}$, using the equation $\tau_{intra} = \tau_{intra}^{Diff.} + \tau_{s.-barr.}$. Since in the NMR measurements a wide range of temperature and pressure may be covered (see Ref. 84), the relevant data may be determined under conditions quite close to the reaction conditions, provided that the conditions for a successful application of PFG NMR (see Sec. III.B.2.) are given.

E. PFG NMR Measurements under Nonequilibrium Conditions

As a noninvasive method, NMR spectroscopy provides ideal conditions for the in situ measurement of the dynamical and structural properties of systems undergoing internal changes. This is true in particular for the observation of catalytic reactions in beds of zeolite crystallites [204–208] and, very recently, of the space and time dependence of molecular diffusion during adsorption by combining PFG NMR with NMR imaging [209]. By applying Fourier transform PFG NMR it is possible to monitor simultaneously the diffusivities of various reactant and product molecules during the process of chemical reaction, provided that there is at least one characteristic line in the spectra of the relevant components. This procedure allows a separation into the different components. The first PFG NMR measurements during chemical reaction have been carried out during the conversion of cyclopropane to propene over zeolite NaX [208]. These measurements were made within sealed NMR sample tubes at a reaction temperature of 200°C. Under such conditions, the intracrystalline concentrations of the reactant and product molecules remain spatially constant during the reaction. The measured quantities, though determined under reaction conditions, are therefore self-diffusivities. This is in contrast to the situation in conventional flow reactors, where the concentrations of the individual components are space dependent and one has to deal with net mass transfer processes within the sample being described by coefficients of transport and/or counterdiffusion.

Figure 33 shows the time dependence of the diffusivities of the reactant molecules (cyclopropane) and of the product molecules (propene) as well as of their relative amount, as determined by an analysis of the contributions of both molecular species to the NMR signal following the first $\pi/2$ pulse (free induction decay) under reaction conditions [208]. The diffusivities of these reactant and product molecules happen to be quite close to each other, so there is no essential change in the diffusivity of either of the components with increasing reaction time. It is remarkable that this result may be predicted already on the basis of PFG NMR measurements at lower temperatures (i.e., far away from reaction conditions), where both diffusivities are also found to be nearly identical and independent of the composition [208]. In such a case the self-diffusivity should apparently coincide with the coefficient of counterdiffusion.

From Eq. (2), the measured diffusivities may be used to determine the mean lifetime of the reactant and product molecules within the individual crystallites under the assumption that the molecular exchange is exclusively controlled by intracrystalline diffusion. These values, being of the order of 30 ms, are found to agree with the real intracrystalline mean lifetime directly determined by NMR tracer desorption studies [208], so that any influence of crystallite surface barriers may be excluded. From an analysis of the time dependence of the intracrystalline concentration of the reactant and product molecules, the intrinsic reaction time constant is found to be on the order of 10^4 s. This value is much larger than the intracrystalline mean lifetimes determined by PFG NMR, and thus any limiting influence of mass transfer for the considered reaction may be excluded. In agreement with this conclusion, the size of the applied crystallites was found to have no influence on the conversion rates in measurements with a flow reactor [208].

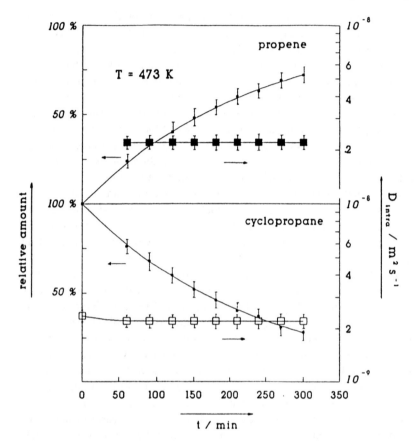

Figure 33 Time dependence of the relative amount of cyclopropane and propene during the conversion of cyclopropane to propene in NaX, together with values for their self-diffusion coefficients (☐ cyclopropane; ■ cyclopropene) at 473 K. (From Ref. 208.)

REFERENCES

1. G. F. Froment and K. B. Bischoff, *Chemical Reactor Analysis and Design*, Wiley, New York, 1979.
2. J. J. Carberry, *Chemical and Catalytic Reaction Engineering*, McGraw-Hill, New York, 1976.
3. C. N. Satterfield, *Mass Transfer in Heterogeneous Catalysis*, MIT Press, Cambridge, MA, 1970.
4. D. M. Ruthven, *Principles of Adsorption and Adsorption Processes*, Wiley, New York, 1983.
5. J. Kärger and D. M. Ruthven, *Diffusion in Zeolites and Other Microporous Solids*, Wiley, New York, 1992.

6. C. Förste, J. Kärger, H. Pfeifer, L. Riekert, M. Bülow, and A. Zikánová, *J. Chem. Soc. Faraday Trans.* 86: 881 (1990).

7. R. M. Barrer, *Zeolites and Clay Minerals as Adsorbents and Catalysts*, Academic Press, London, 1978.

8. R. Ash, R. M. Barrer, and R. J. B. Craven, *J. Chem. Soc. Faraday Trans. 1 74*: 40 (1978).

9. M. Kočiřik and A. Zikánová, *Z. Phys. Chem.*, *Leipzig, 250*: 250 (1972).

10. M. M. Dubinin, I. T. Erashko, O. Kadlec, V. I. Ulin, A. M. Voloshchuk, and P. P. Zolotarev, *Carbon 13*: 193 (1975).

11. A. Zikánová, M. Bülow, and H. Schlodder, *Zeolites 7*: 115 (1987).

12. H. G. Karge and W. Niessen, *Catal. Today 8*: 451 (1991).

13. P. T. Callaghan, *Principles of Nuclear Magnetic Resonance Microscopy*, Clarendon, Oxford, 1991.

14. B. Blümich and W. Kuhn, eds., *Magnetic Resonance Microscopy*, VCH, Weinheim, 1992.

15. P. Lauterbur, *Nature 242*: 190 (1973).

16. P. Mansfield and P. G. Morris, *NMR Imaging in Biomedicine*, Academic Press, New York, 1982.

17. E. R. Andrew, *Acc. Chem. Res. 16*: 114 (1983).

18. K. Roth, *NMR-Tomographie und Spektroskopie*, Springer, Berlin, 1984.

19. W. Heink, J. Kärger, and H. Pfeifer, *Chem. Eng. Sci. 33*: 1019 (1978).

20. G. Guillot, A. Trokiner, L. Darrasse, and H. Saint-Jalmes, *J. Phys. D: Appl. Phys. 22*: 1646 (1989).

21. G. Guillot, G. Kassab, J. P. Hulin, and P. Rigord, *J. Phys. D: Appl. Phys. 24*: 763 (1991).

22. E. G. Smith, J. W. Rockliffe, and P. I. Riley, *J. Coll. Interf. Sci. 131*: 29 (1989).

23. R. L. Armstrong, A. Tzalmona, M. Mensinger, and C. Lemaire, in *Magnetic Resonance Microscopy* (B. Blümich and W. Kuhn, eds.), VCH, Weinheim, 1992, p. 309.

24. S. P. Zhdanov, S. S. Kvoshchov, and N. N. Samulevich, *Synthetic Zeolites*, Khimia, Moscow, 1981.

25. D. T. Hayhurst, A. Nastro, R. Aiello, F. Crea, and G. Giordano, *Zeolites 8*: 416 (1988).

26. U. Müller and K. K. Unger, *Zeolites 8*: 154 (1988).

27. G. Finger and J. Kornatowski, *Zeolites 10*: 615 (1990).

28. T. Ito and J. Fraissard, in *Proceed. 5th Intern. Conf. on Zeolites* (L. V. C. Rees, ed.), Heyden, London, 1980, p. 510.

29. J. Ripmeester, *J. Am. Chem. Soc. 104*: 289 (1982).

30. J. Fraissard and T. Ito, *Zeolites 9*: 350 (1989).

31. J. Demarquay and J. Fraissard, *J. Chem. Phys. Lett. 136*: 314 (1987).

32. T. Ito and J. Fraissard, *J. Chem. Soc. Faraday Trans. 1 83*: 451 (1987).

33. J. Fraissard, T. Ito, and L. C. de Menorval, *Proceedings 8th International Congress on Catalysis*, Berlin, 1984, pp. 111–125.

34. Q. J. Chen and J. Fraissard, *Chem. Phys. Lett. 169*: 595 (1990).

35. J. Fraissard, A. Gedeon, Q. Chen, and T. Ito, Proc. *Intern. Symp. on Zeolite Chem. and Catal.*, Prague, 1991.

36. T. Ito, L. C. de Menorval, E. Guerrier, and J. Fraissard, *J. Chem. Phys. Lett. 111*: 3271 (1984).
37. R. Shoemaker and T. Apple, *J. Phys. Chem. 91*: 4024 (1987).
38. B. F. Chmelka, R. Ryoo, S. B. Liu, and L. C. de Menovral, *J. Am. Chem. Soc. 110*: 4465 (1988).
39. A. Gedeon, T. Ito and J. Fraissard, *Zeolites 8*: 376 (1988).
40. N. Bansal and C. Dybowski, *J. Magn. Res. 89*: 21 (1990).
41. J. Kärger, *J. Magn. Res. 93*: 184 (1991).
42. B. F. Chmelka, J. V. Gillis, E. E. Petersen, and C. J. Radke, *AIChE-J. 36*: 1562 (1988).
43. J.-F. Wu, T.-L. Chen, L.-J. Ma, M.-W. Lin, and S.-B. Liu, *Zeolites 11*: 86 (1991).
44. B. F. Chmelka, D. Raftery, A. V. McCormick, L. C. de Menorval, R. D. Levine, and A. Pines, *Phys. Rev. Lett. 66*: 580 (1991).
45. J. Kärger, H. Pfeifer, T. Wutscherk, S. Ernst, and J. Weitkamp, *J. Phys. Chem. 96*: 5059 (1992).
46. W. Heink, J. Kärger, H. Pfeifer, and F. Stallmach, *J. Am. Chem. Soc. 112*: 2175 (1990).
47. C. Förste, A. Germanus, J. Kärger, H. Pfeifer, J. Caro, W. Pilz, and A. Zikánová, *J. Chem. Soc. Faraday Trans. 1 1987*: 2301 (1987).
48. J. Kärger, *AIChE-J. 28*: 417 (1982).
49. K. Beschmann, G. T. Kokotailo, and L. Riekert, *Chem. Eng. Process. 22*: 223 (1987).
50. D. B. Shah, D. T. Hayhurst, G. Evanina, and C. J. Guo, *AIChE-J. 34*: 1713 (1988).
51. H. Pfeifer, in *Fundamental Aspects of Heterogeneous Catalysis Studied by Particle Beams*, NATO ASI Series B *265*: 151 (1991).
52. H. Pfeifer, *Physics Reports (Series C of Physics Letters) 26*: 293 (1976).
53. H. Winkler, M. Nagel, D. Michel, and H. Pfeifer, *Z. Phys. Chem., Leipzig 248*: 17 (1971).
54. D. Geschke and H. Pfeifer, *Z. Phys. Chem., Leipzig, 257*: 365 (1976).
55. H. Pfeifer, A. Gutsze, and S. P. Zhdanov, *Z. Phys. Chem., Leipzig, 257*: 721 (1976).
56. H. W. Spiess, *NMR—Basic Principles and Progress 15*: 55 (1978).
57. H. C. Torrey, *Phys. Rev. 92*: 962 (1953).
58. G. J. Krüger, *Z. Naturforsch. 24a*: 560 (1969).
59. M. Nagel, H. Pfeifer, and H. Winkler, *Z. Phys. Chem., Leipzig, 255*: 283 (1974).
60. H. Pfeifer and H. Winkler, *Proceedings 4th Specialized Colloque Ampere*, Leipzig, 1979, p. 31.
61. J. Kärger, H. Pfeifer, M. Rauscher, and A. Walter, *J. Chem. Soc. Faraday Trans. 1 76*: 717 (1980).
62. A. Germanus, J. Kärger, H. Pfeifer, N. N. Samulevich, and S. P. Zhdanov, *Zeolites 5*: 91 (1985).
63. J. Caro, M. Bülow, W. Schirmer, J. Kärger, W. Heink, H. Pfeifer, and S. P. Zhdanov, *J. Chem. Soc. Faraday Trans. 1*: 2541 (1985).
64. B. Boddenberg and B. Beerwerth, *J. Phys. Chem. 93*: 1440 (1989).

65. B. Boddenberg and G. Neue, *Mol. Phy. 67*: 385 (1989).
66. B. Zibrowius, J. Caro, and H. Pfeifer, *J. Chem. Soc. Faraday Trans. 1 84*: 2347 (1988).
67. N. Boden, S. M. Hanlon, Y. K. Levine, and M. Mortimer, *Mol. Phys. 36*: 519 (1978).
68. B. Boddenberg and R. Burmeister, *Zeolites 8*: 480 (1988).
69. B. Boddenberg and R. Burmeister, *Zeolites 8*: 488 (1988).
70. E. O. Stejskal and J. E. Tanner, *J. Chem. Phys. 42*: 288 (1965).
71. P. Stilbs, *Progr. NMR Spectrosc. 19*: 1 (1987).
72. P. T. Callaghan, *Aust. J. Phys. 37*: 359 (1984).
73. J. Kärger, H. Pfeifer, and W. Heink, *Adv. Magn. Res. 12*: 1 (1988).
74. J. Kärger and W. Heink, *J. Magn. Res. 51*: 1 (1983).
75. D. G. Cory and A. N. Garroway, *Magn. Reson. Med. 14*: 435 (1990).
76. P. Mansfield and P. K. Grannell, *Phys. Rev. B 12*: 3618 (1975).
77. P. T. Callaghan, D. MacGowan, K. J. Packer, and F. O. Zelaya, *J. Magn. Res. 90*: 177 (1990).
78. P. T. Callaghan, A. Coy, D. MacGowan, K. J. Packer, and F. O. Zelaya, *Nature 351*: 467 (1991).
79. R. M. Cotts, *Nature 351*: 443 (1991).
80. J. E. Tanner, *J. Chem. Phys. 52*: 2523 (1970).
81. P. T. Callaghan. J. Lelievre, and J. A. Lewis, *Carbohydrate Research 162*: 33 (1987).
82. G. Fleischer, *Coll. Polymer Sci. 265*: 89 (1987).
83. W. Heink, J. Kärger, H. Pfeifer, and G. Fleischer, *J. Magn. Res.*, in preparation,
84. W. Heink, J. Kärger, H. Pfeifer, P. Salverda, K. P. Datema, and A. Nowak, *J. Chem. Soc. Faraday Trans. 88*: 515 (1992).
85. R. Blinc, J. Pirš, and I. Zupančič, *Phys. Rev. Lett. 33*: 1192 (1973).
86. M. Silva-Crawford, B. C. Gerstein, A.-L. Kuo, and C. G. Wade, *J. Am. Chem. Soc. 102*: 3728 (1980).
87. A. Germanus, H. Pfeifer, W. Heink, and J. Kärger, *Ann. Phys.* (Leipzig) *40*: 161 (1983).
88. D. Zax and A. Pines, *J. Chem. Phys. 78*: 6333 (1983).
89. G. J. Krüger, H. Spiesecke, R. von Steenwinkel, and F. Noack, *Mol. Cryst. Liq. 40*: 103 (1977).
90. D. Canet, B. Diter, A. Belmajdoub, J. Brondeau, C. J. Boubel, and K. Elbayed, *J. Magn. Res. 81*: 1 (1989).
91. R. Dupeyre, Ph. Devoulon, D. Bourgeois, and M. Decorps, *J. Magn. Res. 95*: 589 (1991).
92. J. Stepišnik, *Physica B* (Amsterdam) *104B*: 350 (1981).
93. C. Förste, W. Heink, J. Kärger, H. Pfeifer, N. N. Feoktistova, and S. P. Zhdanov, *Zeolites 9*: 299 (1989).
94. J. Kärger and H. Jobic, *Colloids Surf. 58*: 203 (1991).
95. W. Heink, J. Kärger, and H. Pfeifer, *J. Chem. Soc., Chem. Commun.*: 1454 (1991).
96. W. Heink, J. Kärger, and H. Pfeifer, *Z. Phys. Chem. 170*: 199 (1991).
97. B. Gross and R. Kosfeld, *Meβtechnik 7/8*: 171 (1969).

98.   J. Kärger and W. Heink, *Exp. Techn. Phys.* *19*: 454 (1971).
99.   M. I. Hrovat and C. G. Wade, *J. Magn. Res.* *44*: 62 (1981).
100. P. T. Callaghan, *J. Magn. Res.* *88*: 493 (1990).
101. F. Stallmach, J. Kärger, and H. Pfeifer, *J. Magn. Res.* *102A*: 270 (1993).
102. L. E. Drain, *Proc. Phys. Soc.* *80*: 1380 (1962).
103. R. C. Wayne and R. M. Cotts, *Phys. Rev.* *151*: 264 (1966).
104. K. J. Packer, C. Rees, and D. J. Tomlinson, *Mol. Phy.* *18*: 421 (1970).
105. W. D. Williams, E. F. W. Seymour, and R. M. Cotts, *J. Magn. Res.* *31*, 271 (1978).
106. R. F. Karlicek and I. J. Lowe, *J. Magn. Res.* *37*: 75 (1980).
107. R. M. Cotts, M. J. R. Hoch, T. Sun, and J. T. Markert, *J. Magn. Res.* *83*: 252 (1989).
108. J. Kärger, H. Pfeifer, and S. Rudtsch, *J. Magn. Res.* *85*: 381 (1989).
109. J. Kärger, *Z. Phys. Chemie, Leipzig,* *248*: 27 (1971).
110. C. Förste, A. Germanus, J. Kärger, G. Möbius, M: Bülow, S. P. Zhdanov, and N. N. Feoktistova, *Isotopenpraxis 25*: 48 (1989).
111. R. Ash and R. M. Barrer, *Surface Sci.* *8*: 461 (1967).
112. L. Riekert, *Adv. Catalysis 21*: 281 (1970).
113. J. Kärger, *Surface Sci.* *36*: 797 (1973).
114. J. Kärger, *Surface Sci.* *57*: 749 (1976).
115. J. Kärger and J. Caro, *J. Chem. Soc. Faraday Trans.* *73*: 1363 (1977).
116. H.-J. Doelle and L. Riekert, *Am. Chem. Soc. Symp. Ser.* *40*: 401 (1977).
117. L.-K. Lee and D. M. Ruthven, *J. Chem. Soc. Faraday Trans. 1 75*: 2406 (1979).
118. D. M. Ruthven and L.-K. Lee, *AIChE Journal 26*: 654 (1981).
119. M. Bülow, J. Kärger, M. Kočiřík, and A. M. Voloshchuk, *Z. Chemie 21*: 175 (1981).
120. J. Caro, J. Kärger, G. Finger, H. Pfeifer, and R. Schöllner, *Z. Phys. Chem., Leipzig,* *257*: 903 (1976).
121. E. Cohen de Lara, R. Kahn, and F. Mezei, *J. Chem. Soc. Faraday Trans. 1 79*: 1911 (1983).
122. D. M. Ruthven, M. Eic, and Z. Xu, in *Catalysis and Adsorption by Zeolites* (G. Öhlmann, H. Pfeifer, and R. Fricke, eds.), Elsevier, Amsterdam, 1991, p. 233.
123. J. Kärger and D. M. Ruthven, *J. Chem. Soc. Faraday Trans. 1 77*: 1485 (1981).
124. H. Yucel and D. M. Ruthven, *J. Chem. Soc. Faraday Trans. 1 76*: 71 (1980).
125. H. Jobic, M. Beé, J. Caro, M. Bülow, J. Kärger, and H. Pfeifer, in *Catalysis and Adsorption by Zeolites* (G. Öhlmann, H. Pfeifer, and R. Fricke, eds.), Elsevier, Amsterdam, 1991, p. 445.
126. M. Bülow, P. Lorenz, W. Mietk, and P. Struve, *J. Chem. Soc. Faraday Trans. 1 79*: 1099 (1983).
127. A. Germanus, J. Kärger, and H. Pfeifer, *Zeolites 4*: 188 (1984).
128. C. J. Wright and C. Riekel, *Mol. Phys.* *36*: 695 (1978).
129. H. Jobic, M. Beé, J. Kärger, H. Pfeifer, and J. Caro, *J. Chem. Soc., Chem. Commun.*: 341 (1990).
130. M. Bülow, W. Mietk, P. Struve, and P. Lorenz, *J. Chem. Soc Faraday Trans. 1 79*: 2457 (1983).

131. M. Eic, M. Goddard, and D. M. Ruthven, *Zeolites 8*: 327, (1988).
132. J. Caro, M. Bülow, W. Schirmer, J. Kärger, W. Heink, H. Pfeifer, and S.P. Zhdanov, *J. Chem. Soc. Faraday Trans. 1 85*: 4201 (1989).
133. H. Jobic, M. Beé, J. Caro. J. Kärger, and M. Bülow, *J. Chem. Soc. Faraday Trans. 1 85*, 4201 (1989).
134. P. Demontis, E. S. Fois, G. B. Suffritti, and S. Quartieri, *J. Phys. Chem. 94*: 4329 (1990).
135. D. T. Hayhurst and A. R. Paravar, *Zeolites 8*: 27 (1988).
136. S. D. Pickett, A. K. Nowak, J. M. Thomas, B. K. Peterson, J. F. P. Swift, A. K. Cheetham, C. J. J. den Ouden, B. Smit, and M. F. M. Post, *J. Phys. Chem. 94*: 1233 (1990).
137. R. L. June, A. T. Bell, and D. N. Theodorou, *J. Phys. Chem. 94*: 8232 (1990).
138. M. Bülow, U. Härtel, U. Müller, and K. K. Unger, *Ber. Bunsenges. Phys. Chem. 94*: 74 (1990).
139. J. Kärger and H. Pfeifer, *J. Chem. Soc. Faraday Trans. 87*: 1989 (1991).
140. C. Förste, J. Kärger, and H. Pfeifer, in *Zeolites: Facts, Figures, Future* (P. A. Jacobs and R. A. van Santen, eds.), Elsevier, Amsterdam, 1989, p. 907.
141. C. Förste, J. Kärger, and H. Pfeifer, *J. Am. Chem. Soc. 112*: 7 (1990).
142. A. Germanus, J. Kärger, H. Pfeifer, N. N. Samulevich, and S. P. Zhdanov, *Zeolites 5*: 91 (1985).
143. S. Rudtsch, diploma thesis, Leipzig University, 1988.
144. Ph. Grenier, F. Meunier, P. G. Gray, L. M. Sun, S. Rudtsch, J. Kärger, Z. Xu, and D. M. Ruthven, *Zeolites*, submitted.
145. J. Kärger and H. Pfeifer, *Zeolites 7*: 90 (1987).
146. K. J. Laidler, *Chemical Kinetics*, McGraw-Hill, London, 1965.
147. D. M. Ruthven and R. I. Derrah, *J. Chem. Soc. Faraday Trans. 1 68*: 2332 (1972).
148. J. Kärger, H. Pfeifer, and R. Haberlandt, *J. Chem. Soc. Faraday Trans. 1 76*: 1569 (1980).
149. A. Tiselius, *Nature 133*: 212 (1933).
150. A. Tiselius, *J. Phys. Chem. 40*: 223 (1936).
151. R. M. Barrer, *Adv. Chem. Ser. 101*: 101 (1971).
152. D. M. Ruthven, M. Eic, and E. Richard, *Zeolites 11*: 647 (1991).
153. U. Hong, J. Kärger, R. Kramer, H. Pfeifer, G. Seiffert, U. Müller, K. K. Unger, H.-B. Lück, and T. Ito, *Zeolites 11*: 816 (1991).
154. B. Zibrowius, J. Caro, and J. Kärger, *Z. Phys. Chem., Leipzig, 269*: 1101 (1988).
155. J. Kärger, *J. Phys. Chem. 95*: 5558 (1991).
156. U. Hong, J. Kärger, H. Pfeifer, U. Müller, and K. K. Unger, *Z. Phys. Chem., Leipzig, 173*: 225 (1991).
157. D. N. Theodorou, private communication.
158. S. J. Goodbody, K. Watanabe, D. MacGowan, J. P. B. Walton, and N. Quirke, *J. Chem. Soc. Faraday Trans. 87*: 1951 (1991).
159. H. Pfeifer, D. Freude, and J. Kärger, in *Catalysis and Adsorption by Zeolites* (G. Öhlmann, H. Pfeifer, and R. Fricke, edts.), Elsevier, Amsterdam, 1991, p. 89.
160. W. Keller, diploma thesis, Leipzig University, 1992.

161. J. Kärger, *Phys. Rev. A 45*: 4173 (1992).
162. J. Kärger, M. Petzold, H. Pfeifer, S. Ernst, and J. Weitkamp, *J. Catal. 136*: 283 (1992).
163. J. Kärger, M. Bülow, and P. Lorenz, *J. Coll. Interf. Sci. 65*: 181 (1978).
164. P. Lorenz, M. Bülow, and J. Kärger, *Izv. Akad. Nauk SSSR, Ser. Khim.*: 1741 (1980).
165. P. Lorenz, M. Bülow, and J. Kärger, *J. Colloids Surf. 11*: 353 (1984).
166. O. H. Tezel, D. M. Ruthven, and D. L. Wernick, in *Proc. 6th Intern. Zeolite Conf.* (D. H. Olson and A. Bisio, eds.), Butterworths, Guildford, 1984, p. 232.
167. D. L. Wernick and E. J. Osterhuber, in *Proc. 6th Internat. Zeolite Conf.* (D. H. Olson and A. Bisio edits.), Butterworths, Guildford, 1984, p. 122.
168. T. L. James and G. G. McDonald, *J. Magn. Res. 11*: 58 (1973).
169. P. Stilbs and M. E. Moseley, *Chem. Scr. 13*: 26 (1978–79).
170. P. Stilbs, M. E. Moseley, and B. Lindman, *J. Magn. Res. 40*: 401 (1980).
171. K. P. Datema, C. J. J. den Ouden, W. D. Ylstra, H. P. C. E. Kuipers, M. F. M. Post, and J. Kärger, *J. Chem. Soc. Faraday Trans. 87*: 1935 (1991).
172. U. Hong, J. Kärger, and H. Pfeifer, *J. Amer. Chem. Soc. 113*: 4812 (1991).
173. R. Richter, thesis (Promotion A), Leipzig University, 1983.
174. R. Richter, R. Seidel, J. Kärger, W. Heink, H. Pfeifer, H. Fürtig, W. Höse, and W. Roscher, *Z. Phys. Chem., Leipzig 267*: 841 (1986).
175. J. Kärger and H. Pfeifer, *ACS Sympos. Ser. 368*: 376 (1988).
176. J. Kärger and D. M. Ruthven, *Zeolites 9*: 267 (1989).
177. G. T. Kerr, *J. Catal. 15*: 200 (1969).
178. F. Stallmach, thesis, Leipzig University, 1993.
179. J. Kärger, H. Pfeifer, F. Stallmach, N. N. Feoktistova, and S. P. Zhdanov, *Zeolites 13*: 50 (1993).
180. J. Kärger, M. Bülow, G. R. Millward, and J. M. Thomas, *Zeolites 6*: 146 (1986).
181. E. F. Kondis and J. S. Dranoff, *Ind. Eng. Chem. Process Design Develop. 10*: 108 (1971).
182. J. Kärger, H. Pfeifer, R. Seidel, B. Staudte, and T. Gross, *Zeolites 7*: 282 (1987).
183. J. Kärger, H. Pfeifer, R. Richter, H. Fürtig, W. Roscher, and R. Seidel, *AIChE-J.*: 1185 (1988).
184. R. Richter, R. Seidel, J. Kärger, W. Heink, H. Pfeifer, H. Fürtig, W. Höse, and W. Roscher, *Z. Phys. Chem., Leipzig, 267*: 1145 (1986).
185. J. Kärger, H. Pfeifer, F. Stallmach, M. Bülow, P. Struve, R. Entner, H. Spindler, and R. Seidel, *AIChE-J. 36*: 1500 (1990).
186. R. Seidel and J. Kärger, *AIChE-J.*, in preparation.
187. J. Kärger, H. Pfeifer, J. Caro, M. Bülow, H. Schlodder, R. Mostowicz, and J. Völter, *Appl. Catal. 29*: 21 (1987).
188. J. Kärger, H. Pfeifer, J. Caro, M. Bülow, J. Richter-Mendau, B. Fahlke, and L. V. C. Rees, *Appl. Catal. 24*: 187 (1986).
189. J. Völter, J. Caro, M. Bülow, B. Fahlke, J. Kärger, and M. Hunger, *Appl. Catal. 42*: 15 (1988).
190. D. H. Olson and W. O. Haag, *ACS Symp. Ser. 248*: 275 (1984).
191. J. A. Lercher and G. Rumplmayr, *Appl. Catal. 25*: 215 (1986).

192. J. Caro, M. Bülow, M. Derewinski, J. Haber, M. Hunger, J. Kärger, H. Pfeifer, W. Storek, and B. Zibrowius, *J. Catal. 124*: 367 (1990).
193. J. Kärger, M. Kočiřík, and A. Zikánová, *J. Coll. Interf. Sci. 84*: 240 (1981).
194. J. E. Tanner and E. O. Stejskal, *J. Chem. Phys. 49*: 1768 (1968).
195. R. M. Barrer, in *Diffusion in Polymers*, Academic Press, London, 1968, p. 165.
196. J. Crank, *The Mathematics of Diffusion*, Clarendon Press, Oxford, 1975, p. 266.
197. R. E. Meredith and C. W. Tobias, *J. Appl. Phys. 31*: 1279 (1960).
198. S. C. Cheng and R. I. Vachon, *Int. J. Heat Mass Transfer 12*: 149 (1969).
199. S. H. Jury, *J. Franklin Inst. 305*: 79 (1978).
200. H. Stach, thesis (Prom. B), Berlin, Academy of Sciences, 1977.
201. J. Kärger, A. Zikánová, and M. Kočiřík, *Z. Phys. Chem.*, *Leipzig*, *265*: 587 (1984).
202. H. Thamm, thesis (Prom. A), Berlin, Academy of Sciences, 1975.
203. M. Bülow, A. Grossmann, and W. Schirmer, *Z. Chem. 12*: 161 (1972).
204. D. Michel, W. Meiler, and H. Pfeifer, *J. Mol. Catal. 1*: 85 (1975/76).
205. H. Pfeifer, W. Meiler, and D. Deininger, *Ann. Rep. NMR Spectrosc. 15*: 292 (1983).
206. M. W. Anderson and J. Klinowski, *Nature 339*: 200 (1989).
207. H. Ernst and H. Pfeifer, *J. Catalysis 136*, 202 (1992).
208. U. Hong, J. Kärger, B. Hunger, N. N. Feoktistova, and S. P. Zhdanov, *J. Catalysis 137*: 243 (1992).
209. J. Kärger, G. Seiffert, and F. Stallmach, *J. Magn. Res. A 102*: 327 (1993).

3

In Situ NMR

James F. Haw

Texas A&M University, College Station, Texas

I. INTRODUCTION

In 1992, the National Research Council published *Catalysis Looks at the Future* [1], a report from the Panel on New Directions in Catalytic Science and Technology. The Executive Summary of that report stated that:

> Substantial progress and scientific breakthroughs have been made in recent years in several fields, including atomic resolution of metal surfaces, in situ observation of an olefin complexed to zeolite acid sites by nuclear magnetic resonance (NMR) spectroscopy, and in situ characterization of several reaction intermediates by a variety of spectroscopic techniques. Theoretical modeling is now ready for substantial growth as a result of progress in computer technology and in theory itself. For these reasons, it is desirable to focus on areas in which the extensive scientific and technological resources of academe and industry may lead to the fastest practical results. In order of priority, these areas are
>
> 1. in situ studies of catalytic reactions;
> 2. characterization of catalytic sites (of actual catalysts) at atomic resolution (metals, oxides);
> 3. synthesis of new materials that might serve as catalysts or catalyst supports; and
> 4. theoretical modeling linked to experimental verification.

Various forms of spectroscopy have been applied to in situ studies of catalysis, and it is appropriate to cite a few examples. FT-IR is frequently employed for in situ investigations. The experimental configurations used can be either transmission studies of free-standing catalyst wafers [2] or diffuse reflectance measurements on samples in "catalytic reaction chambers" [3]. In situ Raman spectroscopy has also been applied [4]. X-rays have been used to study catalysts in situ, either by powder diffraction methods [5,6] or XAFS [7]. In situ imaging techniques are beginning to be applied to the measurement of spatial distributions and residence times in catalytic reactors. A recent example of this method employed positron-emission tomography [8].

No single spectroscopic technique can possibly be expected to solve all or even most of the mechanistic problems in heterogeneous catalysis. In situ NMR, however, has several compelling characteristics that suggest that it should receive particular attention for further development and application.

1. In situ NMR experiments can be performed on actual catalyst samples using increasingly realistic temperatures, pressures, and reactant loadings.
2. In the absence of high concentrations of paramagnetic sites or metal particles with undesirable magnetic properties, high resolution magic-angle spinning spectra of species on catalysts are relatively easy to obtain and interpret.
3. Suitable NMR experiments [9–11] can probe the structures of the organic adsorbates (^{13}C), Bronsted acid sites (^{1}H), other nuclei associated with active sites (e.g., ^{27}Al), other framework sites (e.g., ^{29}Si), or cations ion-exchanged into the catalyst or present as promoters. Probes such as ^{129}Xe [12–14] or bases [15] can also be used to characterize zeolites.
4. NMR can measure spatial distributions and rates of diffusion [16,17].
5. Chemists have developed a considerable amount of expertise in the use of conventional NMR methods for the elucidation of reaction mechanisms in solution, and much of this experience is directly transferable to in situ studies of reaction mechanisms or catalysts.
6. Suitable NMR experiments are able to characterize equilibrium structures and the frequencies and amplitudes of molecular motion. Such information is essential for the experimental verification of theoretical calculations on the structure and molecular dynamics of catalytic systems.

The term *in situ* literally means "in place." The ultimate in situ spectroscopic experiment would be conducted in place inside of a standard catalytic reactor. This extreme is rarely practical or even desirable, though, and an in situ experiment is generally understood to be one in which an attempt has been made to recreate some characteristics of the industrial process in an environment suitable for spectroscopy. Since reactor conditions such as temperature, pressure, and flowrate or reactor design (e.g., fluidized bed) may not be optimal for spectroscopy, some compromise is inevitable.

For the purposes of this chapter, an in situ experiment is considered to be one in which a chemical reaction occurs during the course of a series of spectroscopic measurements. In many in situ NMR experiments, the observations are made while the reaction is in progress at a suitably high temperature. Alternatively, the procedure might be divided into reaction periods at higher temperatures and observation periods at a lower temperature.

This chapter begins with a description of instrumentation specific for in situ NMR experiments, namely, in situ flow probes and sample preparation devices suitable for magic-angle spinning (MAS) experiments. We also review some of the literature through mid-1992 on the application of different in situ experimental modes to various catalytic problems. Since the magnetic properties of small metal particles create unique problems for the characterization of adsorbates on supported metals by NMR, such studies are treated in depth in a separate chapter in this volume. Only cursory mention is made of them here. The next section presents several brief examples of in situ MAS experiments and motivates a section on techniques for assigning resonances in in situ MAS experiments and the types of NMR experiments that are useful in these investigations. That discussion concludes with an example of some of the problems that can be encountered in attempting to reconcile in situ NMR and other types of in situ experiments performed under different conditions. The penultimate section illustrates several approaches to the use of NMR to test reaction mechanisms, and the chapter ends with suggested directions for future progress using in situ NMR.

II. INSTRUMENTATION AND TECHNIQUES FOR IN SITU NMR EXPERIMENTS

The basic NMR spectrometer for an in situ study differs little from a standard solids instrument. The heavy emphasis on variable temperature studies generally suggests a wide-bore magnet, but the specifications of the transmitting and receiving components are dictated by the particular type of NMR experiment. In situ ^{13}C MAS techniques, for example, typically involve narrow spectral widths and fairly sharp lines, and these requirements can be fulfilled using a modified solutions console with power amplifiers providing approximately 100 W on the ^{1}H decoupler channel and 400 W on the X-nucleus channel. On the other hand, nonspinning ^{1}H or ^{2}H measurements in flow probes or involving supported metal catalysts often result in line shapes determined by dipolar, quadrupolar, or Knight shift interactions. Such studies require consoles and amplifiers designed for general-purpose solids experiments.

Instrumentation and techniques unique to in situ studies are associated with sample preparation and probe design. Figure 1 depicts an in situ flow probe developed by Haddix et al. [18], a design that sacrifices the ability to rotate the sample in return for conditions that very closely mimic a bench-top flow reactor.

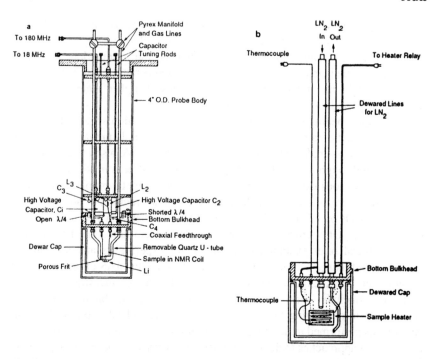

Figure 1 In situ NMR probe developed by Haddix, Reimer, and Bell: (a) gas flow system and the resonant circuit; (b) heating and cooling system components. (From Ref. 18.)

A typical experiment with this probe might involve in situ activation of the catalyst in a flowing gas stream, followed by the establishment of steady-state reaction conditions at elevated temperature through the introduction of one or two reactants into the flow stream. At this point in the experiment it is possible to monitor the product distribution exiting the NMR probe by means of a sampling valve and gas chromatograph. In principle, this probe could be used to monitor adsorbed species at high temperature during reactant flow. In practice, though, the temperature is reduced and the flow is halted once the desired steady state is established. Spectroscopy is then performed at a suitable temperature—usually ambient, although the probe is capable of operating at temperatures as low as 77 K.

Figure 2 shows broad-line ^1H spectra of ammonia on γ-Mo_2N obtained at 298 K following evacuation at progressively higher temperatures [18]. This study is an NMR analog of a temperature-programmed desorption (TPD) experiment. The probe design in Fig. 1 has been applied to chemistry of H_2 [19], NH_3 [20], and acetonitrile [21] on γ-Mo_2N, and a similar in situ flow probe has been used by Went and Reimer to characterize ammonia adsorption and decomposition on

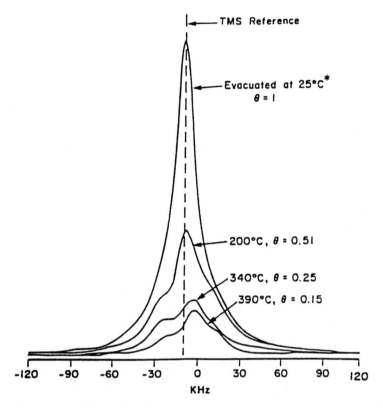

Figure 2 Temperature-programmed desorption ^{1}H NMR spectra of NH$_3$ on γ-Mo$_2$N. (From Ref. 18.)

titania-supported vanadium catalysts [22]. Iggo and coworkers have developed a somewhat different in situ flow probe, which they have applied to studies of reactions of methanol or hydrogen on a Cu/ZnO/Al$_2$O$_3$ methanol synthesis catalyst [23].

The ^{13}C resonances observed in the latter study were very broad, as expected for overlapping powder patterns. Many catalytic reactions involve the conversion of one or more reactants into a distribution of products by way of one or more long-lived intermediates. Once the reaction is initiated, the catalyst bed actually may contain a dozen or more species, each with varying concentrations over time. To obtain spectral resolution sufficient to establish the identities of such species, additional line narrowing is required. Magic-angle spinning is typically used to improve resolution under these circumstances.

An important difference between in situ MAS experiments and experiments using flow probes is that catalyst activation and reagent loading for in situ MAS occur outside the probe. These steps are most conveniently performed in a vacuum line. The sample must then be sealed and placed in a rotor and transferred to the spectrometer without any unintentional reactions. Two general experimental protocols are common. If the sample is sealed in a glass ampule, it is possible to alternate sample heating steps outside of the probe with spectroscopic characterization steps at room temperature. This protocol avoids the instrumental complexities associated with variable temperature MAS NMR, and it has been applied by Anderson, Klinowski, and coworkers in several studies of methanol chemistry on zeolites and other molecular sieves [24–26]. The same strategy has also been used by Nosov and coworkers to study CH_3OH and H_2S on HZSM-5 [27] and by Ernst and Pfeifer and to investigate synthesis of methylamines on a variety of catalysts [28].

The second general protocol uses variable-temperature MAS probes to observe the reaction while in progress. These experiments became feasible owing to the continued evolution of probes and experimental techniques for variable temperature operation [29–32]. The reliability of variable-temperature MAS probes has improved considerably in recent years. Commercial probes are available from several vendors with demonstrated temperature ranges of ca. 77 K to 523 K or higher. Using specially designed probes, Yannoni has performed MAS studies at temperatures as low at 5 K [33], and Stebbins has reported MAS spectra of a mineral sample at 773 K [34].

The conflicting requirements of handling a catalyst sample in a vacuum line and preparing it for MAS have been addressed by a variety of clever devices for sample handling and sealing. Figure 3 shows a design used by Gerstein and coworkers [35]. The catalyst sample can be treated at high temperature in the lower tube. Following cooling, the sample is transferred to what will become the rotor by rotating the Cajon ultra-torr tee. After the catalyst has fallen into place, an adsorbate is introduced and the sample is sealed. The rotor is formed by collapsing the tapered neck with a small glassblower's torch. An advantage of the design in Fig. 3 is that adsorption and sealing can be performed at cryogenic temperatures, and the sample can be transferred to a precooled NMR probe for in situ MAS studies or reactions at subambient temperatures.

A possible disadvantage of glass ampules is that they cannot be reopened for the subsequent addition or removal of reagents in more elaborate in situ NMR protocols. Figure 4 is a line drawing of another simple preparation device, the CAVERN apparatus [36], which permits the rotor to be sealed or unsealed in the vacuum line to facilitate multiple reagent adsorption stages. The rotor design depicted in Fig. 4 is based on a zirconia tube, which tolerates temperatures between 77 and 673 K, and plastic components on either end that are kept close to ambient temperature during high- or low-temperature experiments. Volatile adsorbates are introduced into the CAVERN from the vacuum line, but nonvolatile adsorbates can also be introduced by crushing a small glass capsule

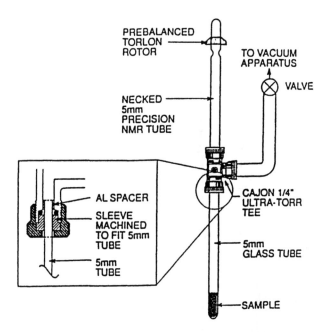

PREBALANCED
TORLON
ROTOR

TO VACUUM
APPARATUS

VALVE

NECKED
5mm
PRECISION
NMR TUBE

AL SPACER

SLEEVE
MACHINED
TO FIT 5mm
TUBE

5mm
TUBE

CAJON 1/4"
ULTRA-TORR
TEE

5mm
GLASS TUBE

SAMPLE

Figure 3 Two-part sample preparation cell. Insert shows sleeve and spacer that are used on both upper and lower 5-mm glass tubes. (From Ref. 35.)

above the catalyst bed during the sealing step. These procedures can be performed at cryogenic temperatures if necessary for subambient studies of very reactive species. Variations of the CAVERN apparatus have been used for a number of in situ studies including olefin oligomerization [37–39], hydrocarbon cracking [40], the chemistry of acetylene on acidic [41] and supported metal catalysts [42], methanol-to-gasoline chemistry on zeolite HZSM-5 [43–45], and chemistry relating to methanol synthesis on $Cu/ZnO/Al_2O_3$ [46].

One potential drawback of devices in which adsorption occurs on a "deep bed" of catalyst packed in an MAS rotor is the occurrence of a nonuniform distribution of reactant. This problem is illustrated by the ^{133}Cs MAS spectra of zeolite CsZSM-5 in Fig. 5 [47]. Without adsorbates, the ^{133}Cs chemical shift was -157 ppm at 298 K. An amount of methanol-^{13}C equivalent to 1 methanol molecule for every Cs^+ in the zeolite was then introduced using the apparatus in Fig. 4, and the spectrum in Fig. 5b was obtained. The spectrum shows the consequences of a deep-bed adsorption of a slowly diffusing adsorbate. Rather than a single ^{133}Cs resonance reflecting the adsorption of one equivalent of the methanol, there are two signals: one at -157 ppm originating from the bottom of the catalyst bed and a second at -82 ppm from the top of the catalyst bed. The

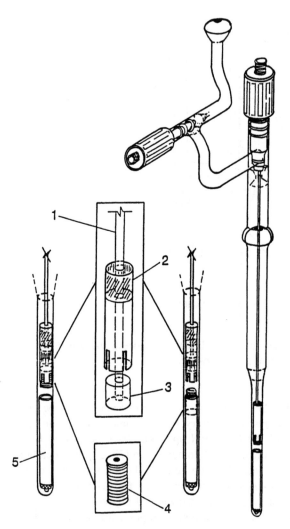

Figure 4 Diagram of the CAVERN apparatus used for both capping rotors following adsorption of reactants and the capping and uncapping of rotors to allow the sequential adsorption of reactants onto catalysts. Inset: Expanded views of the lower section of the CAVERN apparatus that depict the capping procedure. The diagrams on the left and right show the lower section of the CAVERN prior to capping and immediately after capping, respectively, and the middle sections show enlarged views of the plunger mechanism and Kel-F cap used. The parts are as follows: (1) aluminum rod; (2) Kel-F sleeve; (3) plunger; (4) Kel-F cap; (5) zirconia rotor.

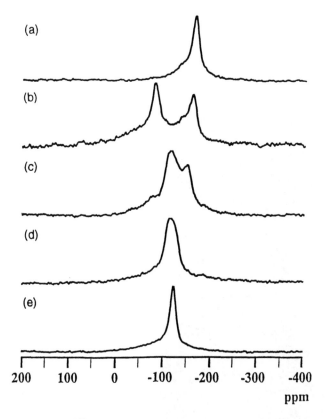

Figure 5 ^{133}Cs MAS NMR spectra of zeolite CsZSM-5 comparing the effects of sample preparation using a deep-bed CAVERN and a shallow-bed CAVERN: (a) zeolite without adsorbate; (b) zeolite after deep-bed absorption of methanol at 298 K; (c) the above sample after heating at 423 K for 15 min; (d) after 30 min at 423 K; (e) a second sample prepared by shallow-bed adsorption using the device in Figure 6.

former received no adsorbate, while the latter was nearly saturated. As shown in Fig. 5c and 5d, a uniform distribution of methanol could be obtained by heating the sample rotor in the MAS probe to a moderate temperature. In this example, the ^{13}C MAS spectrum (not shown) confirmed that no reaction took place, but in many cases the reactivity of the adsorbate precludes heating the sample in order to obtain uniform distribution. The photographs in Fig. 6 depict a device that can be used to achieve a shallow bed adsorption. The catalyst sample is initially distributed in a shallow layer on a glass "trap door" for activation at high temperature followed by adsorption at ambient temperature. The trap door is then opened and the catalyst falls into the MAS rotor and is sealed. This device was tested by the adsorption of methanol on CsZSM-5. As suggested by the ^{133}Cs

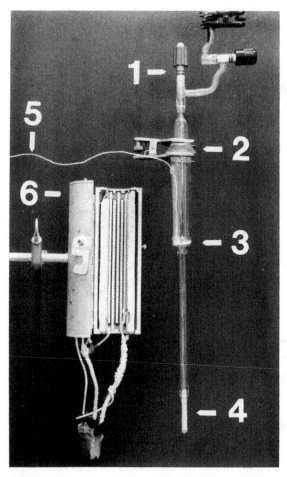

(a)

Figure 6 (a) Photograph of a shallow-bed CAVERN apparatus that permits catalyst activation at high temperature before adsorption at room temperature or above: 1. Teflon stopcock for extending and retracting stainless steel rod; 2. 35/25 ball-and-socket joint secured with a metal clamp; 3. activation chamber and catalyst bed; 4. Kel-F cap and zirconia MAS rotor; 5. thermocouple wire for monitoring catalyst temperature during activation; and 6. heater used to activate catalyst. (b) An enlarged view of the internal components of the activation CAVERN: 1. stainless steel rod; 2. bell-shaped glass trap door to prevent catalyst from falling into rotor during activation; 3. glass tubing that lifts trap door when the rod is retracted; 4. Kel-F cap; 5. zirconia MAS rotor.

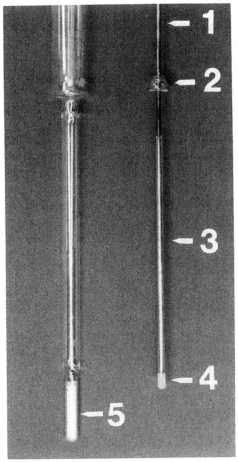

(b)

MAS spectrum in Fig. 5e, a uniform distribution of adsorbates was achieved by shallow-bed adsorption at room temperature.

A second major advantage of the device in Fig. 6 is that activation of the catalyst occurs in place prior to adsorption. This feature eliminates the need to handle the activated catalyst in a glove box.

III. SPECTROSCOPIC CAVEATS

A. Spectral Acquisition

Many of the examples in this chapter involve observation of ^{13}C, which is perhaps not surprising since catalytic processes are often concerned with the synthesis of organic products and because the spectroscopic properties of ^{13}C are favorable. ^{13}C MAS NMR is customarily performed with ^{1}H decoupling to remove line broadening arising from heteronuclear dipolar coupling. For very mobile physisorbed species, however, it may be possible to observe scalar coupling patterns in the absence of decoupling, and this approach is sometimes a useful way to assign resonances. Indeed, some catalyst/adsorbate systems are sufficiently liquidlike in their dynamical properties that a variety of solution state 1-D and 2-D NMR experiments is applicable with little or no modification [48].

One characteristic of in situ NMR experiments is that there is typically a wide range of correlation times characterizing molecular motion. Some species will be essentially immobile as a result of strong chemisorption to the catalyst surface or physical entrapment, as in the case of a coke molecule. Other species may reside exclusively in the gas phase or else be partitioned into adsorbed and gas phase populations in slow exchange on the NMR time scale owing to diffusional constraints. Figure 7 shows an example. At high temperatures, methanol and dimethyl ether are partitioned between the gas phase and adsorbed phase on zeolite HZSM-5 [31]. For many adsorbates on zeolites, especially at reaction

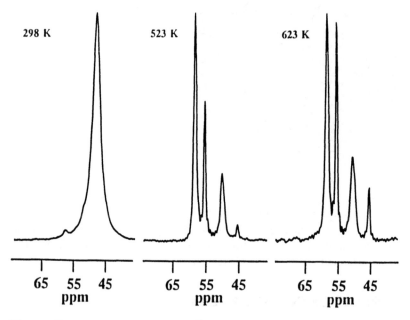

Figure 7 Expanded views from a ^{13}C in situ MAS study of methanol-to-gasoline chemistry showing the signals from methanol and dimethyl ether. In the high-temperature spectra two peaks are resolved for both species: gas phase (48.0 ppm) and adsorbed (53.1 ppm) methanol, and gas phase (58.0 ppm) and adsorbed (61.0 ppm) dimethyl ether.

temperature, diffusion will be rapid enough to make the ^{13}C NMR properties intermediate between solutionlike and solidlike behavior.

One of the practical consequences of this wide variation in dynamics is that cross-polarization (CP) [49] spectra may reflect a corresponding distribution in enhancements, and thus may be only semiquantitative. This problem is illustrated by the graphs of relaxation data in Fig. 8. ^{13}CH$_3$OH was adsorbed onto a sample of zeolite HZSM-5 and sealed into a magic-angle spinning rotor, which was then heated to 373 K to form a mixture of methanol, dimethyl ether, and water. The CP/MAS behavior of this system was then investigated at 298 K. The time constants characterizing polarization transfer (T_{CH}) were relatively long for both methanol and dimethyl ether, as expected for mobile species. These species, however, also exhibited fairly rapid and nonexponential ^{13}C $T_{1\rho}$ relaxation. Thus any arbitrary but reasonable choice of a contact time is likely to produce a quantitation error up to a factor of two or so. Fortunately, the signal-to-noise typically encountered in in situ MAS studies employing moderate loadings of ^{13}C-labeled adsorbates is more than sufficient, and the theoretical enhancement in sensitivity by a factor of four with efficient CP is not needed. Indeed, in the example shown in Fig. 8, cross-polarization is inefficient owing to motion, and

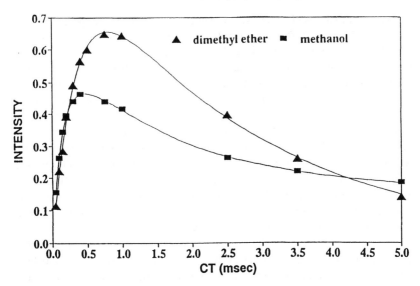

Figure 8 Plots of ^{13}C signal intensities versus cross-polarization contact time for methanol and dimethyl ether on zeolite HZSM-5. Intensities were normalized by division by the intensity obtained for each species in a direct 90° flip-observe experiment. Note that in this case the cross-polarization signals are less intense than the Bloch decay signals for all choices of the contact time (CT).

the CP signals are less intense than fully relaxed Bloch decays for all choices of contact time.

The other possible advantage of CP is to avoid a bottleneck caused by long ^{13}C T_1 values in rigid solids. The ^{13}C T_1s of these carbons are quite short, again owing to motion. The available reports in the literature suggest that ^{13}C T_1s for organic species on zeolites or amorphous oxide catalysts are rarely longer than several seconds. Even in the case of rigid coke deposits formed during cracking reactions, the ^{13}C T_1s are short by virtue of paramagnetic sites [50]. Exceptions to this rule can no doubt be found, but one is probably better off acquiring ^{13}C MAS spectra using 90° pulses (Bloch decays) and pulse delays of several seconds or more as needed to ensure quantitation. The modest CP enhancement or time savings achieved by neglecting to use these safeguards may not be worth the risk of semiquantitative results. The value of cross-polarization for in situ NMR experiments is more as an assignment technique or simply as a means to emphasize resonances from less mobile adsorbates. In practice, one can usually have it both ways. If the temperature is raised in small increments to avoid fast kinetics, it is possible to measure not only Bloch decay and CP spectra at each temperature, but also T_1s or other relaxation data as needed.

B. Chemical Shift Interpretation

The most commonly used spectroscopic parameter in ^{13}C MAS spectrometry is the isotropic chemical shift. This quantity can be difficult to interpret in the case of adsorbates on supported-metal catalysts as a result of possible Knight shifts and inhomogeneous broadening due to the magnetic anisotropy of the metal particle. It must be emphasized that these are not problems in the absence of metals, and they are often not problems for metals with favorable properties (a closed d-shell configuration, for instance, as with Ag) or for weakly bound, mobile adsorbates. The ^{13}C MAS linewidths of species bound to the frameworks of zeolites are usually no more than 5 ppm. Weakly adsorbed species typically have linewidths of 1 ppm or less, and the resolution of gas phase species may be limited by magnet homogeneity. The accuracy of ^{13}C chemical shifts may be established to within 1 ppm or better by external referencing, and this figure may be further improved sometimes by adding a suitable internal standard, such as methane, which is not strongly adsorbed by the catalyst.

The ^{13}C isotropic chemical shifts of adsorbates in zeolites are typically within 3 ppm of their corresponding values in solution. Small downfield shifts attributable to hydrogen bonding are often seen, mirroring the familiar solvent effects observed with changes in concentration or solvent. For example, Fig. 7 shows that gas-phase methanol has a chemical shift of 48.0 ppm vs 53.1 ppm in zeolite HZSM-5 at the same temperature [31]. Thus, in many cases, one can use ^{13}C shifts from solution data to help assign MAS spectra of products formed during in situ experiments. Because of the small shifts mentioned above, however, it is often advisable to verify the assignment by determining the chemical shift of an authentic sample of that compound in the zeolite.

An exception to the straightforward correspondence between ^{13}C shifts in zeolites (or other catalysts) and solution values occurs when the structure of the compound is significantly perturbed on the catalyst. The most common example is protonation equilibria on acidic catalysts. Indeed, there have been a number of reports of the use of protonation shifts of amines [51,52], phosphines [15], and phosphine oxides [53] as probes of catalyst acidity. Similar effects are occasionally encountered in in situ experiments when a basic molecule is formed as an intermediate or product. An interesting case is the conversion of acetone to hydrocarbons on zeolites, which may involve the intermediacy of diacetone alcohol, mesityl oxide, phorone, and isophorone—all ketones. The chemical shifts of the carbonyl carbons of all these species in acidic zeolites were found to be up to 10 ppm downfield of the corresponding values in reference compilations. Furthermore, although the chemical shifts of the olefinic carbons α to the carbonyl were in reasonable agreement with values for CDCl$_3$ solutions, the resonances of the olefinic carbons β to the carbonyl were very broad and shifted 20–30 ppm downfield [54].

Early in the development of ^{13}C NMR, it was recognized that the ^{13}C shifts of carbonyls are very sensitive to protonation equilibria [55,56].

Acetone, for example, has a chemical shift of 205 ppm in neat solution and a shift of 244 ppm in sulfuric acid. This equilibrium also explains the olefinic carbon shifts, as illustrated for the case of mesityl oxide.

Farcasiu has suggested that shifts of probe molecules like mesityl oxide be used in an acidity scale based on NMR measurements [57]. Farcasiu and coworkers showed that whereas the α olefinic carbon shifted only 2–3 ppm going from CDCl$_3$ to 100% H$_2$SO$_4$, the β olefinic carbon moved downfield by nearly 50 ppm. Thus, in the in situ study of acetone conversion, the apparent chemical shift discrepancies are accounted for by protonation equilibria.

These large apparent discrepancies between observed ^{13}C shifts and literature values are a double-edged sword. If one is not aware of such effects, they can confound all attempts at spectral assignment. But if one does recognize them, they are a source of information about the reaction mechanisms and perhaps some property of the catalyst such as acidity. The effects are not overly common, but given a working knowledge of physical organic chemistry they can usually be anticipated. Olah et al. have provided a very extensive literature of ^{13}C shifts of ions in strong acid and superacid solutions [58], and this work is very useful for in situ studies of reactions on acidic zeolites. Observed chemical shift discrepancies are usually discrepant only in appearance, the trick being to compare measurements in acidic zeolites to measurements in the appropriate solution.

^{1}H MAS NMR has also been applied to in situ work [59]. The familiar problem with proton NMR of solids, ^{1}H-^{1}H dipolar coupling, is usually not a major obstacle if the spins are dilute (as in the case of most zeolites and many amorphous oxides) and if any protonated adsorbates are sufficiently mobile. Thus the CRAMPS experiment [60] is not needed for line narrowing, and ^{1}H MAS spectra of zeolites with reasonable resolution may be obtained using MAS with modest spinning speeds, as demonstrated for zeolite HZSM-5 in Fig. 9. Nevertheless, linewidths of approximately 1 ppm cause more of a problem for ^{1}H studies than for ^{13}C studies owing to the modest chemical shift range of ^{1}H. In some cases it might be necessary to selectively deuterate either the catalyst or the adsorbate to resolve the feature of interest, and one must be aware of the possibility of ^{1}H chemical exchange in acidic media. The proton spectra of common zeolites (with adsorbates in some instances) have been at least partially assigned by Pfeifer and coworkers [61], Klinowski and coworkers [62], and

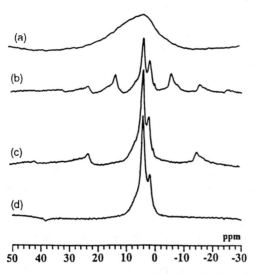

Figure 9 Proton MAS spectra of activated zeolite HZSM-5 at various spinning speeds: (a) no sample spinning; (b) 2.5 kHz; (c) 6 kHz; (d) 11 kHz.

others. Protonation or hydrogen bonding shifts of several ppm are often observed.

IV. EXAMPLES OF IN SITU MAS EXPERIMENTS

We now consider selected results that illustrate several of the in situ MAS protocols and caveats discussed above.

A. Variable-Temperature ^{13}C MAS—Catalytic Cracking

The catalytic cracking of crude oil into gasoline-range hydrocarbons on zeolite catalysts is responsible for a large portion of world petroleum production. Figure 10 shows the ^{13}C MAS NMR spectra of the reactions of ethylene oligomers on zeolite HZSM-5 [31]. The room-temperature spectrum consists of a broad peak centered at 32 ppm assigned to $(-CH_2-)_n$ chains and a narrower peak at 13 ppm assigned to $-CH_3$ groups. Heating of the sample to 523 K resulted in the cracking of the oligomers to propane, butanes, and higher aliphatic hydrocarbons (10–40 ppm). A small amount of aromatics was observed (130–140 ppm), as well as a peak at 250 ppm due to cyclopentenyl cations (vide infra). A general shift toward lower molecular weight hydrocarbons (propane and butanes) and more aromatics occurred when the sample temperature was increased to 573 K. At 623 K, the products consisted almost entirely of methane (-10 ppm), ethane (5 ppm), propane (15 ppm), and methyl-substituted benzenes (125–140 ppm).

These results are consistent with the expected cracking chemistry of ethylene oligomers in zeolite catalysts [63]. At 523 K, cracking reactions produced aliphatic hydrocarbons and carbenium ions. No ethane or methane was generated at 523 K, suggesting that β-elimination to form C_3 and higher hydrocarbons was the predominant cracking reaction. When the sample was heated 573 K, the average carbon chain length decreased as secondary cracking reactions began. Methane and ethane did not appear until the sample was heated to 623 K. The observation of these species demonstrates that different cracking chemistry was occurring on the zeolite at 623 K than at 523 K. The formation of methane can be explained by mechanisms that involve the formation of penta-coordinated carbonium ions, which can eliminate methane.

The peak at 250 ppm is indicative of three-coordinate carbenium ions [64]. A resonance at 250 ppm, observed in an in situ study of the oligomerization of propene on zeolite HY at low temperatures, was assigned to alkyl-substituted cyclopentenyl cations [37]. In that paper, the 250 ppm resonance was assigned to

I   II

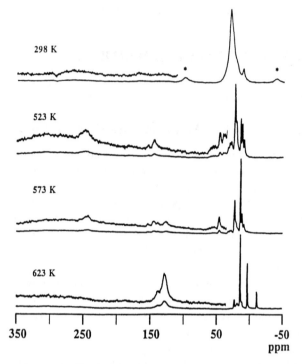

Figure 10 ^{13}C MAS NMR spectra showing the cracking of ethylene-^{13}C$_2$ oligomers on zeolite catalyst HZSM-5. As the sample temperature was increased, the hydrocarbon product distribution shifted from medium-range hydrocarbons (C$_3$–C$_6$) to light hydrocarbons (C$_1$–C$_3$) and coke. Highlighted regions show the presence of methyl-substituted cyclopentenyl carbenium ions (250, 148, and 48 ppm) in the spectra at 523 and 573 K. * denotes spinning sideband.

the C$_1$ and C$_3$ carbons in 1,2,3-trimethylcyclopentenyl cation (**I**), along with a peak at 158 ppm due to the C$_2$ carbon. In the cracking study there was no resonance at 158 ppm, but rather one at 148 ppm that correlated with the appearance and disappearance of the peak at 250 ppm. The ratio of the integrated intensity of the peak at 250 ppm to the peak at 148 ppm is approximately 2:1, which suggests that the resonances are due to 1,3-dimethylcyclopentenyl cation (**II**). The solution-state ^{13}C chemical shifts of 249 ppm for the C$_1$ and C$_3$ carbons and 148 ppm for the C$_2$ carbon for **II** agree with this assignment [65]. Further evidence for the existence of the cation is the resonance at 48 ppm, which lies outside the normal chemical shift region for simple aliphatic hydrocarbons but is in very good agreement with the reported value of 48.7 ppm for the C$_4$ and C$_5$ carbons of **II**. It is known that cyclopentenyl cations are very stable carbenium ions [66], and these species have long been proposed to form in the reactions of ethylene on zeolites. From studies of the reactions of ethylene on rare-earth exchanged X-type zeolites performed in the late 1960s, Venuto and Habib proposed a mechanism that involved cyclopentenyl cations as intermediates in the production of aromatic hydrocarbons [63]. The above observation is consistent with that proposal. Similar ions have also been proposed to form on the basis of in situ IR studies [67].

B. Off-Line Heating—Methanol Conversion Chemistry

One of the most successful routes to synthetic fuels is the conversion of methanol to gasoline (MTG) using the zeolite catalyst HZSM-5. This process, developed by Mobil, now produces one third of New Zealand's gasoline supply. Although this conversion has long been studied, the reaction mechanism is still poorly understood. Key questions in MTG chemistry include the mechanism of initial C-C bond formation, the identity of the "first" olefin, and the reason for an observed induction period preceding hydrocarbon synthesis [68,69].

Proposed mechanisms for C-C bond formation can be organized into the following classifications: carbene [70,71], carbocation [72], oxonium ylide [73,74], and free radical [75]. Some of these mechanisms [72–74] invoke a framework-bound methoxy species as a methylating agent or intermediate in the reaction. Dybowski and coworkers reported a ^{13}C NMR study that supported the formation of methoxy groups in HZSM-5 [76], but the existence of these species is still controversial [77]. Framework-bound alkoxy species clearly do form on other molecular sieve catalysts. For example, Anderson and Klinowski have shown that when methanol is heated to 573 K on the silicoaluminophosphate catalyst SAPO-5, a framework-bound methoxy species forms, which is readily seen in a ^{13}C MAS spectrum obtained after heating [77].

In situ NMR experiments of various types have been applied by several groups in an attempt to better understand MTG chemistry on HZSM-5 and other materials. Figure 11 shows results from a study by Anderson et al. on the reactions of methanol on the zeolite Offretite [24]. The catalyst was activated and loaded with 30 wt % methanol-^{13}C and sealed in a Pyrex capsule. The in situ experiment was performed by alternating between off-line heating steps and spectral acquisition at room temperature. Methanol was converted to dimethyl ether and then, after heating at 573 K for 10 minutes, to a mixture of hydrocarbons.

C. Natural Abundance ^{13}C Protocol

In most in situ ^{13}C MAS studies reported in the literature thus far, ^{13}C enrichment has been used to greatly increase sensitivity. A number of simple ^{13}C-labeled compounds are indeed available commercially at moderate cost, but overdepen-

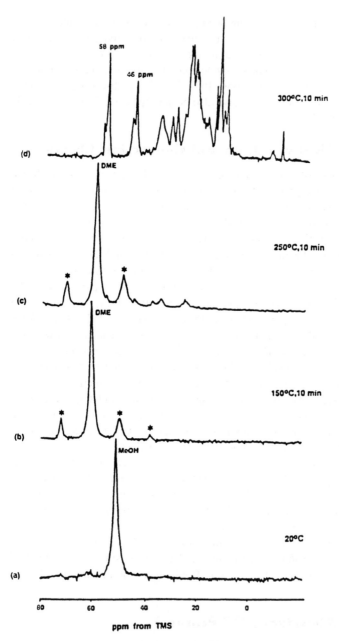

Figure 11 ^{13}C MAS NMR spectra of the products of the conversion of methanol over zeolite H-TMA-OFF at (a) room temperature, (b) 150°C, 10 min, (c) 250°C, 10 min, (d) 300°C, 10 min. * denotes spinning sidebands. (From Ref. 48.)

dence on labeled compounds can result in delays of weeks or months for the synthesis of a custom-labeled compound that may cost several thousand dollars per gram.

Modern ^{13}C MAS probes are sufficiently sensitive to bring in situ studies at natural abundance within reach. Figure 12 shows an in situ ^{13}C CP/MAS study of natural abundance diacetone alcohol, $CH_3COCH_2COH(CH_3)_2$, on zeolite HZSM-5. Each spectrum was the result of 20,000 scans taken at 298 K following 30 minutes of heating in the MAS probe at progressively higher temperatures. The experiment in Fig. 12 required 48 hours from start to finish, but the quality of the results was high enough to make it possible to sort out some of the observations in an in situ study of the acid-catalyzed aldol chemistry of acetone, which proceeds through diacetone alcohol.

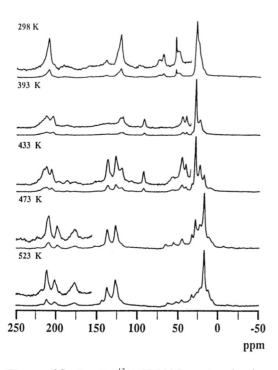

Figure 12 In situ ^{13}C CP/MAS spectra showing reactions of natural abundance diacetone alcohol on zeolite HZSM-5. Each spectrum was the result of 20,000 scans taken at 298 K following 30 minutes of heating in the MAS probe at progressively higher temperatures.

D.   ¹H MAS Study—Isotope Exchange

Labeled compounds are used in in situ NMR studies for reasons other than selectivity. Selective labeling can be used to avoid spectral overlap or to identify a reaction pathway by following the fate of a label. Figure 13 shows a case in point [59]. Benzene-d_6 was adsorbed on zeolite HZSM-5 at cryogenic temperature and transferred to an NMR probe, and ¹H MAS spectra were acquired as the temperature was increased. The spectrum taken at 173 K shows signals due exclusively to the zeolite: the external silanol peak at 2.0 ppm and the Bronsted acid site at 4.3 ppm. The latter site is the locus of catalytic activity for the zeolite. Several studies have shown that hydrogen-bonded adsorption complexes can form between acid sites in zeolites and electron-rich adsorbates. In the experiment depicted in Fig. 13, benzene-d_6 diffused into the zeolite as the temperature was raised to 268 K, forming a hydrogen-bonded adsorption complex with the Bronsted site. Interaction with the benzene molecules shifted the Bronsted site

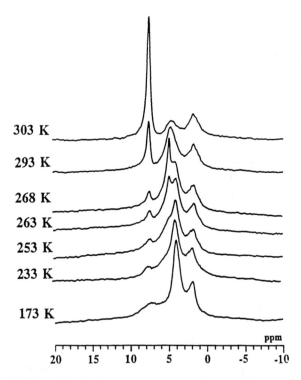

Figure 13   Proton MAS spectra showing the formation of an adsorption complex followed by proton-deuterium exchange as a sample of frozen benzene-d_6 on H-ZSM5 was heated from 173 K to ambient.

[1]H resonance from 4.3 to 5.1 ppm. At 293 K and above, label exchange between the Bronsted sites and benzene occurred at a measurable rate, and a ^{1}H signal for the former grew in at 7.5 ppm at the expense of the signal from the latter.

V. ASSIGNMENT PROBLEMS

A general problem in spectroscopic studies of catalytic systems is that of assigning a spectroscopic signal to a proposed species whose existence is not universally accepted. The problem can lead to controversies regarding the existence and significance of proposed intermediate species. An example of such a controversy familiar to the surface science community was the debate arising from studies in different laboratories using different surface spectroscopic techniques that culminated with the identification of ethylidyne on Pt(111). This section illustrates ways in which assignments can be developed in in situ MAS NMR with two detailed examples. The first deals with framework-bound alkoxys, which have figured prominently in a number of in situ studies. The second is a case history of the reconciliation of apparently discrepant results from in situ NMR and FTIR studies of the same system under somewhat different conditions.

In situ NMR [37,78] and infrared studies [2] in several laboratories have found evidence for the function of framework-bound alkoxy species such as **III–V** during the reaction of the parent alcohols or olefins on acidic zeolites. One

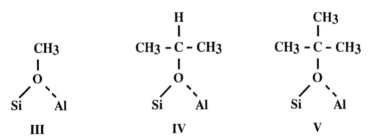

significant feature of the framework-bound methoxy species **III** is that it is proposed to be an important intermediate in several of the mechanisms proposed for MTG chemistry. One way to think of the formation of framework-bound alkoxy species is that the "incipient carbenium ion" forming during the dehydration of an alcohol or protonation of an olefin is captured by the nucleophilic framework oxygen.

Figure 14 shows an in situ ^{13}C MAS study of the reaction of methyl iodide on the basic zeolite CsX [79]. When [13]CH$_3$I was adsorbed on CsX zeolite at 298 K, two isotropic ^{13}C signals were seen in the MAS NMR spectrum. Unreacted methyl iodide adsorbed in the zeolite was observed at −15 ppm, a value 5 ppm downfield from its value in neat solution. The second ^{13}C signal in Fig. 15 is at a

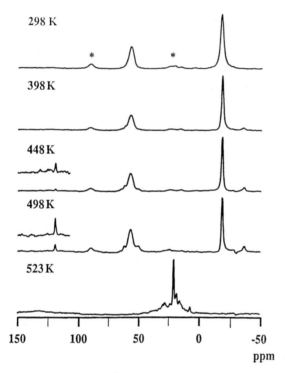

Figure 14 In situ ^{13}C MAS NMR study of the reactions of methyl iodide-^{13}C on zeolite CsX. Methyl iodide (-15 ppm) was partially converted to a framework-bound methoxy (58 ppm) upon adsorption at 298 K. Ethylene (120 ppm) formed between 448 and 498 K, and oligomerized and cracked to a mixture of hydrocarbons at 523 K. * denotes spinning sideband.

chemical shift of 58 ppm, which is apparently due to the framework-bound methoxy species **III**. This species was proposed to form by nucleophilic attack of the conjugate base site of the zeolite framework of methyl iodide with the assistance of Cs$^+$ in the removal of the leaving group, I$^-$. When the temperature was raised to the range of 448–498 K, ethylene began to form as indicated by its single, characteristic ^{13}C resonance. As will be shown later, when larger steady-state quantities of ethylene were generated, two ^{13}C signals due to ethylene were seen at 298 K, one at 125 ppm (strong adsorption site) and the other at 122 ppm (weak adsorption site). These assignments were confirmed through an exhaustive investigation in which different loadings of ethylene were adsorbed directly on CsX and their NMR properties were determined as the sample temperature was cycled. Whereas at room temperature all of the ethylene acted like an adsorbed species (cross-polarized efficiently, for example), a second peak, which cross-

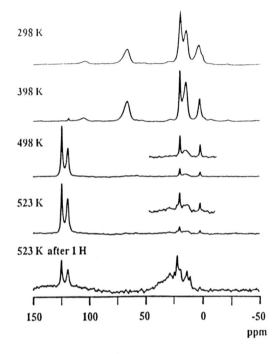

Figure 15 In situ ^{13}C MAS NMR study of the reactions of ethyl iodide-1,2-^{13}C on zeolite CsX. A framework-bound ethoxy formed upon adsorption at 298 K, and this was quantitatively converted to ethylene at 498 K. This product was partitioned between strong and weak adsorption sites (see text). Extended heating at 523 K led to further reactions.

polarized much less efficiently and showed a well-defined scalar multiplet without decoupling, grew in at 120 ppm when the temperature was raised. These effects were reversible. Evidence for two environments was also seen with high loadings of alkyl halides.

The MAS NMR probe temperature was raised to 523 K, and the final spectrum in Fig. 14 was obtained. This spectrum shows that the methyl iodide and framework-bound methoxy were converted quantitatively to a mixture of light aliphatic hydrocarbons (12–37 ppm) and possibly a small amount of aromatics or coke (broad signal ca. 125–150 ppm).

For the purposes of this chapter, the central claim of the in situ experiment depicted in Fig. 14 is that the framework-bound methoxy species **III** forms to an appreciable extent when methyl iodide is adsorbed on CsX zeolite at room temperature. The only evidence for this assignment in Fig. 14 is that the isotropic ^{13}C chemical shift (58 ppm) is consistent with a methyl group bound to an oxygen, and the small spinning sidebands suggest a species with restricted mobility. We will now consider how one might further support the assignment of the 58 ppm resonance to species **III**. For simplicity, this evidence will be restricted to chemical reactivity arguments and measurements of ^{13}C NMR properties, but corroborating results from other forms of spectroscopy or from nonspectroscopic physical methods might be essential to firmly establish a particularly controversial assignment.

If methyl iodide reacts readily to form a framework-bound methoxy species on CsX, then an analogous reaction should result for other sterically unhindered alkyl iodides. This process in fact occurs, as illustrated by the in situ ^{13}C MAS study of ethyl iodide-1,2-^{13}C on CsX depicted in Fig. 15. Here adsorbed ethyl iodide (C_1, 6 ppm; C_2 22 ppm) formed an appreciable amount of the framework-bound ethoxy species (**VI**) at room temperature, and the isotropic ^{13}C shifts of **VI** (C_1 68 ppm; C_2 17 ppm) are suggestive of an ethyl group bound to oxygen. For example, C_1 of diethyl ether is at 66 ppm in solution. Upon heating, **VI** was converted to ethylene, which was partitioned between strong adsorption sites at 125 ppm and weak sites at 122 ppm.

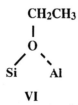

VI

Additional chemical evidence for the assignment of the 58 ppm resonance to the methoxy species **III** was the observation that it also formed from methyl bromide and methyl chloride in relative yields consistent with the leaving group stability $I^- > Br^- > Cl^-$. Methyl iodide was adsorbed on several zeolites with different Si/Al ratios, and the intensity of the 58 ppm resonance correlated with the Al content, as it must for a framework-bound alkoxy. The final example of chemical evidence for the assignment regards the expected chemistry of species such as **III** and **VI** upon exposure to moisture. The Si-O-C linkage is easily hydrolyzed on a silica gel surface to form alcohols and/or ether. As demonstrated in Fig. 16, the species assigned to **III** readily hydrolyzes to methanol and dimethyl ether, whereas the proposed ethoxy species formed from ethyl iodide-^{13}C hydrolyzed to ethanol upon exposure to atmospheric moisture.

The use of straightforward NMR techniques to further support the proposal that the species formed from alkyl halides on CsX are the framework-bound structures **III** and **VI** is illustrated in Figs. 17 and 18. A number of NMR properties is strongly dependent on the frequency and amplitude of molecular motion, the most familiar example being relaxation times. ^{13}C T_1 values are sometimes significantly higher for chemisorbed than for physisorbed moieties; in

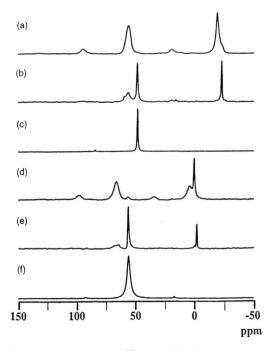

150   100   50   0   -50
ppm

Figure 16 Selected ¹³C MAS NMR results showing the hydrolysis of framework-bound alkoxy species to alcohols and ethers upon exposure to atmosphere. Methyl iodide-¹³C on CsX: (a) before exposure, showing a methoxy signal at 58 ppm; (b) after exposure, showing methanol at 51 ppm and a trace of dimethyl ether at 61 ppm; (c) ¹³C MAS spectrum of an authentic sample of methanol-¹³C on CsX. Ethyl iodide-1-¹³C on CsX: (d) before exposure; (e) after exposure, showing ethanol at 58 ppm; (f) ¹³C MAS spectrum of an authentic sample of ethanol-1-¹³C on CsX. All spectra were obtained at 298 K. Unreacted alkyl iodides partially diffused out of the catalysts during exposure.

the case of methyl iodide on CsX, the physisorbed CH_3I and the 58 ppm resonance had comparable rates.

A second general approach to probing dynamics is to look for changes in lineshape or signal intensity due to dipolar coupling or chemical shift anisotropy. Although these properties are averaged by rapid isotropic motion, they are scaled in a predictable manner by anisotropic motions such as uniaxial rotation [80]. In the case of species **III**, free methyl group rotation about the O-C bond would be expected, but no other large amplitude motion would be possible for a nonexchanging, framework-bound species. The spectra in Fig. 17 are consistent with the dynamics described above. Figure 17a is a Bloch decay spectrum obtained using a pulse delay $> 5\ T_1$ in order to ensure quantitation. Figure 17b was obtained with cross-polarization using a 2-ms contact time. Even weakly ad-

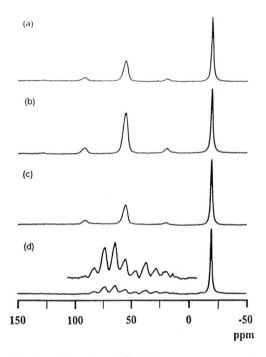

Figure 17   Selected ^{13}C MAS NMR results supporting the assignment of the methoxy resonance at 58 ppm: (a) Bloch decay spectrum; (b) cross-polarization spectrum; (c) spectrum with cross-polarization and interrupted decoupling; (d) slow-speed spectrum obtained at 830 Hz. All spectra were acquired at 298 K.

sorbed species in zeolites show some cross-polarization response, but relative cross-polarization efficiencies tend to be much higher for immobilized forms. Comparing Fig. 17a and b, one notes a significant cross-polarization enhancement for the 58 ppm resonance relative to the unreacted methyl iodide. An analogous result was seen for the signals assigned to the framework-bound ethoxy species (Fig. 18).

The most commonly applied assignment technique in ^{13}C MAS NMR is interrupted decoupling (also called dipolar dephasing) [81]. This experiment involves generating transverse magnetization and allowing it to evolve for a short period (typically 40–50 μs) in the absence of ^{1}H decoupling prior to observing the FID with the decoupler on. In a typical crystalline organic solid, short-range ^{1}H-^{13}C dipolar coupling results in near complete dephasing of ^{13}C magnetization from CH$_2$ and CH groups, while only attenuating signals from quaternary carbons and typical (unhindered) CH$_3$ groups. The interpretation of an interrupted decoupling experiment can be tenuous for a physisorbed species, but the

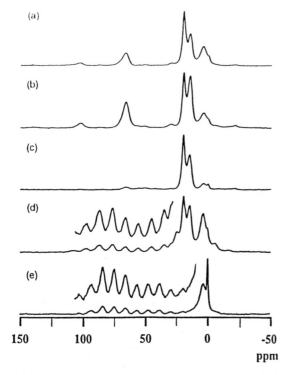

Figure 18 Selected ^{13}C MAS NMR results for ethyl iodide-1,2-^{13}C on CsX supporting the assignment of the ethoxy CH$_2$ resonance at 68 ppm and CH$_3$ resonance at 17 ppm: (a) Bloch decay spectrum; (b) cross-polarization spectrum; (c) spectrum obtained with cross-polarization and interrupted decoupling showing suppression of the methylene resonance of the ethoxy group; (d) slow-speed spectrum obtained at 930 Hz; (e) same as in d except that the ethoxy was prepared from ethyl iodide-1-^{13}C and the spinning speed was 820 Hz. All spectra were acquired at 298 K.

complete dephasing of ^{13}C magnetization is usually definitive for a rigid CH or CH$_2$ group. Figure 17c shows that the signal assigned to the methoxy species was attenuated by a factor of about 2 after 50 μs of interrupted decoupling, while the signal due to physisorbed CH$_3$I was unaffected. The resonance assigned to the CH$_2$ of the ethoxy species (Fig. 18) was almost completely eliminated by interrupted decoupling, whereas the CH$_3$ signal of the ethoxy (17 ppm) was attenuated by a factor of nearly two. These interrupted decoupling results are consistent with the proposed assignments.

Figures 17d and 18d and e show slow-speed MAS spectra from experiments with ^{13}CH$_3$I, ^{13}CH$_3$^{13}CH$_2$I, and CH$_3$^{13}CH$_2$I, respectively. These spectra demonstrate that the signals assigned to framework alkoxys give rise to four or more orders of spinning sidebands, a result consistent with species immobilized in some manner. The intensities of the sideband patterns in Fig. 17d were analyzed by the method of Herzfeld and Berger [82], and the following principal components of the ^{13}C chemical shift were calculated for the 58 ppm resonance: $\delta_{11} = 82$ ppm, $\delta_{22} = 81$ ppm, and $\delta_{33} = 9$ ppm. These values are in the range for methoxy groups in organic solids as well as for analogous species on other catalytic surfaces. The chemical shift anisotropy (73 ppm) and asymmetry parameters (0.01) calculated from the principal components are consistent with a nearly axially symmetric methoxy group undergoing rapid rotation about the O-C bond, but not other large amplitude motions. In the case of the ethoxy group, the large chemical shift anisotropy (92 ppm) and asymmetry parameter (0.56) determined from Fig. 18e suggests that the bulk of the methyl group precludes rapid motion about the O-C bond.

In recent years, several new experiments have been developed or proposed that show promise for facilitating the assignments of species formed during in situ magic angle spinning. Burum has described so-called WIMSE experiments, for example, which can differentiate unambiguously between carbon classes even in the presence of partial averaging of ^{1}H-^{13}C dipolar coupling by molecular motion [83]. For catalyst/adsorbate systems, the technique should have a significant advantage over interrupted decoupling. MAS also is usually thought to remove homonuclear dipolar coupling between pairs of ^{13}C and heteronuclear coupling between ^{13}C and other low-γ spins like ^{15}N. Schaefer and coworkers, however, have developed a family of related methods, termed REDOR, for the selective reintroduction of heteronuclear dipolar couplings [84–86]. These experiments should be readily applicable to in situ assignment problems in which the adsorbate contains both ^{13}C and ^{15}N or ^{31}P, and it is interesting to speculate about a REDOR experiment involving one spin on the adsorbate and a second spin on the catalyst. In a recent series of papers, Griffin and coworkers have described a procedure in which the ^{13}C-^{13}C homonuclear dipolar coupling is selectively reintroduced by setting the MAS spinning speed equal to the frequency difference between two resonances (or a submultiple thereof) [87–89]. This effect is called rotational resonance, and it can be used to establish that two ^{13}C nuclei are within several Å of each other. Figure 19 shows MAS spectra of $^{13}CH_3{}^{13}CH_2I$ on CsX at 298 K and various spinning speeds. When the spinning speed was set to match rotational resonance between C_1 and C_2 of the proposed ethoxy species (4646 Hz), the ^{13}C-^{13}C dipolar coupling between these spins was reintroduced, resulting in a characteristic splitting pattern. The rotational resonance experiment thus established a spatial correlation between the two carbons and further supports the proposed assignments.

A number of NMR experiments can be carried out to further verify the assignments considered in this example. One might adsorb CD_3I on CsX and characterize the ^{2}H lineshape, which should consist of a nearly isotropic compo-

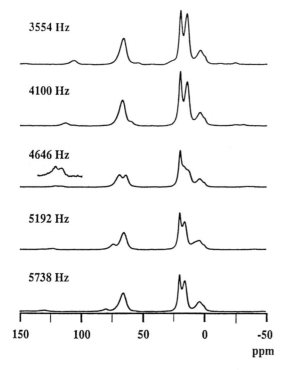

Figure 19 Demonstration of the rotational resonance effect for the framework-bound ethoxy species obtained on CsX from ethyl iodide-1,2-^{13}C. The rotational resonance condition was achieved when the spinning speed was set to match the frequency difference (4646 Hz) between the signals at 68 and 17 ppm. Rotational resonance reintroduced the ^{13}C-^{13}C dipolar coupling for the rigid ethoxy species, resulting in a splitting pattern that reflects the short internuclear distance.

nent from unreacted CD$_3$I centered on a Pake pattern with a width of approximately 50 kHz due to species **III**. The linewidth of the latter is characteristic of a methyl rotor, and a similar ^{2}H lineshape analysis has been used previously to identify ethylidyne on a supported Pt catalyst [90]. ^{2}H methods are generally useful for supported metal catalysts for which the susceptibility broadening often precludes the measurement of high-resolution chemical shift spectra.

Controversies in surface science and catalysis have arisen when different groups have applied different spectroscopic techniques to similar (but not necessarily identical) systems. Disagreements can even develop when different NMR techniques are applied to similar samples. We now consider an example of a (largely successful) attempt to reconcile a series of in situ ^{13}C MAS studies with an in situ infrared investigation of a similar system. This history highlights the advantages of applying multiple spectroscopic methods in catalytic studies.

In 1987, Forester and Howe published a detailed in situ FTIR study of methanol and dimethyl ether on HZSM-5 [2]. Their experiments were performed in transmission mode using a very thin free-standing catalyst wafter. In a typical experiment, 1–5 μL of either CH_3OH or CH_3OCH_3 was pulsed onto the catalyst at an appropriate temperature. These conditions result in fairly low loadings of adsorbate on the catalyst wafer.

The problems of peak assignment in infrared spectroscopy are probably greater than those faced in analogous NMR experiments. Furthermore, infrared transition probabilities are highly variable owing to symmetry considerations and other molecular properties. In the absence of calibration, relative intensities of infrared absorption bands do not accurately reflect relative concentrations. By characterizing the infrared properties of appropriate model compounds and

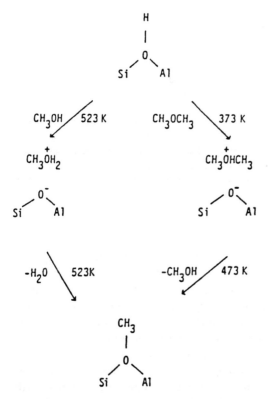

Figure 20 Schematic of the formation of protonated species and the framework-bound methoxy deduced from an in situ FTIR experiment. (From Ref. 2.)

drawing from other assignments in the literature, however, Forester and Howe were able to assign their in situ FTIR spectra with confidence; their chemical conclusions are summarized in Fig. 20. The adsorbates did not react at room temperature. Protonated dimethyl ether was observed at 373 K, and this species formed a framework-bound methoxy at 473 K. The behavior of methanol was analogous, although a temperature of 523 K was required in that case. One key negative result of the in situ FTIR study was the failure to observe trimethyloxonium, **VII**. The significance of this finding is that trimethyloxonium is a key

$$
\begin{array}{c}
CH_3 \\
| \\
O^{\,+} \\
\diagup \quad \diagdown \\
H_3C \qquad CH_3
\end{array}
$$

VII

intermediate in two related mechanisms that have been proposed for MTG chemistry on HZSM-5 or analogous reactions on other catalysts (Scheme I). In 1991, an in situ ^{13}C MAS study of dimethyl ether-$^{13}C_1$ on HZSM-5 was published [44]. The principal conclusion of that work was that dimethyl ether disproportionates on an acid site in the vicinity of room temperature to form trimethyloxonium (80 ppm) and a stoichiometric amount of methanol (Scheme II). The results of that as well as a subsequent study [45] (see Fig. 21 for a representative result) seemed to argue against the reaction mechanism in Scheme I. For example, trimethyloxonium was not observed in the presence of a large excess of methanol, and the ion decomposed at temperatures below the onset of hydrocarbon formation.

Nevertheless, the conflicting conclusions regarding the formation of trimethyloxonium were problematic since different forms of spectroscopy should ultimately reveal the same truth. There were several possible sources of this discrepancy. The first could be that the assignments in either the ^{13}C NMR or the FTIR work were incorrect. The second possibility was that there was some subtle difference in the zeolite samples or activation procedures that might account for the difference. Finally, both studies could be correct and the apparent discrepancy could reflect differences in the experimental protocols used in the respective in situ experiments.

The spectral assignments in the two investigations seemed clear enough. The interested reader is referred to the paper of Forester and Howe [2] for the arguments regarding the infrared work. As for NMR, the ^{13}C chemical shift of trimethyloxonium (80 ppm) formed in situ was identical to that observed in superacid solutions, as well as that observed for an authentic sample of trimethyloxonium-ZSM-5 prepared by direct ion exchange from a chilled solution. Other evidence for the assignment was the stoichiometric formation of methanol

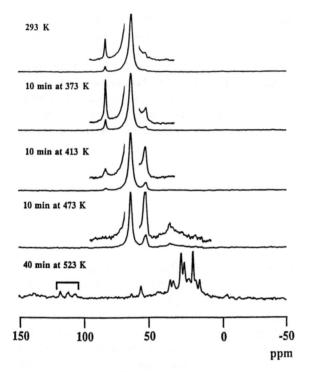

Figure 21 In situ ^{13}C CP/MAS NMR spectra showing the reactions of dimethyl ether-1-^{13}C on zeolite HZSM-5 as the sample was heated from 293 to 523 K. Highlighted regions show the increase and decrease of trimethyloxonium upon heating. Bracket denotes background signals from Kel-F caps.

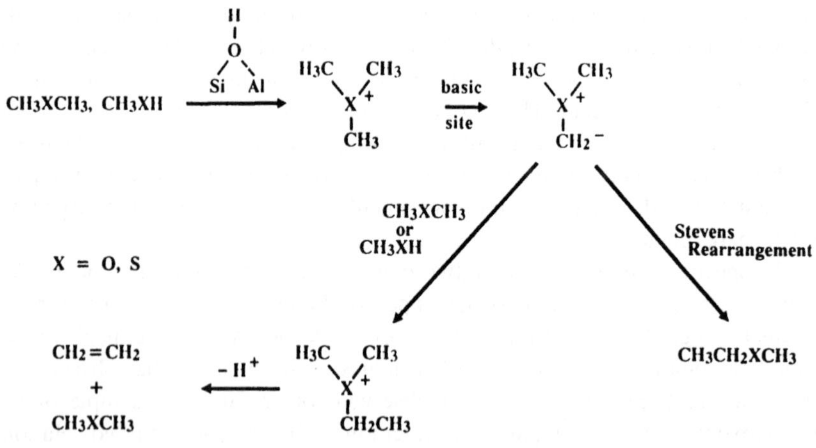

Scheme I

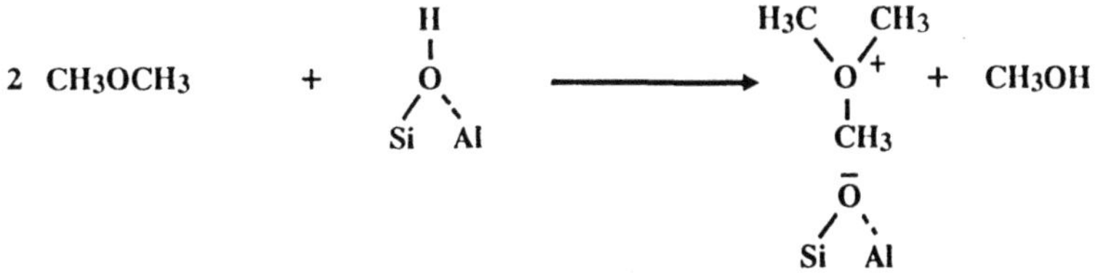

$$2\ CH_3OCH_3\quad +\quad \underset{\underset{Si\quad Al}{\diagup\ \diagdown}}{\overset{\overset{H}{|}}{O}}\quad \longrightarrow\quad \underset{\underset{\underset{Si\quad Al}{\diagup\ \diagdown}}{\overset{|}{\overline{O}}}}{\underset{\overset{|}{CH_3}}{\overset{\overset{H_3C\quad CH_3}{\diagdown\ \diagup}}{O^+}}}\quad +\quad CH_3OH$$

Scheme II

and consumption of an acid site, as predicted by Scheme II. The 80 ppm resonance did not form on NaZSM-5. These results offer compelling but perhaps not definitive proof that the 80 ppm resonance is in fact due to trimethyloxonium. Further evidence in support of this assignment included the formation and study of other oxonium species such as dimethylethyloxonium and diethylmethyloxonium ions [45]. These species also had chemical shifts and reactivities consistent with oxonium ions and readily formed on HZSM-5 from the corresponding ethers. The best evidence as obtained by generalizing the observations to include the formation of trimethylsulfonium, **VIII**, and trimethylselenium, **IX**, from the

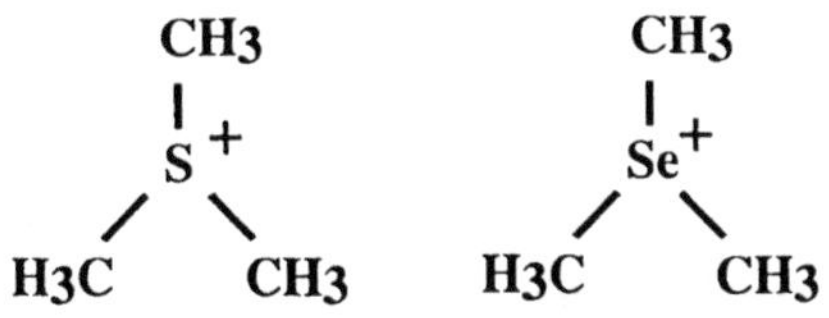

$$\underset{\underset{H_3C}{\diagup}\quad \underset{CH_3}{\diagdown}}{\overset{\overset{CH_3}{|}}{S^+}}\qquad\qquad \underset{\underset{H_3C}{\diagup}\quad \underset{CH_3}{\diagdown}}{\overset{\overset{CH_3}{|}}{Se^+}}$$

$$\textbf{VIII}\qquad\qquad\qquad \textbf{IX}$$

reaction of dimethyl sulfide and dimethyl selenide, respectively, on HZSM-5. Representative results are shown in Fig. 22. Dimethylsulfide disproportionates by a reaction analogous to Scheme II to form **VIII** and CH_3SH, and the chemical shift of **VIII** (28 ppm) is identical to that obtained for a sample of trimethylsulfonium-ZSM-5. In the reaction of dimethyl selenide, formation of trimethylselenonium was signaled by the ^{13}C spectrum, with an unambiguous resonance at 245 ppm in the corresponding ^{77}Se spectrum. A fairly comprehensive onium ion chemistry on zeolites was thus elaborated, and the trimethyloxonium assignment was verified by a wealth of circumstantial evidence.

The second possible explanation of the discrepancy—subtle differences in the catalyst used—was partly ruled out by reproducing the NMR results on a variety of well-characterized HZSM-5 samples from different industrial sources. A more definitive check of this hypothesis would have involved sample exchanges between the two groups, but this step was deemed not to be necessary.

The apparent discrepancy regarding the formation of trimethyloxonium was due instead to a difference in the experimental protocols used in the two in situ investigations. Figure 23 shows two in situ ^{13}C studies of dimethyl ether-$^{13}C_1$ on

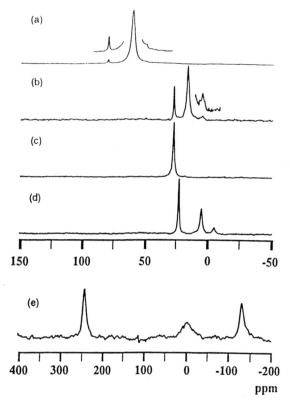

Figure 22 ^{13}C and ^{77}Se MAS NMR spectra showing the formation of trialkylonium ions from the disproportionation of dimethyl chalcogenides on zeolite HZSM-5. (a) ^{13}C spectrum of dimethyl ether adsorbed on HZSM-5; (b) ^{13}C spectrum of dimethyl sulfide on HZSM-5; (c) ^{13}C spectrum of an authentic sample of trimethylsulfonium-ZSM-5; (d) ^{13}C spectrum of dimethyl selenide on HZSM-5; (e) ^{77}Se spectrum of dimethyl selenide and methanol on HZSM-5.

HZSM-5 that were identical in all respects except for the loading of the adsorbate. Trimethyloxonium formed in a high loading experiment with 3.0 mmol dimethyl ether per gram of catalyst, but did not form when the loading was reduced by about one order of magnitude [91]. Those samples were prepared using shallow-bed adsorption; trimethyloxonium also formed when low loadings of dimethyl ether were introduced to a deep bed of catalyst. The formation of trimethyloxonium at high local loadings of ether apparently reflects solvation of the ion by excess ether as well as mass action. The experimental procedure used in the in situ FTIR study resulted in lower loadings and hence no trimethyloxonium.

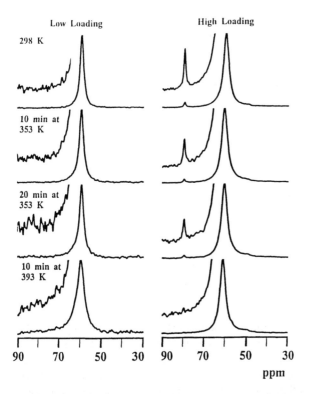

Figure 23 ¹³C MAS NMR spectra of the reactions of high (1.8 Eq/acid site) and low (0.2 Eq/acid site) loadings of dimethyl ether-1-¹³C on HZSM-5. Trimethyloxonium formed only in the high-loading sample and disappeared as the sample was heated to 393 K.

Variable temperature ¹³C MAS measurements were performed on several samples in order to find NMR evidence for the protonation of dimethyl ether at high temperatures. Plots of chemical shift vs. temperature are shown in Fig. 24. An important difference (and not always a good one) between NMR and other forms of spectroscopy is the time scale of the measurement. Rapid exchange of the proton at high temperature will lead to a single resonance at the population-weighted average shift of the protonated and free ether. Thus no discrete signal is observed for the protonated ether, and its presence could be masked by an excess of the nonprotonated form. The time scale of an infrared measurement is much shorter than that of most chemical exchange processes.

The chemical shift of dimethyl ether on HZSM-5 at 298 K was between 60 and 61 ppm, depending on loading. For comparison, Olah et al. reported that $CH_3OH^+ CH_3$ had a shift of 69.9 ppm in superacid solution [92]. Examining the

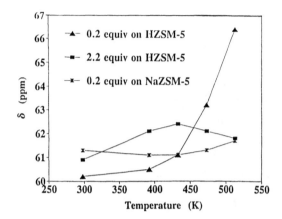

Figure 24 ^{13}C chemical shift change of high (2.2 equiv/acid site) and low (0.2 equiv/ acid site) loadings of dimethyl ether on HZSM-5 and a low loading of dimethyl ether on NaZSM-5 as the samples were heated from 393 to 513 K. Protonation shifts were observed at higher temperatures for low loadings of dimethyl ether on HZSM-5.

data in Fig. 25, one notes that the ^{13}C chemical shift of a low loading (less than one equivalent) of dimethyl ether approached that of the protonated form at high temperatures. A smaller shift was seen with a much higher loading, and essentially no shift was seen on NaZSM-5. These observations are consistent with protonation of dimethyl ether at high temperature, as was concluded from the in situ FTIR study. A further observation of the infrared study, formation of a framework-bound methoxy, could not be established from the temperature dependence of the chemical shift; a second exchange process might be operative at high temperature.

One conclusion from the above "detective story" is that two or more forms of spectroscopy or other evidence should be applied whenever possible, and one should keep in mind the possible consequences of differences in experimental protocols. The time scale of NMR chemical shift measurements is often useful for characterizing intermediate exchange rates, but it can be an obstacle to studies of complex systems in fast exchange.

VI. PROBING REACTION MECHANISMS WITH IN SITU NMR

The account given above illustrates the first caveat with any in situ study: the concentrations of intermediates and products may be strongly dependent on loading, pressure, steady-state vs. transient behavior, etc. Unimolecular processes may be replaced with bimolecular reactions at higher loadings, and

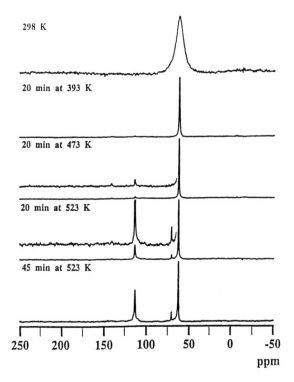

Figure 25 ^{13}C MAS NMR spectra of the reactions of allyl alcohol-1-^{13}C on zeolite CsHX. Label exchange between the 1 (64 ppm) and 3 (115 ppm) positions were observed as well as the formation of allyl ether (72 ppm).

parallel studies at various reactant loadings can be a prudent measure. In situ flow probes provide a much closer approximation to steady-state reactor conditions, but in many cases the complexity of the product distribution challenges the resolution attainable without sample rotation. A typical in situ MAS experiment involves a much longer residence time at a significantly lower temperature than the actual catalytic process. Whenever possible, the conclusions of an in situ NMR experiment should be compared to flow reactor studies or other forms of spectroscopy.

Despite the above cautions, some interesting mechanistic work has already been accomplished, such as a detailed characterization of framework-bound alkoxys by in situ MAS NMR. These species, a subject of much current research including theoretical investigations [93], have previously been proposed to be intermediates in several mechanisms for carbon-carbon bond formation. For example, Chang and coworkers suggested the mechanism in Scheme III as an

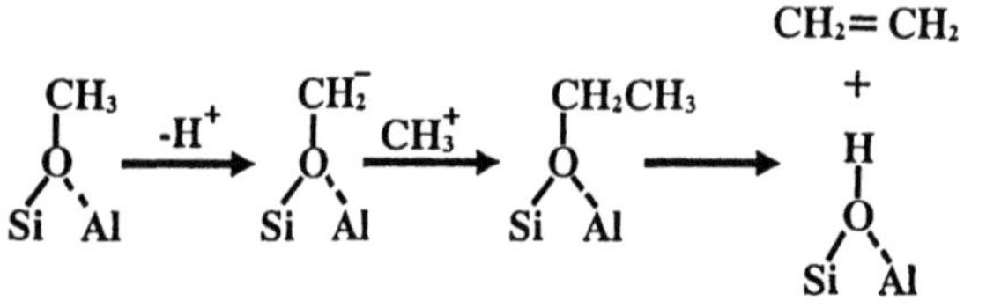

$$CH_2 = CH_2$$

Scheme III

alternative to the onium-ylide routes (Scheme II) for MTG chemistry, and an analogous mechanism might account for the formation of ethylene from methyl iodide on CsX.

Three-coordinate carbenium ion and five-coordinate carbonium ion intermediates satisfactorily account for many of the acid-catalyzed reactions of hydrocarbons at high temperatures. Yannoni et al. have characterized the structure and dynamics of several carbenium ions trapped in (noncatalytic) solids at low temperatures [32,94,95], but lifetimes of such ions on active surfaces at higher temperatures would preclude NMR observation in all but special cases. Maciel observed triphenyl carbenium ion on alumina [96]. The alkyl-substituted cyclopentenyl ions discussed earlier are also special ions; they are commonly observed products in "conjunct polymerization" reactions of olefins in acidic solutions. The five member ring cannot easily rearrange to an aromatic structure, and ions like **I** and **II** are apparently too hindered to be captured by the framework to form alkoxy species.

In the case of carbenium ions that are too reactive to be observed easily, it may still be possible to infer their existence from appropriate experiments. For example, Hutchings and coworkers proposed that allyl alcohol reacts on HZSM-5 by several pathways including protonation and dehydration to form the allylic carbenium ion (**X**), which goes on to form hydrocarbons [97]. Several in

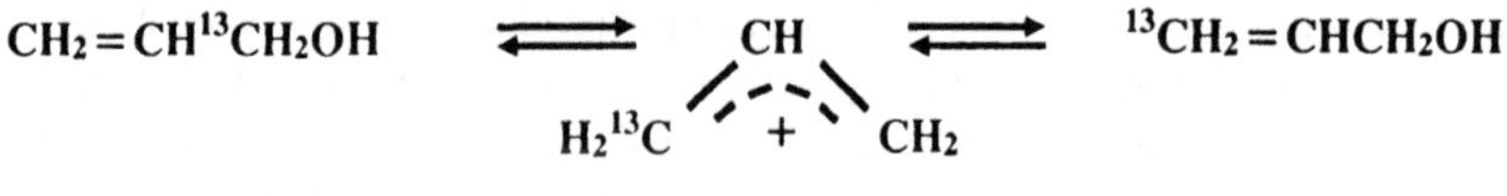

$$CH_2 = CH^{13}CH_2OH \quad \rightleftharpoons \quad {}^{13}CH_2 = CHCH_2OH$$

X

situ ^{13}C MAS experiments were performed on various zeolites to examine the reactions of $CH_2 = CH^{13}CH_2OH$ [98]. No evidence of label scrambling was seen on HZSM-5, presumably because the high ratio of adsorbate to acid sites favored further reaction of X as soon as it was generated. On X zeolites (Fig. 25), the formation of $^{13}CH_2 = CHCH_2OH$ was signaled by the resonance at 115 ppm. The simplest explanation of this 1-3 scrambling reaction is formation of the symmetrical allylic carbenium ion. In another study, the intermediacy of the symmetrical trimethyl carbenium ion in the oligomerization of isobutylene was inferred by an analogous scrambling reaction [39].

When label scrambling does not occur, it may be necessary to perform parallel experiments with labels in different positions in order to observe signals from all environments. For acetone on HZSM-5, label scrambling does not occur as demonstrated by the parallel in situ studies of acetone-2-^{13}C and acetone-1,3,-^{13}C in Fig. 26. The NMR spectra are complementary at each temperature.

One way to test a reaction mechanism is to incorporate a proposed intermediate into an in situ experiment and either look for a rate effect or follow the fate of a label. For example, mechanisms were proposed for MTG chemistry that involved CO as either an intermediate [25] or catalyst [99]. These arguments were reasonable since CO often forms in low concentrations during MTG reactions. These ideas were investigated by performing in situ experiments with added CO, which was generated in situ in the zeolite by thermal decomposition of formic acid [43]. Parallel experiments with or without CO exhibited similar rates; there was no catalytic effect. The experiments in Fig. 27 address the proposed intermediary role. The top four ^{13}C MAS spectra show the results of heating methanol-^{13}C and formic acid-[13]C. CO formation from formic acid was quantitative after 5 minutes at 523 K. The methanol and dimethyl ether reacted to form hydrocarbons while the CO was partially converted to CO_2 by the water-gas shift reaction. The total integrated intensity of CO and CO_2 remained constant during hydrocarbon synthesis, suggesting that CO was not incorporated into the products. A more definitive experiment was one identical to that described above, with the exception that formic acid-^{13}C was adsorbed with unlabeled methanol. The final spectrum at 523 K from that experiment is shown at the bottom of Fig. 27. Clearly, no ^{13}C label was incorporated into the hydrocarbon products. A recent mass spectrometry study has arrived at a similar conclusion for flowing reactors [100].

In some cases it may be useful to reduce the catalyst activity level to facilitate the observation of early reaction steps. An example of this approach is shown by the two in situ studies illustrated in Fig. 28 [101]. When acetaldehyde-1,2-^{13}C was heated on a zeolite sample activated to 673 K, a complex product distribution was formed, which decomposed to CO, CO_2, and other products at higher temperatures. If a small amount of water was first adsorbed uniformly on the zeolite, acetaldehyde was converted almost quantitatively to crotonaldehyde by a similar in situ protocol. It seems that water levels the acidity of the zeolite in a manner analogous to that seen in nonaqueous acid-base chemistry. As an aside, note that the ^{13}C chemical shifts of the carbonyl and β olefinic carbons are shifted downfield owing to the protonation equilibrium. This effect was discussed previously as a caveat to chemical shift interpretation.

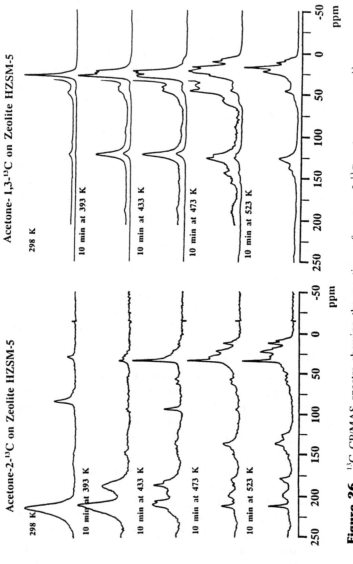

Figure 26 ^{13}C CP/MAS spectra showing the reactions of acetone-2-^{13}C and acetone-1,3-^{13}C on zeolite HZSM-5. No ^{13}C label scrambling was observed, and the two studies provide complementary information on the reaction mechanism.

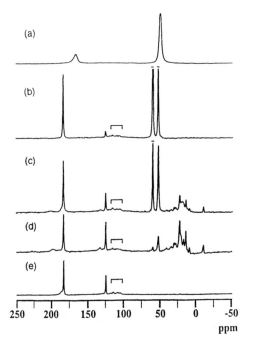

Figure 27 In situ ^{13}C MAS NMR spectra showing the chemistry of formic acid-^{13}C and methanol-^{13}C (a–d) and of formic acid-^{13}C and unlabeled methanol (e) on HZSM-5. (a) At 298 K prior to raising the probe temperature; (b) after 5 min at 523 K, showing conversion of formic acid to CO; (c) after 150 min at 523 K; (d) after 210 min at 523 K; (e) same as (d) except that unlabeled methanol was used, demonstrating that the label from ^{13}CO is not incorporated into the hydrocarbon products. Background signals from Kel-F endcaps, which were used to seal the sample rotor, are denoted by brackets.

VII. FUTURE DIRECTIONS

In situ NMR will be applied to a number of systems in the next few years. There is a clear need to further develop MAS methods so they will better simulate reactor conditions.

A. Pressure

Most MAS rotor designs tend to explode at 10 atm or less, and little attention has been given to pressurization techniques. High pressures are required for many catalytic reactions. For example, methanol synthesis is carried out at approximately 50 atm and 523 K. At low pressures, the reaction is spontaneous in the reverse reaction. A recent in situ study of the Cu/ZnO/Al$_2$O$_3$ catalyst concen-

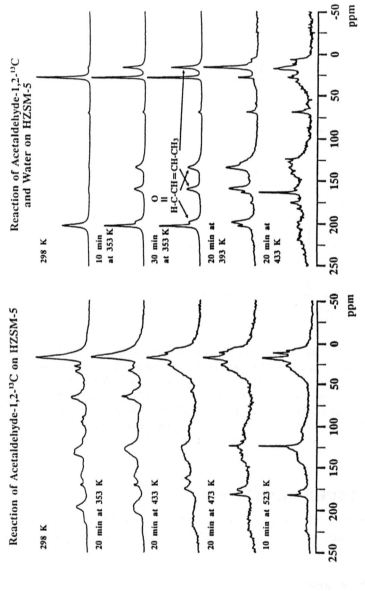

Figure 28 Demonstration of reaction tuning in zeolites: acetaldehyde undergoes complex aldol chemistry on activated HZSM-5 but selectively forms crotonaldehyde at 353 K when water is coadsorbed.

trated on methanol decomposition rather than synthesis [46]. It should be possible to develop high-pressure MAS rotors analogous to high-pressure solution NMR tubes.

B. Temperature

The temperature range of currently available VT MAS probes is probably adequate for most in situ experiments, although some improvement in reliability is needed for probes that operate at the highest temperatures. The most intriguing opportunity for advancement in this area is a temperature-jump experiment. Recently, Taulelle and coworkers reported the use of lasers to achieve very high sample temperatures in NMR experiments [102]. A laser heater might be used to take an MAS rotor from a low temperature to a much higher temperature and back in a few seconds to tens of seconds. This heating protocol might simulate the equivalent of short contact times and could possibly quench moderately reactive species.

C. Reagent Flow

The Achilles heel of an in situ MAS experiment is the absence of flow. Several existing strategies partly compensate for this problem. For example, the device in Fig. 4 permits the rotor to be cycled between treatment steps in the vacuum line and observation periods in the NMR probe. Continuous removal of volatiles from an MAS rotor with a porous cap has been demonstrated [103], but it is difficult to imagine a satisfactory way of continuously introducing a reagent to a spinning rotor. Once contrived approach to this problem would be to position an ampule of reagent at one end of the rotor which could be burst at the start of an experiment, perhaps by a jump in spinning speed. A better approach might be to develop sample cartridges that form part of a flow reactor and then demount for loading into an MAS rotor. This procedure would be, in a sense, a spinning version of a protocol in which a steady state was established with flow before the reactor tube was sealed off and transferred to a probe.

D. Spectroscopy

An area that might be explored is correlating the spectroscopy of the framework or cations with that of the adsorbates. Fyfe et al. have shown that a phase transformation is induced in ZSM-5 by the introduction of a small amount of adsorbate and that this change is reflected in the ^{29}Si MAS spectrum [104]. The ^{27}Al spectra of zeolites are greatly affected by the introduction of water or other adsorbates. The possibility of polarization transfer experiments between spins in the adsorbate and framework should be investigated. A related topic would be in situ double rotation (DOR) experiments on, for example, ^{17}O or ^{27}Al [105–107]. At least one vendor has advertised a VT DOR probe.

Another productive area would be in situ imaging and/or spatially localized spectroscopy of model flow reactors. Pulsed-field gradient measurements could conceivably be used to characterize diffusion and pore blockage due to coking while the reaction is in progress. ^{129}Xe might be similarly useful. This technique has already been used to probe adsorbate distributions in catalyst beds [108].

E. Theoretical Modeling and Experimental Verification

As stated in the introduction, theoretical modeling linked to experimental verification has been identified as a priority area in which the focus of scientific resources may lead to rapid progress in catalysis. NMR studies of adsorbates on catalysts are well suited for determining dynamics, diffusion, and equilibrium structures and may be one of the best ways to experimentally verify the results of simulations.

Sauer has reviewed theoretical work on solids and surfaces [109], and van Santen et al. recently published an introduction to zeolite theory and modeling [110]. A concise tutorial on simulation of molecular dynamics has been provided by van Gunsteren and Berendsen [111]. Such calculations are already having an impact on protein and nucleic acid research, but analogous progress in catalysis research depends in part on the development of more accurate force fields and additional studies linking calculations and experimental results. Nevertheless, some intriguing calculations of diffusion [112–119] and other phenomena [120–126] in zeolite science have already appeared in the literature. Quantum mechanical methods are required to model chemical reactions and properties for which the explicit treatment of electrons is required. Ab initio calculations [127] have usually been restricted to cluster models of acid sites such as $(HO)_3Si-OH-Al(OH)_3$ [128–130]. The strengths and weaknesses of such models have been pointed out [131].

One of the biggest limitations of current molecular dynamics calculations is that trajectories can only be followed for around 100 ps with reasonable computation times. This short time scale complicates the observation of activated phenomena and makes the method sensitive to the choice of initial conditions, since the simulation may not run long enough to adequately sample configuration space. 100 ps is far too short to model dynamics that affect the NMR lineshape, for which the time scale is typically on the order of a ms. As demonstrated by Thomas and coworkers [132] and others, hybrid approaches employing Monte Carlo methods and other methods in addition to molecular dynamics can extend these simulations to longer effective time scales.

A simple example of how NMR and simulations might be combined for a better understanding of zeolite-adsorbate interactions is provided by a study of acetylene on CsZSM-5 [79]. Here the interaction between Cs^+ and acetylene was first modeled using van der Waals and coulombic nonbonded interactions. For Cs^+ the initial parameters were approximated by those of Xe with an assigned charge of +1. Existing van der Waals parameters for carbon and hydrogen were used together with partial charge assignments obtained from semi-empirical molecular orbital calculations. The carbon and hydrogen atoms had small negative and positive charges, respectively. A series of energy calculations and minimizations were performed on several models to check and modify slightly the van der Waals parameters of Cs^+.

Configurations were first calculated for "gas phase" complexes. There was room around the Cs^+ for six or seven acetylene molecules to complex in the first coordination sphere; additional molecules formed a second layer. Figures 29 and 30 show computational results for 4 and 10 acetylenes, respectively. Not surprisingly, the acetylenes in the first coordination sphere have a perpendicular geometry as their most stable configuration. A crystalline ZSM-5 lattice with one conjugate base site was constructed, and one Cs^+ with 13 acetylenes were introduced into reasonable initial positions near the base site. Figure 31 shows a wire frame model after energy minimization. A different representation of this system with the Cs^+ and acetylenes shown as van der Waals spheres and the lattice as tetrahedra is shown in Fig. 32. Six or seven of the acetylene molecules

Figure 29   Energy-minimized structure calculated for Cs^+ and 4 acetylenes.

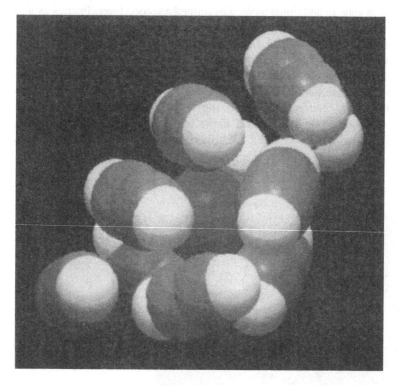

Figure 30   Energy-minimized structure calculated for Cs^+ and 10 acetylenes.

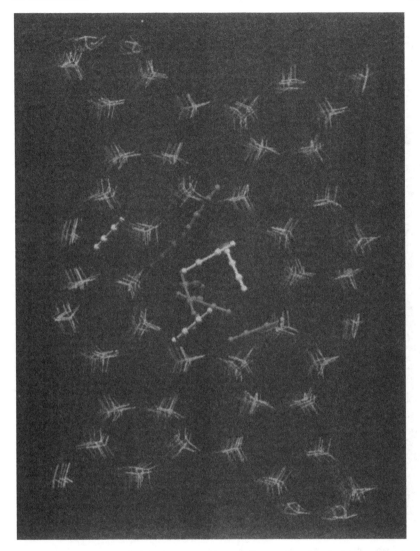

Figure 31 Energy-minimized structure calculated for Cs$^+$ and 13 acetylenes after insertion into the ZSM-5 lattice.

continued to be coordinated to the Cs$^+$, but steric constraints forced the remaining adsorbate molecules into other parts of the zeolite.

If six or seven acetylene molecules can coordinate Cs$^+$ in CsZSM-5, it should be reflected in the ^{133}Cs chemical shift of that material. Spectra obtained as a function of acetylene loading are shown in Figure 33. It is apparent that at least several acetylenes can strongly interact with the cation in CsZSM-5.

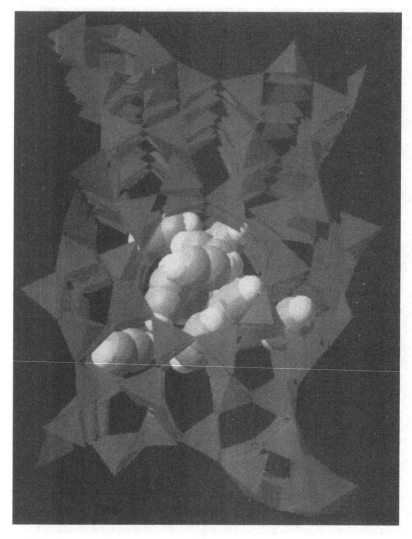

Figure 32 Same as Fig. 31 except for different representation.

F. The Use of Zeolites as Catalysts in Organic Synthesis

Organic chemists have discovered zeolites and other microporous catalysts. Holderich et al. have published a review of the use of zeolites as catalysts for organic synthesis with 455 references prior to 1988 [133]. In a recent review of *New Directions in Zeolite Catalysis*, Weitkamp suggested that zeolites could be viewed as reaction vessels with appropriate molecular dimensions and thus act as

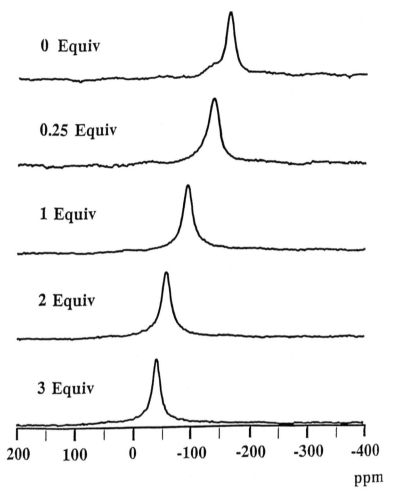

Figure 33 ^{133}Cs MAS NMR spectra of various loadings of acetylene on CsZSM-5 confirming that the Cs cation in ZSM-5 can be "solvated" by several acetylenes.

a template directing synthetic reactions to the desired product [134]. A large number of organic reactions has been carried out or attempted with the aid of zeolites, either as gas phase reactions or with the zeolite added to refluxing solvent. In many cases, the product distribution is very different from that obtained using conventional catalysts. Also, the products and yields can depends very strongly on the zeolite framework or cationic form. Further progress in the use of zeolites in synthesis will depend in part on a better understanding of the reactions of simple compounds in zeolites including the nature of framework-bound intermediates. Synthetic chemists are very familiar with the use of solution-state NMR as a tool in the discovery of new reactions. Perhaps in situ NMR will be useful in the discovery of new organic syntheses.

ACKNOWLEDGMENT

This work was supported by the National Science Foundation (Grant No. CHE 92-21406).

REFERENCES

1. National Research Council, Catalysis Looks to the Future, National Academy Press, Washington, DC, 1992.
2. T. R. Forester and R. F. Howe, *J. Am. Chem. Soc. 109*: 5076 (1987).
3. C. Schild, A. Wokaun, and A. Baiker, *J. Mol. Catal. 63*: 223 (1990).
4. J.-M. Jehng and I. E. Wachs, *J. Phys. Chem. 95*: 7373 (1991).
5. M. J. Annen, D. Young, M. E. Davis, O. B. Cavin, and C. R. Hubbard, *J. Phys. Chem. 95*: 1380 (1991).
6. D. R. Corbin, L. Abrams, G. A. Jones, M. M. Eddy, W. T. A. Harrison, G. S. Stucky, and D. E. Cox, *J. Am. Chem. Soc. 112*: 4821 (1990).
7. J. W. Couves, J. M. Thomas, C. R. A. Catlow, G. N. Greaves, G. Baker, and A. J. Dent, *J. Phys. Chem. 94*: 6517 (1990).
8. G. Jonkers, K. A. Vonkeman, S. W. A. van der Wal, and R. A. van Santen, *Nature 355*: 63 (1992).
9. B. F. Chmelka and A. Pines, *Science 246*: 71 (1989).
10. J. Klinowski, *Chem. Rev. 91*: 1459 (1991).
11. G. Engelhardt, in *Introduction to Zeolite Science and Practice* (H. Van Bekkum, H. E. M. Flanigen, and J. C. Jansen, eds.), Elsevier, Amsterdam, 1991.
12. C. Tsiao, D. R. Corbin, V. Durante, D. Walker, and C. Dybowski, *J. Phys. Chem. 94*: 4195 (1990).
13. C. J. Tsiao, J. S. Kauffman, D. R. Corbin, L. Abrams, E. E. Carroll, and C. Dybowski, *J. Phys. Chem. 95*: 5586 (1991).
14. R. Grosse, R. Burmeister, B. Boddenberg, A. Gedeon, and J. Fraissard, *J. Phys. Chem. 95*: 2443 (1991).
15. J. H. Lunsford, W. P. Rothwell, and W. Shen, *J. Am. Chem. Soc. 107*: 1540 (1985).
16. H. Pfeifer, D. Freude, and J. Karger, in *Catalysis and Adsorption by Zeolites* (G. Ohlmann, H. Pfeifer, and R. Fricke, eds.), Elsevier, Amsterdam, 1991, p. 365.
17. J. Karger and H. Spindler, *J. Am. Chem. Soc. 113*: 7571 (1991).
18. G. W. Haddix, J. A. Reimer, and A. T. Bell, *J. Catal. 106*: 111 (1987).
19. G. W. Haddix, J. A. Reimer, and A. T. Bell, *J. Catal. 108*: 50 (1987).
20. G. W. Haddix, D. H. Jones, J. A. Reimer, and A. T. Bell, *J. Catal. 112*: 556 (1988).
21. G. W. Haddix, J. A. Reimer, and A. T. Bell, *J. Phys. Chem. 93*: 5859 (1989).

22. M. S. Went and J. A. Reimer, *J. Am. Chem. Soc. 114*: 5768 (1992).
23. A. Bendata, G. Chinchen, N. Clayden, B. T. Heaton, J. A. Iggo, and C. S. Smith, *Catal. Today 9*: 129 (1991).
24. M. W. Anderson, M. L. Occelli, and J. Klinowski, *J. Phys. Chem. 96*: 388 (1992).
25. M. W. Anderson and J. Klinowski, *J. Am. Chem. Soc. 112*: 10 (1990).
26. M. W. Anderson, B. Sulikowski, P. J. Barrie, and J. Klinowski, *J. Phys. Chem. 94*: 2730 (1990).
27. A. V. Nosov, V. M. Mastikhin, and A. V. Mashkina, *J. Mol. Catal. 66*: 73 (1991).
28. H. Ernst and H. Pfeifer, *J. Catal. 136*: 202 (1992).
29. J. R. Lyerla, C. S. Yannoni, and C. A. Fyfe, *Acc. Chem. Res. 15*: 208 (1982).
30. J. F. Haw, G. C. Campbell, and R. C. Crosby, *Anal. Chem. 58*: 3272 (1986).
31. F. G. Oliver, E. J. Munson, and J. F. Haw, *J. Phys. Chem. 96*: 8106 (1992).
32. J. F. Stebbins, *Chem. Rev. 91*: 1353 (1991).
33. P. C. Myhre, G. G. Webb, and C. S. Yannoni, *J. Am. Chem. Soc. 112*: 8991 (1990).
34. J. F. Stebbins, I. Farnam, E. H. Williams, and J. Roux, *Phys. Chem. Miner. 16*: 763 (1989).
35. M. Pruski, D. K. Sanders, T. S. King, and B. C. Gerstein, *J. Magn. Reson. 96*: 574 (1992).
36. E. J. Munson, D. B. Ferguson, A. A. Kheir, and J. F. Haw, *J. Catal. 136*: 504 (1992).
37. J. F. Haw, B. R. Richardson, I. S. Oshiro, N. D. Lazo, and J. A. Speed, *J. Am. Chem. Soc. 111*: 2052 (1989).
38. B. R. Richardson, N. D. Lazo, P. D. Schettler, J. L. White, and J. F. Haw, *J. Am. Chem. Soc. 112*: 2886 (1990).
39. N. D. Lazo, B. R. Richardson, P. D. Schettler, J. L. White, E. J. Munson, and J. F. Haw, *J. Phys. Chem. 95*: 9420 (1991).
40. J. L. White, N. D. Lazo, B. R. Richardson, and J. F. Haw, *J. Catal. 125*: 260 (1990).
41. N. D. Lazo, J. L. White, E. J. Munson, M. Lambregts, and J. F. Haw, *J. Am. Chem. Soc. 112*: 4050 (1990).
42. M. J. Lambregts, E. J. Munson, A. A. Kheir, and J. F. Haw, *J. Am. Chem. Soc. 114*: 6875 (1992).
43. E. J. Munson, N. D. Lazo, M. E. Moellenhoff, and J. F. Haw, *J. Am. Chem. Soc. 113*: 2783 (1991).
44. E. J. Munson and J. F. Haw, *J. Am. Chem. Soc. 113*: 6303 (1991).
45. E. J. Munson, A. A. Kheir, N. D. Lazo, and J. F. Haw, *J. Phys. Chem. 96*: 7740 (1992).
46. N. D. Lazo, D. K. Murray, M. L. Kieke, and J. F. Haw, *J. Am. Chem. Soc. 114*: 8552 (1992).
47. E. J. Munson, D. K. Murray, and J. F. Haw, *J. Catal. 141*: 733 (1993).
48. M. W. Anderson, and J. Klinowski, *Chem. Phys. Lett. 172*: 275 (1990).
49. A. Pines, M. G. Gibby, and J. S. Waugh, *J. Chem. Phys. 59*: 569 (1973).

50. B. R. Richardson and J. F. Haw, *Anal Chem. 61*: 1821 (1989).
51. G. E. Maciel, J. F. Haw, I.-S. Chuang, B. L. Hawkins, T. E. Early, D. R. McKay, and L. Petrakis, *J. Am. Chem. Soc. 105*: 5529 (1983).
52. J. F. Haw, I.-S. Chuang, B. L. Hawkins, G. E. Maciel, *J. Am. Chem. Soc. 105*: 7206 (1983).
53. L. Baltusis, J. F. Frye, and G. E. Maciel, *J. Am. Chem. Soc. 108*: 7119 (1986).
54. T. Xu and J. F. Haw, unpublished results.
55. G. E. Maciel and G. C. Ruben, *J. Am. Chem. Soc. 85*: 3903 (1963).
56. G. E. Maciel and J. J. Natterstad, *J. Chem. Phys. 42*: 2752 (1965).
57. D. Farcasiu, A. Ghenciu, and G. Miller, *J. Catal. 134*: 118 (1992).
58. G. A. Olah, G. K. Surya Prakash, and J. Sommer, *Superacids*, Wiley-Interscience, New York, 1985.
59. J. L. White, L. M. Beck, and J. F. Haw, *J. Am. Chem. Soc. 114*: 6182 (1992).
60. B. C. Gerstein and C. R. Dybowski, *Transient Techniques in NMR of Solids*, Academic Press, San Diego, 1985.
61. E. Brunner, H. Ernst, D. Freude, T. Frohlich, M. Hunger, and H. Pfeifer, *J. Catal. 127*: 34 (1991).
62. M. W. Anderson, P. J. Barrie, and J. Klinowski, *J. Phys. Chem. 95*: 235 (1991).
63. D. B. Venuto and E. T. Habib, *Fluid Catalytic Cracking with Zeolite Catalysts*, Marcel Dekker, New York, 1979, p. 124.
64. R. N. Young, *Prog. NMR Spectrosc. 12*: 261 (1979).
65. J. C. Rees, and D. Whittaker, *J. Chem. Soc., Perkins Trans. II*: 948 (1980).
66. N. C. Deno, in *Carbonium Ions*, Vol. 2 (G. A. Olah and P. V. R. Schleyer, eds.), Wiley-Interscience, New York, 1970.
67. R. F. Howe, in *Methane Conversion* (D. M. Bibby, C. D. Chang, R. F. Howe, and S. Yurchak, eds.), Elsevier, Amsterdam, 1988, p. 157.
68. C. D. Chang, *Catal. Rev.—Sci. Eng. 25*: 1 (1983).
69. C. D. Chang, in *Perspectives in Molecular Sieve Science* (W. E. Flank, and T. E. Whyte, eds.), ACS Symp. Ser. 368, 1988, p. 596.
70. C. D. Chang and A. J. Silvestri, *J. Catal. 47*: 249 (1977).
71. C. S. Lee and M. M. Wu, *J. Chem. Soc., Chem. Commun.*: 250 (1985).
72. Y. Ono and T. J. Mori, *J. Chem. Soc., Faraday Trans. 77*: 2209 (1981).
73. G. A. Olah, A. Doggweiler, J. D. Feldberg, S. Frohlich, M. J. Grdina, R. Karpeles, T. Keumi, S. Inabe, W. M. Ip, K. Lammertsma, G. Salem, and D. C. Tabor, *J. Am. Chem. Soc. 106*: 2143 (1984).
74. J. P. van den Berg, J. P. Wolthuizen, and J. H. C. van Hooff, in *Proceedings of the 5th International Conference on Zeolites* (L. V. Rees, ed.), Heyden, London, 1980.
75. J. K. Clarke, R. Darcy, B. F. Hegarty, E. O'Donoghue, V. Amir-Ebrahimi, J. R. Rooney, *J. Chem. Soc., Chem. Commun.*: 425 (1986).
76. C. Tsiao, D. R. Corbin, and C. Dybowski, *J. Am. Chem. Soc. 112*: 7140 (1990).
77. M. W. Anderson and J. Klinowski, *J. Chem. Soc., Chem. Commun.*: 918 (1990).
78. M. T. Aronson, R. J. Gorte, W. E. Farneth, and D. J. White, *J. Am. Chem. Soc. 111*: 840 (1989).
79. D. K. Murray, J.-W. Chang, and J. F. Haw, *J. Am. Chem. Soc. 115*: 4732 (1993).

80. M. Mehring, *Principles of High-Resolution NMR in Solids*, 2nd ed., Springer-Verlag, New York, 1983.
81. S. J. Opella and M. H. Frey, *J. Am. Chem. Soc. 101*: 5854 (1979).
82. J. Herzfeld and A. E. Berger, *J. Chem. Phys. 73*: 6021 (1980).
83. D. P. Burum and A. Bielecki, *J. Magn. Reson. 95*: 184 (1991).
84. T. Gullion and J. Schaefer, *J. Magn. Reson. 81*: 196 (1989).
85. Y. Pan, T. Gullion, and J. Schaefer, *J. Magn. Reson. 90*: 330 (1990).
86. Y. Pan and J. Schaefer, *J. Magn. Reson. 90*: 341 (1990).
87. M. H. Levitt, D. P. Raleigh, F. Creuzet, and R. G. Griffin, *J. Chem. Phys. 92*: 6347 (1990).
88. D. P. Raleigh, M. H. Levitt, and R. G. Griffin, *Chem. Phys. Lett. 146*: 71 (1988).
89. A. E. McDermott, F. Creuzet, and R. G. Griffin, *Biochemistry 29*: 5767 (1990).
90. C. A. Klug, C. P. Slichter, and J. H. Sinfelt, *J. Phys. Chem. 95*: 2119 (1991).
91. E. J. Munson, A. A. Kheir, and J. F. Haw, *J. Phys. Chem. 97*: 7321 (1993).
92. G. A. Olah, A. Doggweiler, J. D. Felberg, and S. Frohlich, *J. Org. Chem. 50*: 4847 (1985).
93. V. B. Kazansky, *Acc. Chem. Res. 24*: 379 (1991).
94. P. C. Myhre, K. L. McLaren, and C. S. Yannoni, *J. Am. Chem. Soc. 107*: 5294 (1985).
95. P. C. Myhre, G. G. Webb, and C. S. Yannoni, *J. Am. Chem. Soc. 112*: 8992 (1990).
96. G. E. Maciel, in *Heterogeneous Catalysis* (B. L. Shapiro, ed.), Texas A&M University Press, College Station, 1984.
97. G. J. Hutchings, D. F. Lee, and C. D. Williams, *J. Chem. Soc., Chem. Commun.*: 1475 (1990).
98. E. J. Munson, T. Xu, and J. F. Haw, *J. Chem. Soc., Chem. Commun.*: 75 (1993).
99. J. E. Jackson and F. M. Bertsch, *J. Am. Chem. Soc. 112*: 9085 (1990).
100. G. J. Hutchings and P. Johnston, *Appl. Catal. 57*: L5 (1990).
101. E. J. Munson and J. F. Haw, *Angew. Chem. 32*: 615 (1993).
102. F. Taulelle, J. P. Coutures, D. Massiot, and J. P. Rifflet, *Bull. Magn. Reson. 11*: 318 (1989).
103. R. C. Crosby, D. L. Lewis, and J. F. Haw, *Anal. Chem. 60*: 2695 (1988).
104. C. A. Fyfe, H. Strobl, G. T. Kokotailo, G. J. Kennedy, and G. E. Barlow, *J. Am. Chem. Soc. 110*: 3373 (1988).
105. Y. Wu, B. Q. Sun, A. Pines, A. Samoson, and E. Lippmaa, *J. Magn. Reson. 89*: 297 (1990).
106. Y. Wu, B. F. Chmelka, A. Pines, M. E. Davis, P. J. Grobet, and P. A. Jacobs, *Nature 346*: 550 (1990).
107. R. Jelinek, B. F. Chmelka, Y. Wu, P. J. Grandinetti, A. Pines, P. J. Barrie, and J. Klinowski, *J. Am. Chem. Soc. 113*: 409 (1991).
108. B. F. Chmelka, J. G. Pearson, S. B. Liu, R. Ryoo, L. C. de Menorval, and A. Pines, *J. Phys. Chem. 95*: 303 (1991).
109. J. Sauer, *J. Chem. Rev. 89*: 199 (1989).
110. R. A. van Santen, D. P. de Bruyn, C. J. J. den Ouden, and B. Smit, in *Introduction to Zeolite Science and Practice* (H. Van Bekkum, E. M. Flanigen, and J. C. Jansen, eds.), Elsevier, Amsterdam, 1991.

111. W. F. van Gunsteren and H. J. C. Berendsen, *Angew. Chem. 29*: 992 (1990).
112. A. K. Nowak, C. J. J. den Ouden, S. D. Pickett, B. Smit, A. D. Cheetham, M. F. M. Post, and J. M. Thomas, *J. Phys. Chem. 95*: 848 (1991).
113. R. L. June, A. T. Bell, and D. N. Theodorou, *J. Phys. Chem. 94*: 8232 (1990).
114. P. Demontis, E. S. Fois, G. B. Suffritti, and S. Quartieri, *J. Phys. Chem. 94*: 4329 (1990).
115. S. Yashonath, P. Demontis, and M. L. Klein, *J. Phys. Chem. 95*: 5881 (1991).
116. S. Yashonath, *Chem. Phys. Lett. 177*: 54 (1991).
117. S. Fritzsche, R. Haberlandt, J. Kaerger, H. Pfeifer, and M. Wolfsberg, *Chem. Phys. Lett. 171*: 109 (1990).
118. T. Inui and Y. Nakazaki, *Zeolites 11*: 434 (1991).
119. S. Yashonath, *J. Phys. Chem. 95*: 5877 (1991).
120. J. B. Nicholas, A. J. Hopfinger, F. R. Trouw, and L. E. Iton, *J. Am. Chem. Soc. 113*: 4792 (1991).
121. M. K. Song, H. Chon, and M. S. Jhon, *J. Phys. Chem. 94*: 7671 (1990).
122. R. G. Bell, R. A. Jackson, and C. R. A. Catlow, *J. Chem. Soc., Chem. Commun.*: 782 (1990).
123. S. M. Tomlinson, R. A. Jackson, and C. R. A. Catlow, *J. Chem. Soc., Chem. Commun.*: 813 (1990).
124. J. O. Titiloye, S. C. Parker, F. S. Stone, and C. R. A. Catlow, *J. Phys. Chem. 95*: 4038 (1991).
125. F. Vigne-Maeder and H. Jobic, *Chem. Phys. Lett. 169*: 31 (1990).
126. M. W. Deem and J. M. Newsam, *Nature 342*: 260 (1989).
127. J. Simons, *J. Phys. Chem. 95*: 1017 (1991).
128. P. Ugliengo, V. R. Saunders, and E. Garrone, *Chem. Phys. Lett. 169*: 501 (1990).
129. M. Allavena, K. Seiti, E. Kassab, G. Ferenczy, and J. G. Angyan, *Chem. Phys. Lett. 168*: 461 (1990).
130. E. Kassab, K. Seiti, and M. Allavena, *J. Phys. Chem. 95*: 9425 (1991).
131. G. J. Kramer, A. J. M. de Man, and R. A. van Santen, *J. Am. Chem. Soc. 113*: 6435 (1991).
132. C. M. Freeman, C. R. A. Catlow, J. M. Thomas, and S. Brode, *Chem. Phys. Lett. 186*: 137 (1991).
133. W. Holderich, M. Hesse, and F. Naumann, *Angew. Chem. 27*: 226 (1988).
134. J. Weitkamp, *Catalysis and Adsorption by Zeolites* (G. Ohlmann, H. Pfeifer, and R. Fricke, eds.), Elsevier, Amsterdam, 1991, p. 21.

4

NMR Spectroscopy of Bulk Oxide Catalysts

Hellmut Eckert

University of California, Santa Barbara, California

I. INTRODUCTION

A. Bulk Oxide Catalysts in Industry

Bulk oxides are used extensively in the chemical industry as catalysts for a wide range of chemically important transformations [1,2]. In general, these applications employ one of the two fundamental properties of the oxide surface: (1) redox activity, which enables hydrogen abstraction from, or oxygen transfer to, adsorbed substrate molecules and (2) acid-base character (Lewis or Bronsted type), which facilitates catalytic cracking or isomerization. Table 1 summarizes the most important catalytic bulk metal oxides currently in use and gives an overview of the processes catalyzed by them. Not all of these are currently exploited industrially.

The large-scale production of sulfuric acid, overall one of the most important products in the chemical industry, exploits the redox activity of vanadium(V) oxide. Vanadia-, molybdena-, and tin oxide–based catalysts are used further for a host of selective hydrocarbon oxidation processes including dehydrogenation, oxidative coupling, and oxygenation [3,4]. Much recent activity in this area has been aimed at the development of more active and more selective catalysts by dispersing oxides in the form of monolayers on bulk oxide supports. This approach has proven extremely successful for tailoring catalyst properties to

Table 1 Oxide Catalysts and Reactions

Reactions	Catalyst composition
Oxygenation	
$SO_2 \rightarrow SO_3 \rightarrow$ sulfuric acid	K-V-O
$CO \rightarrow CO_2$	$LnCoO_3$ (Ln = lanthanide ion)
propene $\rightarrow$ acrolein	Bi-Mo-O, Sn-Sb-O
propene $\rightarrow$ acrylonitrile	Bi-Mo-O, Sn-Sb-O
propene $\rightarrow$ acrylic acid	Cu/O
acrolein $\rightarrow$ acrylic acid	V-Mo-O
butane $\rightarrow$ maleic anhydride	V-P-O
methanol $\rightarrow$ formaldehyde	Fe-Mo-O, MoO_3, Bi-V-O, V_2O_5
ethanol $\rightarrow$ acetaldehyde	$LaMO_3$ (M = Mn, Fe, Co, Ni)
$CH_4 + H_2O \rightarrow CO + H_2$	$Ln_2Ru_2O_7$ (pyrochlores)
Oxidative coupling	
methane $\rightarrow$ ethane	Li-Mg-O
methane $\rightarrow$ ethene	$Ln_2Sn_2O_7$ (Ln = Pr, Sm, Eu, Tb, Tm, Yb) (pyrochlores)
Dehydrogenation	
butane $\rightarrow$ butenes, butadienes	Cr-Al-O, Mo-Al-O
butene $\rightarrow$ 1,3 butadiene	Sn-Sb-O
NO reduction	(La, M)CoO_3 (M = divalent metal)
Dehydration, double bond	Sn-Sb-O, Al-P-O, SiO_2-Al_2O_3,
isomerization, hydrocarbon cracking	γ-Al_2O_3, Zr-P-O

specific applications and has resulted in a whole new class of materials. Such systems are the subject of another chapter of this volume.

Regarding their use as cracking and isomerization catalysts, bulk oxides such as clays and amorphous silica-aluminas have been widely displaced by molecular sieve compounds (e.g., zeolites, aluminophosphates), whose well-defined pore structures generally offer higher selectivity and flexibility. Nevertheless, bulk oxides continue to be used for various cracking and isomerization applications in the petroleum industry.

Specific bulk oxides are also of interest as polymerization catalysts and as precursors to HDN and HDS catalysts. Although the latter by themselves are nonoxidic, the structural characterization of the precursors is an important topic, and corresponding studies are included in this review.

B. Structural Issues and Mechanistic Aspects of Bulk Oxide Catalysts

Many bulk oxide catalysts are multiphase and compositionally rather complex materials. Therefore, an important part of academic catalysis research is concerned with simple model systems. The goals of such studies are to identify key constituents of a catalyst, to clarify the individual roles played by its various components, and to develop an understanding of their functions on a structural and mechanistic basis.

Many selective oxidation processes proceed via the Mars van Krevelen mechanism [2], which involves the following distinct steps: (1) α-hydrogen abstraction from the hydrocarbon by lattice oxide, (2) insertion of the oxygen from the catalyst into the resulting intermediate, (3) diffusion of oxygen from the interior to the vacant site, and (4) regeneration of the catalyst by reaction with gaseous oxygen. These processes, and hence the catalytic behavior of a metal oxide, are critically determined by the solid-state and surface properties of this material. In the same vein, the activity of isomerization and cracking catalysts is controlled by the acidic properties of its surface [5].

To summarize, structural catalyst characterization studies must address the following issues:

1. The composition and the crystallographic identities of the various phases present
2. The local environment of the metal and the oxygen constituents in the bulk and at the surface
3. The redox properties of the catalyst, which bear on its capability for oxygen transfer and on the involvement and distribution of mixed metal oxidation states
4. The defect structure, including the concentration, arrangement, and mobilities of dopants and vacancies
5. Surface characteristics such as acid-base behavior, specifically at the temperatures of catalyst operation (Relevant characteristics include the number and concentration of acidic sites, their types [Bronsted or Lewis], and the correlation of these properties with structure and composition of the bulk catalyst.)

Nuclear magnetic resonance offers unique opportunities for the study of these issues. The atomic nuclei interact with various local magnetic and electric fields that reflect site symmetry, chemical bonding, and interatomic distances. The lineshapes derived from these interactions are often highly site specific and diagnostic of particular structural environments [6–10]. Since the strength of all interactions falls off rapidly with distance, NMR is locally selective, which is advantageous for the investigation of disordered materials and surfaces. As an element-selective method, NMR lends itself easily to the investigation of compositionally very complex materials, and the uniform relation between signal intensity and species concentration opens opportunities for rigorously quantitative applications, or spin counting.

Since the solid-state NMR approach is relatively new, most applications to bulk oxide catalysts so far have focused on introducing a new experimental technique to an existing catalytic system, rather than actually maximizing knowledge about such a system. To achieve the latter goal would require careful attention to details of sample preparation, structural characterization by complementary techniques, and correlation of structural features with catalytic activities. Much remains to be done in this area, and, unavoidably, the present review reflects this state of the literature. Section II presents many of the chief concepts and ideas relating to NMR applications to oxide catalysts, whereas the remaining sections summarize the current state of knowledge in specific areas of bulk oxide catalysis.

II. NMR SPECTROSCOPIC APPROACHES

In a general solid-state NMR experiment, three distinct internal interaction mechanisms must be considered [6–10]: (1) the coupling of the nuclear magnetic moments with local magnetic fields created by surrounding electrons known as the chemical shift interaction, (2) the magnetic dipole-dipole coupling among the nuclei, and (3) the interaction of nuclear electric quadrupole moments with local electric field gradients. Under standard experimental conditions, the resulting NMR lineshapes are rather complex and difficult to analyze unambiguously. Much of the power of NMR spectroscopy, however, comes instead from various selective experiments that allow the experimenter to study each of the above three interactions individually. Such experiments not only increase the overall informational content of individual spectra, but often complement each other nicely. Techniques and approaches that have been important for characterizing bulk oxide catalysts are discussed below.

A. Static (Wideline) NMR

The use of standard NMR spectroscopy without any selective averaging techniques has generally had little importance in the field of catalysis. An exception is high-field ^{51}V NMR, which yields characteristic lineshapes in the solid state that are easily interpreted in terms of the chemical shift anisotropy [11]. Generally, we can distinguish the three situations illustrated in Fig. 1: The spectrum in Fig. 1c is observed for compounds with asymmetric coordination environments. It shows three distinct features, which can be identified with the three cartesian chemical shift components δ_{xx}, δ_{yy}, δ_{zz} in the molecular axis system. Figure 1b corresponds to the case of cylindrical symmetry, where $\delta_{xx} = \delta_{yy} \neq \delta_{zz}$ and hence two distinct lineshape components appear. Finally, for chemical environments with spherical symmetry, the chemical shift is the same in all three directions. The solid-state NMR spectrum then contains only a single symmetric peak (Fig. 1a).

Figure 2 shows representative solid-state ^{51}V NMR spectra of crystalline vanadates. Each model compound typifies a certain local environment with well-defined symmetry as shown. One can see from these data that the solid-state ^{51}V

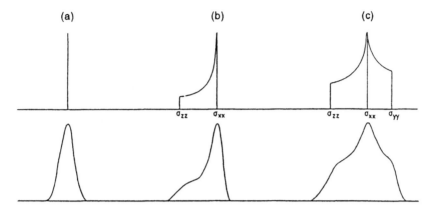

Figure 1 Solid-state NMR powder patterns, dominated by chemical shift anisotropy effects: (a) spherically symmetric chemical shift tensor, (b) axially symmetric chemical shift tensor, (c) asymmetric chemical shift tensor. Top traces: theoretical powder patterns; bottom traces: powder patterns broadened by other anisotropic interactions or chemical shift distribution effects.

chemical shift anisotropies are uniquely well suited for differentiating among the various site symmetries. VO_4^{3-} groups with approximate spherical symmetry yield simple single-peak spectra; dimeric $V_2O_7^{4-}$ groups, which possess a threefold axis and hence cylindrical symmetry, yield spectra resembling Fig. 1b; whereas the spectra of the completely asymmetric $VO_{2/2}O_2^{-}$ groups are of the kind shown in Fig. 1c. Highly characteristic lineshapes are also observed for vanadium in distorted octahedral environments (ZnV_2O_6) and in square-pyramidal environments (V_2O_5).

B. Multinuclear MAS-NMR

A review of the available literature reveals that, among the large number of selective averaging experiments, the technique of magic-angle spinning NMR has found the widest application to bulk oxide catalysts. In brief, the sample is spun around an axis oriented relative to the magnetic field direction at an angle of 54.7°. For spin-1/2 nuclei, this manipulation modulates and averages out the anisotropic components of the internal interactions in the solid state (dipole-dipole couplings and chemical shift anisotropies) [12,13]. The resulting spectra consist of sharp, well-resolved resonances, characterized by the isotropic chemical shift parameter δ_{iso}, on the basis of which assignments to specific coordination environments and bonding characteristics are possible with the help of model compounds. For the nuclei of greatest interest here, the usual reference compounds for published isotropic chemical shift values are tetramethylsilane (^1H, ^{13}C, ^{29}Si), 1 M $Al(NO_3)_3$ in H_2O (^{27}Al), 85% H_3PO_4 (^{31}P), and liquid $VOCl_3$ (^{51}V).

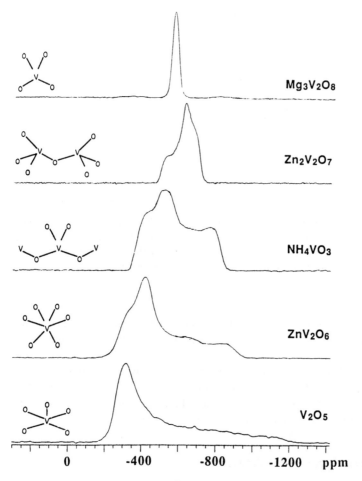

Figure 2 79.0 MHz solid-state [51]V NMR spectra of crystalline metal vanadates. Each compound typifies a certain vanadium environment, as shown.

One example of MAS applications to catalytic materials includes the precise measurement of ^{1}H chemical shifts in the solid state. These data differentiate sensitively among B-OH, Al-OH, Si-OH, and P-OH groups and can also be correlated with the Bronsted acidity. Several applications of this idea to bulk oxide catalysts have been published and reviewed in the literature [5,14–18].

For quadrupolar nuclei ($I > 1/2$), MAS-NMR has proven similarly useful. Line narrowing is incomplete in the presence of strong nuclear electric quadrupole interactions [19], however, and the resonance maximum does not correspond exactly to the isotropic chemical shift. Quantitative aspects of such spectra are more complicated as a result. Nevertheless, ^{27}Al MAS-NMR (I = 5/2) has proven very useful for the assignment of aluminum coordination symmetries. Tetrahedral Al sites resonate within the region 40–80 ppm, whereas six-coordinated Al sites have resonance shifts between -20 and 10 ppm [19]. Sometimes five-coordinate species can be detected at intermediate chemical shift values [20].

Oxide catalysts also frequently contain paramagnetic species, either as the result of redox behavior under process conditions or because these have been added intentionally to modify catalyst properties. In general, the paramagnetic species cause strong resonance shifts, line-broadening effects, and wide spinning sideband patterns [21], which together complicate analysis and interpretation of the spectra considerably. However, such shift effects in turn reveal structural proximity and thus have been used in favorable cases as a source of structural information [22].

C. Quadrupolar Nutation NMR

Nuclei with a spin quantum number of $I > 1/2$ possess nonspherically symmetric nuclear charge distributions (nuclear electric quadrupole moments), which can interact with asymmetric charge distributions (electric field gradients) caused by the coordination environment. This interaction is parametrized by the nuclear electric quadrupole coupling constant, e^2qQ/h, and the asymmetry parameter η. The coupling constant (given in MHz) characterizes the overall magnitude of the interaction, while the asymmetry parameter characterizes the deviation of the electric field gradient from cylindrical symmetry. As a result of this interaction, the $2I + 1$ Zeeman states in the magnetic field are no longer equally spaced, and one thus expects $2I$ energetically distinct transitions, as illustrated in Fig. 3. For nuclei with half-integer spins (the only ones of interest here), the central resonance corresponding to the $m = 1/2 \rightarrow m = -1/2$ transition remains unaffected, while the other transitions form satellite powder patterns symmetrically displaced around the central resonance [23]. For strong nuclear quadrupolar couplings, these satellites are shifted far off resonance into spectral regions lying outside the bandwidth of the radiofrequency pulses used to elicit the NMR signal. Failure to excite the quadrupolar satellites in such cases has an interesting consequence [24,25]. Under these circumstances the magnetization associated with the central transition precesses more rapidly in the applied radio frequency field, resulting in a concomitant shortening of the 90° pulse length t_p (90°). In the limit of completely selective excitation of the central transition (i.e., in the situation where the quadrupole splitting is large compared to the excitation bandwidth) one finds

$$t_p \text{ (90° selective)} = t_p \text{ (90° nonselective)}/(I + 1/2) \tag{1}$$

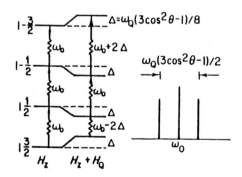

Figure 3 Energy level splittings and transitions for half-integer quadrupolar nuclei, as illustrated for the example of a spin-3/2 nucleus, ω_Q is the NMR transition frequency in the absence of the quadrupolar interaction, ω_Q is the quadrupolar frequency, and θ is the angle between the magnetic field and the principal axis of the electric field gradient tensor.

where t_p (90° nonselective) is the pulse length measured if the quadrupole splitting is either absent as in the liquid state or else so small that all the quadrupole-perturbed Zeeman transitions are uniformly excited. Experimentally, the excitation behavior is mapped out by measuring the signal intensity under systematic incrementation of the pulse length. Such so-called nutation plots [26] provide qualitative information about quadrupole couplings and site symmetries and can be used as fingerprints for compound characterization.

D. Cross-Polarization of Nuclei in the Surface Layer

When selective information about the surface is desired, the MAS experiment is often combined with cross-polarization. Such applications take advantage of the fact that the surface generally contains hydroxyl or molecular H_2O groups, which in the presence of an external magnetic field represent a large reservoir of ^1H magnetization. Cross-polarization employs strong double irradiation at the Hartmann-Hahn condition [27,28]

$$\gamma_H B_{1H} = \gamma_x B_{1X} \qquad (2)$$

where the γ_i and the B_{1i} are the gyromagnetic ratios and radio frequency field strengths for the protons (H) and the heteronuclei (X). This condition enables ^1H magnetization transfer to other nuclei nearby at the surface. The experiment discriminates very well against the more remote nuclei in the bulk, because the cross-polarization rate depends strongly on the magnitude of the dipolar interaction with (and hence the internuclear distance from) the ^{1}H species. Cross-polarization experiments applied to oxide catalysts have sometimes involved quadrupolar nuclei such as ^{17}O ($I = 5/2$), ^{23}Na ($I = 3/2$), ^{27}Al ($I = 5/2$), or ^{95}Mo ($I = 5/2$). In all these cases the central $1/2 \rightarrow -1/2$ transition is spin-locked selectively, and a modified Hartmann-Hahn condition applies:

$$\gamma_H B_{1H} = \gamma_{1X} B_{1X} (I + 1/2) \tag{3}$$

E. Probing Bronsted and Lewis Acid Surface Sites by Adsorbed Molecules

Since the catalytic role of oxide surfaces is often related to the presence of Bronsted and Lewis acidic sites on the surface, it is desirable to characterize the acidic surface properties and the respective concentrations of these species. Information about the Bronsted sites can be obtained from ^{1}H MAS chemical shifts, which are strongly correlated with acidities [4,29]. Complementary information is available from multinuclear MAS-NMR studies of basic probe molecules adsorbed to the surface. Interaction of the probe molecules with acidic sites is expected to cause characteristic shift effects compared to the resonances of the same molecules in either unperturbed or physisorbed states, and hence quantitative information about site populations should be available. Although a number of such investigations have been carried out, it is often difficult to compare results obtained in different laboratories on the same system because experimental details such as sample history, surface coverage, and impurities (frequently water) have large effects on the spectra.

One of the most popular probe molecules has been pyridine. ^{13}C NMR spectra, obtained at high surface coverages under usual high-resolution liquid state conditions, are affected by chemical exchange between physisorbed molecules and molecules bound to acidic sites. As a consequence, the chemical shifts measured depend on surface coverage and are weighted averages [30–32]:

$$\delta_{exp} = f_a \delta_a + (1 - f_a) \delta_n \tag{4}$$

Here f_a is the fraction of base interacting with the acidic sites, and δ_a and δ_n are the chemical shifts of the base adsorbed to the acidic site and the base in the physisorbed state, respectively. Under the assumption that the acidic sites are always fully occupied if sufficient amine is present, f_a is identical with the concentration of acidic sites on the surface and can be determined quantitatively if the total amount of adsorbed base is known.

At low surface coverages molecular motion is more restricted, chemical exchange is slow, and the resonances are generally too broad to be observed by wideline NMR. Use of line-narrowing techniques such as MAS then becomes essential. ^{15}N as the reporter isotope is generally preferable over ^{13}C, because the spectra are simpler and the resonance shifts are larger for directly bonded atoms

[33]. For the same reasons, ^{31}P MAS-NMR studies of adsorbed phosphine molecules have proven very informative [34–36]. While in all these applications only single-pulse spectra yield quantitatively reliable results, the resonances of physisorbed molecules can often be suppressed by CPMAS techniques, thereby allowing the selective observation (and hence the resonance assignment) of Bronsted or Lewis acid site–bound molecules.

III. OXIDATION CATALYSTS

Many industrially important selective oxidation reactions are catalyzed by transition metal oxides. The activity of such catalysts is related to the reducibility of the transition metal ion, which enables the bulk oxide lattice to participate actively in the redox processes present in the Mars van Krevelen mechanism. Unfortunately, NMR spectroscopic investigations are severely limited by the occurrence of paramagnetic oxidation states. As a general rule, NMR signals from atoms bearing unpaired electron spins cannot be detected by conventional methods, and the spectra of atoms nearby are often severely broadened. For this reason, most of the work published in this area has dealt with diamagnetic vanadium(V) oxide–based catalysts.

A. SO$_2$ Oxidation Catalysts

The oxidation of SO$_2$ to sulfuric acid, SO$_2$ + H$_2$O + 0.5 O$_2$ → H$_2$SO$_4$, is catalyzed by potassium vanadium(V) oxide compounds. A typical catalyst preparation sequence involves impregnation of a silica support with a solution containing potassium vanadate (K/V = 3), followed by drying and subsequent calcination at 500°C in air. Under typical operating conditions in SO$_2$/O$_2$/SO$_3$ atmospheres at 400–500°C, the catalytically active species is molten and forms a thin liquid film on the silica support. As such the system functions like a bulk oxide catalyst under operating conditions, and the silica mostly serves as a mechanical support medium.

In a series of [51]V wideline NMR studies, Mastikhin and coworkers have explored the chemical nature of the catalytically active species [37–42]. While the spectra of industrial catalysts from various sources are found to be substantially different, these differences more or less disappear after exposure to the reaction mixture. This result confirms the previously held view that the catalytically active species forms under operating conditions. Figure 4 shows typical spectra recorded at a field strength of 7.0 T, at which the lineshape is dominated by the chemical shift anisotropy. The principal contribution to the spectrum in Fig. 4 arises from an axially symmetric powder pattern with approximate $\delta_\perp$ and $\delta_\parallel$ values of -300 and -1300 ppm, respectively. Based on comparative studies of model preparations, Mastikhin et al. suggest that the key compound formed has the composition K$_3$VO$_2$SO$_4$S$_2$O$_7$. The anisotropic chemical shift parameters of

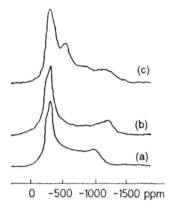

Figure 4 79 MHz solid-state ^{51}V wideline NMR spectra of model preparations and commercial SO_2 oxidation catalysts (a) $K_3 VO_2(SO_4)_2$, (b) $K_3VO_2SO_4S_2O_7$, and (c) the commercial catalyst after exposure to the reaction mixture. (From Ref. 40.)

this material are close to those measured in crystalline V_2O_5, suggesting that vanadium is in a square-pyramidal environment. More insight into multiple lineshape contributions might be expected from MAS NMR studies, which generally yield better-resolved spectra. Such experiments still remain to be performed.

B. Vanadium Phosphorus Oxide Catalysts

Vanadium phosphorus oxide catalysts (with P/V ratios ranging from 0.9 to 1.1) have received widespread industrial attention for the oxidation of butane to maleic anhydride. In these catalysts, vanadium is present in both the oxidation states $+4$ and $+5$, although the active phase is believed to be $(VO)_2P_2O_7$. Despite the favorable NMR properties of ^{51}V and ^{31}P nuclei, though, NMR studies of this system have been scarce [43,44]. Gerstein and coworkers have studied the distribution of vanadium oxidation states by ^{31}P wideline NMR [44]. Since the resonances in this system are extremely broad, spectra were recorded point by point by measuring the spin echo height as a function of carrier frequency, a procedure termed spin echo mapping. All those P atoms near V^{5+} sites give rise to a sharp NMR signal near 0 ppm (corresponding to the resonance of β-$VOPO_4$). In contrast, for P atoms near V^{+4} ions the unpaired electron causes a large downfield shift and line broadening of the ^{31}P resonance, resulting in a very broad absorption centered around 2500 ppm. This feature is exhibited by the compound $(VO)_2P_2O_7$. Only a minor part of the large downfield shift can be accounted for by bulk susceptibility effects; the larger part is most likely attributable to Fermi contact and superexchange interactions. Figure 5 shows that

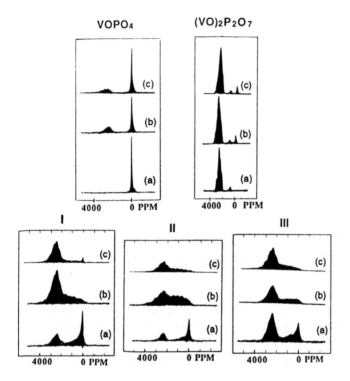

Figure 5 121.5 MHz solid-state ^{31}P wideline NMR spectra of V-O-P catalysts, obtained by spin-echo mapping: (a) fresh, (b) after exposure to 1.5% *n*-butane in air at 450 °C, (c) after exposure to 1.5% 1-butene in air at 450°C. Series I: P/V = 0.9; Series II: P/V = 1.0; Series III: P/V = 1.1. (From Ref. 44.)

exposure of β-VOPO$_4$ or of the commercial catalysts to butane or butene increases the V^{4+}/V^{5+} ratio substantially, and that the distribution of oxidation states is also influenced by the P/V ratio in the catalyst. In contrast, the spectra of $(VO)_2P_2O_7$ remain remarkably constant over the entire course of the reaction. These findings confirm that the latter compound is most likely the active phase in these catalysts.

C. Bismuth Vanadates

Most recently, ^{51}V NMR has been used to characterize vanadium environments in the catalytically active Bi_2O_3-V_2O_5 system [45]. Five crystallographically distinct phases [46] have been identified, with approximately compositions $BiVO_4$ (monoclinic), $Bi_4V_2O_{11}$ (orthorhombic), $Bi_{12}V_2O_{23}$ (triclinic), $Bi_{20}V_2O_{35}$ (fcc), and $Bi_{25}VO_{40}$ (bcc, "sillenite"). There also seem to be extended solid solution ranges, particularly in the $Bi_{25}VO_{40}$-Bi_2O_3 region. Each of the five phases gives rise to characteristic ^{51}V MAS-NMR spectra (Fig. 6) that indicate the numbers of crystallographically distinct sites. In all these compounds the vanadium atoms are four-coordinated, and subtle differences in the coordination symmetry can be identified by ^{51}V nutation NMR. Figure 7 shows typical results. The nutation behavior is shown in the form of stacked one-dimensional Fourier transforms recorded as a function of pulse length. It can be seen that nutation NMR is able to differentiate among the various vanadium environments present in bismuth vanadates. The magnitude of the nuclear electric quadrupole coupling constants, and hence the degree to which the vanadium coordination

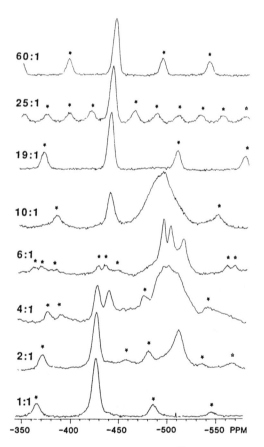

Figure 6 131.4 MHz solid-state ^{51}V MAS-NMR spectra of the bismuth vanadates. The numeral specifies the Bi/V ratio. Spinning sidebands are indicated by asterisks. (From Ref. 45.)

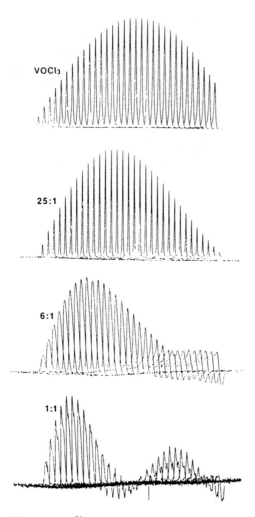

Figure 7 ^{51}V nutation NMR plots of bismuth vanadates and liquid $VOCl_3$ standard. Shown are one-dimensional Fourier transforms as a fraction of pulse length, starting at 0.5 μs and incremented in steps of 0.5 μs. Note that while for liquid $VOCl_3$ the signal maximum is observed at a 90° pulse length around 9 μs, the effective 90° pulse in the solid bismuth vanadates is substantially shorter, indicating partial excitation selectivity owing to quadrupolar splitting. Note the gradual change as a function of the Bi/V ratio. (From Ref. 45.)

environments are distorted from tetrahedral symmetry, decreases with increasing Bi/V content.

D. Other Transition Metal Oxide–Based Catalysts

Bulk V-Mo oxides function as catalysts in the oxidation of acrolein to acrylic acid. ^{51}V NMR has been used to monitor these catalysts through their various preparation steps [47]. Dispersal of the catalysts on a silica support results in increased activities and selectivities. The ^{51}V NMR spectra of such materials differ substantially from those of the bulk catalysts, suggesting that the catalytic behavior of these materials is influenced by the local vanadium coordination environments. It would be interesting to explore whether the molybdenum environment is affected as well, and in principle ^{95}Mo NMR is feasible here owing to the diamagnetic character of these oxides. In general, ^{95}Mo NMR studies are severely handicapped by the low natural abundance, the small magnetic moment, and the quadrupolar character of this isotope. Recently, however, it has been possible to characterize a number of simple binary molybdates and dispersed catalytic systems by ^{95}Mo wideline and MAS NMR [48]. Detailed model compound studies suggest that ^{95}Mo chemical shifts are able to differentiate between four- and six-coordinated molybdenum environments [49]. In view of these developments it can be predicted that ^{95}Mo NMR will find increasing use in a variety of systems, including the industrially important Bi_2O_3-MoO_3 ammonoxidation catalysts.

E. Pyrochlores

Rare earth pyrochlores, possessing the general formula $Ln_2Sn_2O_7$, where Ln denotes a rare earth, are active catalysts for the oxidative coupling of methane [50]. Enhanced conversion to useful hydrocarbons (e.g., ethene) is observed with pyrochlores containing rare earths with mixed valence behavior, particularly Sm, Eu, and Yb. Since the rare earth site thus appears to be crucially linked to the catalytic behavior, the distribution of such rare earth species within the lattice is of intrinsic interest.

Nuclear magnetic resonance spectra of these compounds are highly informative in this regard. The lineshapes of the ^{119}Sn and ^{89}Y resonances are dominated by the coupling of the nuclei with the unpaired electron spins. This interaction results in large resonance shifts arising from both the Fermi contact and the pseudo-contact (dipolar) interaction.

The Fermi contact shift term comes about from unpaired spin density transferred to the nuclei via bonding orbitals as well as screening and polarization effects. This contribution is isotropic and calculated according to

$$\frac{\Delta\omega_c}{h} = a_n \langle S_z \rangle \tag{5}$$

where a_n is the hyperfine coupling constant (in Hz) and $<S_z>$ is the expectation value of the rare earth spin z-component. The pseudo-contact shift term, $\Delta\omega_p$, is orientation-dependent and can be calculated rigorously from the lattice geometry as discussed in Refs. 51 and 52. The experimentally measured isotropic chemical shift is the sum of these two contributions plus a (small) diamagnetic screening effect, $\Delta\omega_d$:

$$\Delta\omega = \Delta\omega_c + \Delta\omega_p + \Delta\omega_d \qquad (6)$$

Figure 8 shows that in $Ln_2Sn_2O_7$ compounds (Ln = Eu, Tm, Yb, Sm, La, Pr, Nd), the experimental isotropic chemical shifts for the ^{119}Sn resonance correlate nicely with theoretical values calculated from Eq. (5) by assuming no variation of a_n with the rare earth cation considered. This result suggests that the Fermi contact shift contribution dominates in such cases. By contrast, the ^{89}Y resonance shifts measured in lanthanide-substituted compounds $Y_{2-x}Ln_xSn_2O_7$ and

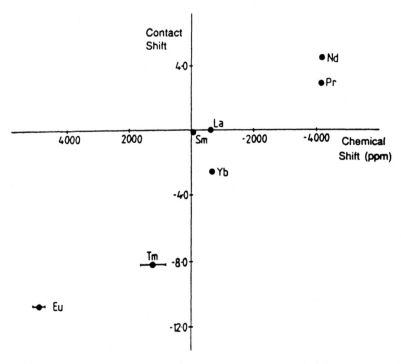

Figure 8 Correlation of the contact shift predicted theoretically with Eq. (5) with the experimentally measured ^{119}Sn isotropic chemical shift [vs. $(CH_3)_4Sn$] in compounds of composition $Ln_2Sn_2O_7$ (Ln = Nd, Pr, La, Sm, Yb, Tm, Eu). Calculations assume a uniform hyperfine coupling constant a_n, and use relative values of $<S_z>$. (From Ref. 51.)

$Y_{2-x}Ln_xTi_2O_7$ are one order of magnitude smaller and are best explained in terms of the pseudo-contact interaction mechanism [52].

As shown by Cheetham and coworkers [51,52], these paramagnetic chemical shift effects can be exploited for site quantitation purposes. Figure 9 provides a nice illustration for a sample with the nominal composition $YSmSn_2O_7$. Figure 9a shows the nearest and next-nearest neighbor coordination environment of the Sn atoms in the pyrochlore structure. The Sn atoms are at the center of an SnO_6

(a)

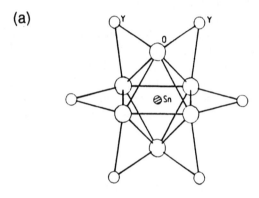

(b)

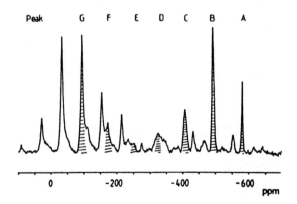

Figure 9 (a) Local environment of the tin atoms in the pyrochlore $Y_2Sn_2O_7$. (b) $YSmSn_2O_7$. The isotropic resonances A–G, shown shaded, correspond to next-nearest-neighbor configurations Y_6, SmY_5, Sm_2Y_4, Sm_3Y_3, Sm_4Y_2, Sm_5Y, and Sm_6 around the central tin atom. The remaining peaks are spinning sidebands. (From Ref. 51.)

octahedron and bonded through oxygen to six rare earth next-nearest neighbors. Thus successive substitution of Y^{3+} by Sm^{3+} cations will result in a total of seven distinct next-nearest neighbor configurations $Y_{6-n}Sm_n$ ($0 \leq n \leq 6$). The ^{119}Sn MAS-NMR spectrum indeed shows seven lines (besides spinning sidebands), which have been marked A–G in Fig. 9b and assigned in order of increasing n. Note the additive effect of the number of the paramagnetic Sm^{3+} ions on the resonance shift.

By careful analysis of the peak intensities and by comparison with random-site occupancy models, one can determine the regions of solid-solution homogeneity in this system and potentially in others. For instance, the spectrum shown in Fig. 9b clearly reveals a very inhomogeneous site distribution, signaling phase separation. Intensity analysis shows that in the system $Y_{2-x}Sm_xSn_2O_7$, solid solutions can only be formed for $x < 0.15$ and $x > 1.97$. Similar results on both ^{119}Sn and [89]Y NMR in related systems indicate that paramagnetically perturbed MAS-NMR spectroscopy is a powerful method for establishing the phase composition and the solid solution range of these catalytically active materials [51,52].

F. Tin Antimony Oxides

Tin oxide-based materials are potent oxidation and isomerization catalysts. Their bulk and surface properties, as well as their presumed mechanism in oxidation catalysis, have been reviewed [53]. Considerable uncertainty remains concerning the phase compositions, solid-solution range, and the redox behavior (Sn^{2+}/Sn^{4+} vs. Sb^{3+}/Sb^{5+}) of these materials. Structural investigations have so far concentrated on the use of [119]Sn and [121]Sb Mossbauer spectroscopy. Surprisingly, no ^{119}Sn solid-state NMR studies have appeared to date on this system, although it was recently demonstrated that isotropic [119]Sn chemical shifts and chemical shift anisotropies give characteristic fingerprints of the various tin coordination environments in Sn(IV) oxide compounds [54]. In situ ^{13}C NMR has been used to study the double bond shift of 1-butene to *cis*-2-butene, and the subsequent *cis-trans* isomerization over tin antimony oxide catalysts [55].

IV. ISOMERIZATION AND CRACKING CATALYSTS

A. Transitional Aluminas

1. The Defect Structure of Transitional Aluminas

A variety of catalytically active aluminas with distinct x-ray powder diffraction patterns are prepared by the controlled dehydration of AlO(OH) (boehmite) or $Al(OH)_3$ (bayerite) according to the following sequences [56]:

$$\text{boehmite (crystalline)} \xrightarrow{450°C} \gamma\text{-Al}_2O_3 \xrightarrow{600\ °C}$$

$$\delta\text{-Al}_2O_3 \xrightarrow{1200°C} \alpha\text{-Al}_2O_3 \qquad (7)$$

$$\text{bayerite} \xrightarrow{230°C} \eta\text{-Al}_2O_3 \xrightarrow{850°C} \theta\text{-Al}_2O_3 \xrightarrow{1200°C} \alpha\text{-Al}_2O_3$$

The η-, γ-, δ-, and θ-forms are known as the transitional aluminas. They have highly disordered structures, high surface areas, and a large number of surface hydroxyl species at exposed surfaces. Transitional aluminas are catalytically active materials and are known to facilitate H-D exchange, C-H activation, skeletal olefin isomerizations, and alcohol dehydrations [57]. In industrial catalytic processes, aluminas are mostly used as catalyst support materials.

All transitional aluminas crystallize in disordered (and distorted) defect spinel lattices. The unit cell contains a cubic close-packed array of 32 O^{2-} ions. Electroneutrality demands that of the 24 sites in the cation sublattice only $21\frac{1}{3}$ are occupied, leaving $2\frac{2}{3}$ vacant positions. In essence, the various transitional aluminas differ in the uniformity of the anionic stacking and in the distribution of aluminum atoms over the octahedral and tetrahedral sites. In contrast to the AlO(OH) and Al(OH)$_3$ precursors and to α-Al$_2$O$_3$, the transitional aluminas show a significant fraction of tetrahedral aluminum sites, which are readily identified and quantitated by ^{27}Al MAS-NMR [58,59]. The fraction of four-coordinate aluminum sites decreases in the order $\eta \rightarrow \gamma \rightarrow \delta \rightarrow \theta$, in accordance with crystallographic predictions.

Recently, these results have been complemented by ^{17}O MAS-NMR investigations on isotopically enriched samples [60]. Two distinct types of O^{2-} sites are expected in the bulk: regular four-coordinated OAl$_4$ sites with moderately weak quadrupolar coupling and three-coordinated OAl$_3$ sites next to a cation vacancy, characterized by a much stronger quadrupolar coupling. These expectations are borne out well by the experimental spectra shown in Fig. 10. Two spectral components indeed appear for all the transitional aluminas. While the ^{17}O NMR spectra are unable to differentiate between the η-, γ-, and δ-aluminas and also reveal comparable disorder in all these phases, the spectrum of θ-Al$_2$O$_3$, which forms the most highly ordered spinel structure among the transitional aluminas, is very distinct. It can be simulated using e^2qQ/h and η of 1.2 MHz and 0, respectively, for the four-coordinated oxygen site, and 4.0 MHz and 0.6, respectively, for the three-coordinated oxygen site.

2. Bronsted Acid Sites

Based on the assumption that undistorted (111), (110), and (100) crystal faces are preferentially exposed at the surface of the crystallites, Knözinger and Ratnasamy have postulated the following five possible types of surface hydroxyl groups on γ- and η-Al$_2$O$_3$ [57]:

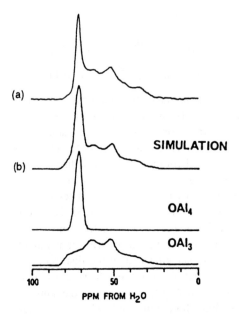

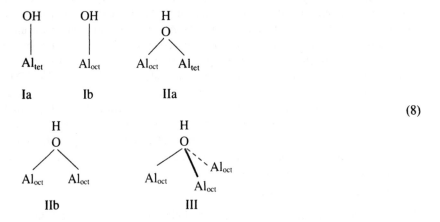

Figure 10 68-MHz ^{17}O MAS-NMR spectrum of θ-alumina (a), and a corresponding simulation (b), consisting of two spectral components in a 1:2 ratio. OAl_4 sites: $e^2qQ/h =$ 1.2 MHz, η = 0.0, δ_{iso} = 72 ppm. OAl_3 sites: e^2qQ/h = 4.0 MHz, η = 0.6, δ_{iso} = 79 ppm. (From Ref. 60.)

These five species have different acidities, which have been predicted to decrease in the order type III → type IIa → type IIb → type Ia → type Ib.

The actual populations of these configurations at the surface depend on the relative contributions of the different crystal faces. If the (111) face dominates, for instance, one expects to find all five species (with type IIa and IIb being preponderant), whereas on top of a layer parallel to the (100) plane only type Ib species can exist. Since the ^{1}H chemical shifts are known to correlate with acidities, it should be possible, in principle, to differentiate among these species by ^{1}H MAS-NMR. Attempts in this direction have indeed identified multiple ^{1}H lineshape components on γ- and δ-alumina [14], but the resolution was not sufficient to allow definite conclusions. The situation might be improved by further experimental developments such as higher field strengths, faster spinning, or multiple pulse line narrowing. Also, low-temperature studies may be required in view of the possibility that the room temperature spectra are influenced by chemical exchange effects.

The ^{1}H species in the surface layer have been utilized to detect the [17]O and [27]Al nuclei at the alumina surface selectively via cross-polarization [60,61]. ^{17}O NMR experiments reveal that the surface sites are spectroscopically quite distinct from the sites in the bulk [60]. In the ^{27}Al NMR studies, essentially the entire signal arises from octahedral aluminum atoms bonded to OH species. Most likely this situation does not mean that tetrahedral surface sites are absent, but that their observation is precluded by unfavorable cross-polarization characteristics such as fast spin-lattice relaxation time in the rotating frame [61].

3. Lewis Acid Sites

Besides the Bronsted acid sites discussed above, transitional aluminas also contain Lewis acid surface sites arising from the partial dehydroxylation of the surface. In this reaction, which is favored at elevated temperatures, an Al-OH bond associated with a type I, II, or III species reacts with a neighboring Bronsted site, resulting in the dissociation of H_2O and a Lewis acid site. Al-OH bonds in type I species are generally thought to be most easily dissociated. To study such Lewis sites, multinuclear NMR experiments have been carried out with adsorbed probe molecules. For example, Morris and Ellis adsorbed pyridine molecules onto the Lewis sites of a previously fully deuterium-exchanged alumina surface and succeeded in selectively cross-polarizing the [27]Al nuclei associated with these Lewis sites from the pyridine ^{1}H magnetization reservoir [61]. Again, mostly octahedral [27]Al sites are detected in this fashion, with spectroscopic parameters quite distinct from those of "regular" octahedral sites in γ-alumina.

A profitable way to quantitate the Lewis acid sites is to study the adsorbed molecules directly. For instance, Fig. 11 shows the ^{13}C NMR results for butylamine adsorbates on γ-alumina [62]. In the adsorbate, there are two distinct resonances each for the α- and the β-carbon atoms of butylamines. Based on the model complex spectra included in Fig. 11, these peaks can be assigned to molecules bound to Bronsted and Lewis acid sites, respectively. In a similar fashion, [13]C [63] and especially ^{15}N CPMAS NMR studies of pyridine adsorbates on γ-alumina are able to differentiate among physisorbed molecules (64

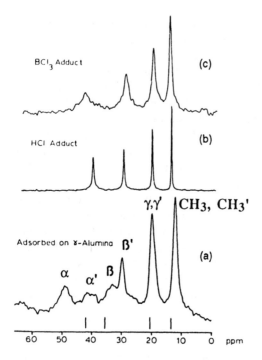

Figure 11 ^{13}C CPMAS NMR spectra of *n*-butylamine (a) adsorbed to the surface of γ-alumina (site assignments are included); (b) solid HCl adduct (model for butylamine bound at a Bronsted-acid site); (c) solid BCl_3 adduct (model for butylamine bound at a Lewis-acid site). (From Ref. 62.)

ppm relative to the nitrate resonance of NH_4NO_3), physisorbed pyridine interacting with water (76 ppm), pyridine molecules interacting with two distinct Lewis acid sites (110 and 138 ppm), and (after air exposure) surface pyridinium ions resulting from the presence of Bronsted acid sites (174 ppm) [33]. In a follow-up study, Majors and Ellis were able to confirm the presence of multiple Lewis acid sites [64]. Two distinct ^{15}N resonances were observed around 105 and 125 ppm and tentatively assigned to Lewis acid sites arising from the dehydroxylation of octahedral and tetrahedral aluminum sites, respectively. The relative intensity ratio and changes observed upon increasing severity of heat treatment have been discussed by the authors in terms of the site statistics at the various index planes. They conclude that the γ-alumina surface is best modeled by assuming equal proportions of (111) and (110) planes.

Qualitatively similar results are observed in ^{31}P CPMAS NMR studies of trimethylphosphine adsorbates [34]. The chemical shifts of Bronsted site–bound, Lewis site–bound, and physisorbed TMP molecules on γ-alumina are −4, −48, and -58 ppm, respectively. Figure 12 shows, however, that the ^{31}P CPMAS-NMR spectra differ dramatically from single-pulse spectra, indicating that the CPMAS technique is not a reliable tool for obtaining quantitative site populations. In the present example, ^{31}P CPMAS discriminates strongly against physisorbed molecules, which are presumably too mobile to be cross-polarized effectively.

Mastikhin et al. propose N_2O as a probe molecule for the selective study of molecular binding to Lewis acid sites [65]. In the gaseous form, the central and terminal N atoms resonate at -147.3 and -235.5 ppm, respectively. N_2O binds to Lewis acidic surfaces at the terminal N position; at room temperature, however, bound and physisorbed N_2O molecules exchange rapidly on the NMR time scale, resulting in averaged chemical shifts that depend on surface coverage. Most notably, the interaction of N_2O with Bronsted sites appears to have virtually no effect on the ^{15}N chemical shifts (see Fig. 13), whereas binding to a Lewis acid site leads to line broadening and characteristic downfield resonance shifts. Using this approach, Mastikhin et al. have selectively measured the number of Lewis acid sites in γ-alumina as a function of heat treatment temperature.

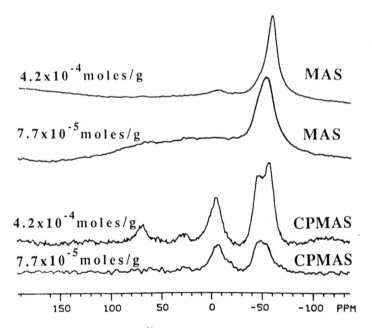

Figure 12 60.7 MHz ^{31}P single-pulse and CPMAS NMR spectra of trimethylphosphine (TMP) on γ-alumina at different surface coverages as indicated. Major resonances at -4, -48, and -58 ppm correspond to Bronsted-site bound, Lewis-site bound, and physisorbed TMP. (From Ref. 35.)

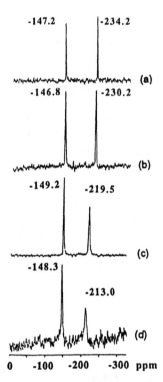

Figure 13 30.42 MHz ^{15}N NMR spectra of N$_2$O adsorbed an (a) 10% H$_3$PO$_4$/SiO$_2$ catalyst, 80 μmole/g (model for Bronsted-site bound N$_2$O); (b) γ-alumina, pretreated at 200°C/4 h in vacuo, 120 μmol/g; (c) γ-alumina, pretreated at 500°C/8 h in vacuo, 120 μmol/g; (d) γ-alumina, pretreated at 600°C/8 h in vacuo, 120 μmol/g. The concentration of Lewis acid sites increases from (b) to (d). Note the effect on the chemical shift (listed with the peaks) and the linewidth of the resonance of the terminal N atom, in contrast to spectrum (a). (From Ref. 65.)

B. Amorphous Silica-Alumina Catalysts

Amorphous silica-aluminas are high-surface-area materials that can be prepared by co-precipitation or sol-gel techniques from solutions containing sodium silicate and sodium aluminate. Because of their high surface acidities they are potent hydrocarbon cracking catalysts, but their activity is strongly dependent on the silica/alumina ratio and on the method of preparation. Solid-state NMR studies have addressed the structural origins of these variations and have served to characterize the surface acidities.

1. Bulk Structure

Depending on the preparation technique, it is possible to generate both single-phase and diphasic materials [66]. Diphasic materials are prepared by mixing sols of silica and boehmite, and ^{27}Al and ^{29}Si MAS NMR studies confirm that the resulting powders can be viewed as intimate mixtures of amorphous silica and γ-alumina. In contrast, preparation from solutions containing both sodium silicate and sodium aluminate results in monophasic materials over a wide compositional region. For low to moderate Al contents, aluminum is thought to substitute into the silica network. Protons balance the resulting negative charge on the network, giving rise to sites of the type which possess high Bronsted acidities.

$$
\begin{array}{c}
\text{H} \\
| \\
\text{O}^+ \\
\diagup \quad \diagdown \\
\text{Si} \qquad \text{Al}^-
\end{array}
\qquad (9)
$$

Charge balancing is also accomplished in part by partially hydrolyzed Al^{3+} cations in octahedral environments, and phase separation occurs for Al/(Al + Si) ratios above 0.8, with the formation of crystalline γ-alumina [67]. ^{27}Al MAS-NMR has proven particularly powerful for structural characterization, owing to its unique ability to differentiate between tetrahedral and octahedral aluminum-oxygen coordination environments. For samples prepared in the same fashion, a compositional trend as shown in Fig. 14 is typical: with increasing Al/(Al + Si) ratio, the fraction of octahedral aluminum increases, and at high alumina content the signal becomes indistinguishable from that of γ-alumina.

A review of the available literature reveals that for samples with comparable Al/(Si + Al) ratios in the low-alumina compositional region, the ratio of octahedral to tetrahedral aluminum can be highly variable [66,68–70]. Even at the lowest aluminum contents, octahedral aluminum is never completely absent. The spectroscopic behavior is also greatly dependent on the state of hydration [71,72] and on details of sample preparation. If the above-described structural model for monophasic silica-aluminas is applicable, there should be an inverse correlation between the number of octahedral aluminum sites and the number of Bronsted sites associated with tetrahedral aluminum. This hypothesis deserves examination by quantitative spin counting studies, and overall there is a general need in this area to address the quantitative aspects of ^{27}Al NMR in more detail. Spin-counting experiments in glasses have shown that often only a fraction of the Al atoms present are detectable by MAS-NMR. The remaining fraction is rendered invisible because of large quadrupolar broadening of Al sites in very

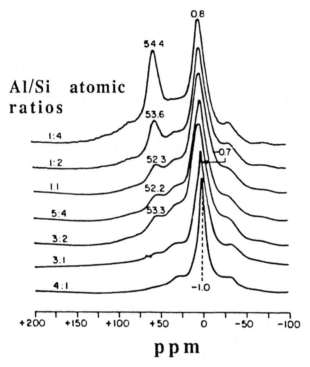

Figure 14 104.22 MHz ^{27}Al MAS-NMR spectra of single-phase amorphous aluminosilicate gels as a function of Al/Si atomic ratio. The numerals indicate resonance maxima assigned to four- and six-coordinate aluminum sites (52–54 and 0–1 ppm, respectively). (From Ref. 66.)

distorted environments [73]. Likewise, in view of the amorphous character of silica-aluminas, more attention to the quantitative aspects of ^{27}Al MAS-NMR is warranted.

^{29}Si MAS-NMR has been used as an auxiliary characterization tool [74]. It is well known that the various $Si(OSi)_n(OAl)_{4-n}$ configurations present in aluminosilicates can be distinguished quantitatively by their ^{29}Si chemical shifts. However, the amorphous nature of silica-aluminas results in chemical shift distribution effects that limit the resolution in the NMR spectra. Furthermore, downfield shift effects not only arise from Al substitution in the silica network but also from the presence of terminal hydroxyl (Si-OH) units. As a result of these problems, ^{29}Si NMR has generally played a lesser role in the characterization of amorphous silica-aluminas.

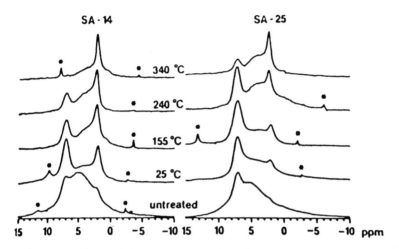

Figure 15 187 MHz high-resolution ¹H solid-state NMR spectra of two silica-aluminas (14 and 25 wt% Al₂O₃, respectively), untreated or evacuated at 10^{-2} torr at the indicated temperature. Spinning sidebands are indicated by asterisks. (From Ref. 77.)

2. Bronsted Acidities

¹H NMR studies have been instrumental in the characterization of the various hydrous species present on the surface and in the bulk of amorphous silica-aluminas [75–77]. Typical results from high-resolution ¹H NMR are shown in Fig. 15 [77]. Four types of hydrogen are present: (1) physisorbed molecular water (broad line at 5 ppm), (2) Si-OH groups (sharp line at 1.8–2.0 ppm), (3) various Al-OH configurations associated with the γ-alumina surface (broad line at 3 ppm), and (4) hydrous species assigned to Bronsted sites (sharp line at 7.0 ppm). The contribution due to physisorbed water (site 1) is easily removed by sample evacuation at 100–200°C. Sites 2 and 3 are best discerned from an analysis of the free induction decays [75,76]: Si-OH groups give rise to exponential decays with T_2 in the vicinity of 150 μs, whereas for Al-bonded hydroxyl species the strong ²⁷Al-¹H heterodipolar interaction results in a much more rapid Gaussian decay. Site 4 shows intermediate behavior.

Most interest has focused on characterization of site 4 (7.0 ppm), the one presumed to be mostly connected with catalytic activity. From their combined analysis of T_2 and high-resolution ¹H NMR data on a set of samples with variable Al contents, Hunger et al. conclude that the concentration of this species reaches a maximum near $Al/(Si + Al) = 0.3$ [75]. This ratio coincides with maximum catalytic activity of these materials, as commonly measured by the rate constant of cumene cracking [75]. The 7.0 ppm peak was originally assigned to the protons in Bronsted acid sites as depicted above. Subsequent studies, however, indicate that the 7.0 ppm peak disappears when the sample is exposed at 340°C, a typical activation temperature where Bronsted sites are presumably generated. To remove this apparent contradiction, Maciel et al. propose the scheme shown below:

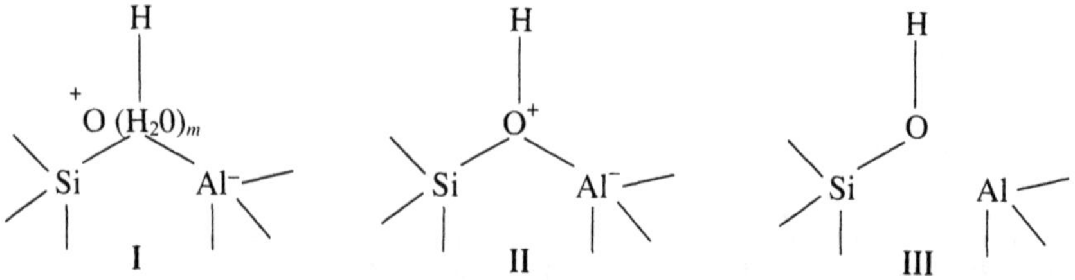

According to this interpretation, the 7.0 ppm peak belongs to hydrated Bronsted acid sites. At elevated temperature this water is lost, resulting in either species II (the bare Bronsted site or species III, a nonacidic SiOH group in the vicinity of a Lewis acid site. Thus far, attempts to detect the bare species have been unsuccessful, and other studies have failed to observe the 7.0 ppm peak altogether [70].

Finally, ^{19}F NMR has been used to characterize the structural features of surface-fluorinated aluminas and amorphous aluminosilicates. Both the ^{19}F NMR chemical shifts and spin-spin relation times differentiate sensitively between Si-F and Al-F bonds. Lineshape and spin echo decay data suggest that the homonuclear ^{19}F-^{19}F dipolar interactions are much weaker than in AlF_3-hydrate, indicating that in samples containing up to 5 wt% fluorine, most of the Al-F species are isolated from each other [78].

3. Surface Characterization by Probe Molecules

Surface acidities and the respective contributions from Lewis and Bronsted sites have been investigated using the ^{13}C, ^{15}N, and ^{31}P resonances of suitable probe molecules such as aniline, pyridine, ethyl-pyridine, trimethyl phosphine, and trimethylphosphine oxide. As in the case of aluminas, well-resolved static ^{13}C spectra are observed at high coverages (above one monolayer), indicating the presence of motion and rapid chemical exchange. Using Eq. (4) Gay and Liang determined total concentrations of surface acidic sites [32]. The distinction between Bronsted and Lewis sites appears to be more difficult using this method. Characteristic chemical exchange effects were also observed in ^{13}C and ^{15}N CPMAS-NMR studies of pyridine on silica-alumina at submonolayer coverages. In these studies, the ^{15}N chemical shift is used as a measure of the relative populations of Bronsted acid site–bound, Lewis acid site–bound, and physisorbed (H-bonded) molecules [79,80].

Other suitable molecules include trialkyl phosphines and trialkyl phosphine oxides [34–36]. Phosphines bound to Bronsted acid and Lewis acid sites and physisorbed molecules yield distinguishable resonances, although the latter two are poorly resolved from each other. There is no indication of exchange broadening in these cases.

Figure 16 shows loading-dependent ^{31}P single pulse NMR spectra of triethylphosphine on silica-alumina, which reflect selective occupation of Bronsted

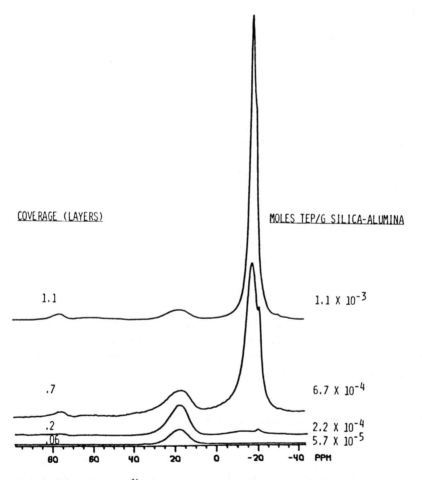

Figure 16 60.7 MHz ^{31}P single-pulse MAS-NMR spectra of triethylphosphine (TEP) adsorbed on silica-alumina (25 wt% Al$_2$O$_3$) at different surface coverages as indicated. Resonances at 20, −17, and −21 ppm correspond to Bronsted site–bound, Lewis site–bound, and physisorbed TEP. (From Ref. 35.)

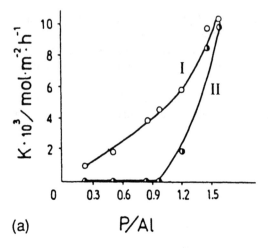

(a)

Figure 17 (a) Dependence of catalytic activities of aluminophosphate catalysts on the P/Al ratio: (I) reaction rate constant for isobutanol dehydration, (II) reaction rate for 1-butene isomerization. (b) ^{1}H MAS-NMR spectra of aluminophosphate catalysts with different P/Al ratios: A: 1.6; B: 1.4; C: 1.0; D: 0.5; E: model compound $Al(H_2PO_4)_2$. Spinning sidebands are indicated by asterisks. Centerband chemical shifts are indicated in the plot. (From Ref. 14.)

sites at low coverages. From such surface titration studies it has been possible to quantitate the number of Bronsted acid sites. For future investigations it would be of interest to correlate the concentration of such sites with the intensity of the 7.0 ppm peak observed in ^{1}H MAS-NMR and also with the Al(4)/Al(6) ratio in the bulk and with the cracking activity of the catalyst. Much remains to be done in this area.

To date, a similarly straightforward determination of Lewis acid sites on amorphous aluminosilicates appears not to be feasible. The binding constants for those probe molecules that are bonded in Lewis acid complexes and those that are physisorbed are too similar, and their ^{31}P resonances are not sufficiently resolved.

C. Aluminophosphates and Related Systems

Aluminium phosphate, $AlPO_4$, is a well-known catalyst and catalyst support, which is generally prepared by neutralizing acidic solutions containing stoichiometric amounts of Al^{3+} and PO_4^{3-} ions. Precipitations conducted off-stoichiometry have resulted in novel aluminophosphates, which have activity in acid-catalyzed isomerization and dehydration reactions. Figure 17a shows that the rate constants of such reactions increase strongly with increasing P/Al ratio, especially as the ratio exceeds the stoichiometric value of 1.0. Figure 17b reveals

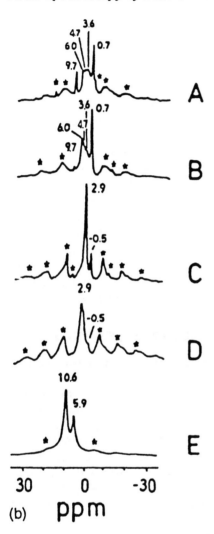

that this increase is correlated with the appearance of downfield shifted resonances in the ^{1}H MAS-NMR spectra (6.0 and 9.7 ppm), signifying the presence of highly acidic P-OH groups at the surface [81]. Enhanced Bronsted acidity associated with such species has also been discussed in conjunction with ^{1}H and ^{31}P MAS-NMR studies of PCl_3- and H_3PO_4-modified silica surfaces [82] and with calcined zirconium phosphate–based catalysts [83].

Phosphorus-deficient aluminophosphates $(0 < P/Al < 1)$ constitute an interesting new class of active polymerization catalysts. Their activity, following promotion with Cr^{3+}, reaches a maximum at $P/Al = 0.8$. The ^{31}P resonance observed in these phases, which does not coincide with that measured in AlPO$_4$ (-32 ppm), shifts downfield with decreasing P/Al ratio. The resonance is assigned to P(OAl$_4$) units, and the gradual shift is attributed to the changing nature of the connecting aluminum sites. The ^{27}Al NMR spectra show the presence of both four- and six-coordinate aluminum environments, with spectroscopic parameters that are quite distinct from those of the bordering phases Al$_2$O$_3$ and AlPO$_4$. In particular the tetrahedral site gives a much broader resonance, and its population is possibly correlated with the catalytic activity. Overall the NMR results demonstrate unambiguously that aluminophosphates are not simply AlPO$_4$-Al$_2$O$_3$ nanocomposites but rather amorphous structures with more or less random distributions of Al-O-Al and Al-O-P connectivities.

The above results can be compared with spectra published more recently on phosphorus-promoted γ-alumina catalysts, prepared via the incipient wetness impregnation techniques [85].The general findings in the ^{31}P and ^{27}Al MAS-NMR results on these materials are quite similar, although a somewhat different interpretation is given for the compositional evolution of the ^{31}P chemical shifts. Compared to the co-precipitates, in these impregnated samples bulk AlPO$_4$ seems to be detectable at much lower phosphorus concentrations.

V. HYDROPROCESSING CATALYST PRECURSORS

Hydrodenitrogenation and hydrodesulfurization processes are carried out over sulfided nickel-molybdena catalysts, supported on γ-alumina. These catalysts also contain small amounts of phosphorus, which acts as a promoter increasing catalytic activity. The role of phosphorus and also other structural questions regarding the interaction of Ni, Mo, and P with the alumina support in the oxidic precursor state have been addressed by ^{27}Al and ^{31}P NMR [85,86]. At high Mo loadings, bulk Al$_2$(MoO$_4$)$_3$ can be identified by a distinct ^{27}Al doublet around -17 ppm. ^{27}Al NMR experiments on P/Mo/Al$_2$O$_3$ preparations offer some evidence that the presence of phosphorus might favor Al$_2$(MoO$_4$)$_3$ formation. This phase is also identifiable in the spectra of some other alumina-supported molybdena samples [48]. In the spectrum of the P/Mo-Ni/Al$_2$O$_3$ catalyst, however, no Al$_2$(MoO$_4$)$_3$ was detectable. This result has to be interpreted with caution, though, in view of the potential interference effects from the paramagnetism of the Ni^{2+} ions.

VI. CONCLUSIONS AND FUTURE PERSPECTIVES

The purpose of this review has been to highlight the innate power of solid-state NMR spectroscopy for investigating detailed questions concerning the solid state and surface chemistry of bulk oxide catalysts. While MAS-NMR has been the mainstay of most applications, more sophisticated selective averaging experiments certainly will be applied in the future [87]. In particular, systems involving abundant spin-1/2 nuclei are expected to benefit from dipolar NMR experiments, which are able to provide information on internuclear distances and atomic distributions [88]. Furthermore, sophisticated sample manipulation techniques such as dynamic angle spinning and double rotation are expected to greatly enhance the state of knowledge in systems containing quadrupolar nuclei such as ^{17}O, ^{27}Al, and ^{51}V [89].

With the inclusion of more sophisticated techniques such as those mentioned above and others, NMR is an excellent quantitative tool for structural catalyst characterization. On the other hand, the question of how this information can be used to understand catalytic mechanisms and to design more potent catalysts often remains unresolved by these studies. Perhaps this is so because to date most NMR applications have sought to correlate catalytic activities with specific structural features present either in the catalyst or on its surface under room temperature conditions. In future studies there should be increasing emphasis on catalyst characterization under operation conditions in situ, including the search for transient adsorbates and reactive intermediates. In fact, such studies are now emerging in other fields of catalysis [90–92].

ACKNOWLEDGMENTS

The author appreciates valuable and insightful discussions with Professor Anthony K. Cheetham (UC Santa Barbara) and Professor Israel E. Wachs (Lehigh University). Support of our experimental research by NSF grant DMR89-13738 is also gratefully acknowledged.

REFERENCES

1. A. W. Sleight, in *Solid State Chemistry*, (A. K. Cheetham, ed.), Oxford Science Publications, Oxford, 1992, p. 166.
2. J. F. Bradzil, in *Characterization of Catalytic Materials* (I. E. Wachs, ed.), Butterworth-Heinemann, Stoneham, MA, 1992.
3. A. Bielanski and J. Haber, *Catal. Rev. Sci. Eng. 19*: 1 (1979).
4. R. K. Grasselli and J. F. Bradzil, eds., *Solid State Chemistry in Catalysis*, American Chemical Society, Washington, DC, 1985.
5. H. Pfeifer, *Colloids Surfaces 45*: 1 (1990).
6. A. Abragam, *Principles of Nuclear Magnetism*, Clarendon Press, Oxford, 1983.
7. M. Mehring, *Principles of High Resolution NMR in Solids*, Springer-Verlag, Berlin, 1983.
8. B. C. Gerstein and C. R. Dybowski, *Transient Techniques in NMR of Solids*, Academic Press Inc., New York, 1985.
9. C. P. Slichter, *Principles of Magnetic Resonance*, Springer-Verlag, New York, 1989.

10. T. M. Duncan and C. R. Dybowski, *Surf. Sci. Rep. 1*: 157 (1981).
11. H. Eckert and I. E. Wachs, *J. Phys. Chem. 93*: 6796 (1989).
12. E. R. Andrew, A. Bradbury, and R. G. Eades, *Nature 183*: 1802 (1959).
13. I. J. Lowe, *Phys. Rev. Lett. 2*: 285 (1959).
14. V. M. Mastikhin, I. L. Mudrakovsky, and A. V. Nosov, *Prog. NMR Spectroscopy 23*: 259 (1991).
15. V. M. Mastikhin, I. L. Mudrakovskii, N. S. Kotsarenko, L. G. Karakchiev, A. G. Pelmenshchikov, and K. I. Zamarev, *React. Kinet. Catal. Lett. 27*: 447 (1985).
16. J. M. Rojo, J. Sanz, J. A. Soria, and J. L. G. Fierro, *Z. Phys. Chem. 152*: 149 (1987).
17. K. I. Zamarev and V. M. Mastikhin, *Colloids Surfaces 12*: 401 (1984).
18. V. M. Mastikhin and K. I. Zamarev, *Z. Phys. Chem. 152*: 59 (1987).
19. J. M. Thomas and J. Klinowski, *Adv. Catal. 33*: 199 (1985).
20. E. Lippmaa, A. Samoson, and M. Magi, *J. Am. Chem. Soc. 108*: 1730 (1986).
21. A. Nayeem and J. P. Yesinowski, *J. Chem. Phys. 89*: 4600 (1988).
22. A. K. Cheetham, C. M. Dobson, C. P. Grey, and R. J. B. Jakeman, *Nature 328*: 706 (1987).
23. M. H. Cohen and F. Reiff, in *Solid State Physics*, Vol. 5, (Seitz and Turnbull, eds.), Academic Press, New York, 1958, p. 321.
24. A. Samoson and E. Lippmaa, *Phys. Rev. B. 28*: 6567 (1983).
25. A. P. M. Kentgens, J. J. M. Lemmens, F. M. M. Geurts, and W. S. Veeman, *J. Magn. Reson. 71*: 62 (1987).
26. P. P. Man, *J. Magn. Reson. 77*: 148 (1988).
27. S. R. Hartmann and E. L. Hahn, *Phys. Rev. 128*: 2042 (1962).
28. A. Pines, M. Gibby, and J. S. Waugh, *J. Chem. Phys. 59*: 569 (1973).
29. H. Pfeifer, *J. Chem. Soc. Faraday Trans. 1 84*: 3777 (1988).
30. I. D. Gay, *J. Catal. 44*: 306 (1976).
31. I. D. Gay, *J. Catal. 48*: 430 (1977).
32. S. Liang and I. D. Gay, *J. Catal. 66*: 294 (1980).
33. J. A. Ripmeester, *J. Am. Chem. Soc. 105*: 2925 (1983).
34. J. H. Lunsford, W. P. Rothwell, and W. Shen, *J. Am. Chem. Soc. 107*: 1540 (1985).
35. L. Baltusis, J. S. Frye, and G. E. Maciel, *J. Am. Chem. Soc. 109*: 40 (1987).
36. L. Baltusis, J. S. Frye, and G. E. Maciel, *J. Am. Chem. Soc. 108*: 7119 (1986).
37. V. M. Mastikhin, O. B. Lapina, and V. F. Lyakhova, *React. Kinet. Catal. Lett. 14*: 317 (1980).
38. V. M. Mastikhin, O. B. Lapina, V. F. Lyakhova, and L. G. Simonova, *React Kinet. Catal. Lett. 17*: 109 (1981).
39. V. M. Mastikhin, V. M. Nekipelov, and K. I. Zamarev, *Kinet. Katal.* (USSR) *23*: 1323 (1982).
40. V. M. Mastikhin, O. B. Lapina, V. N. Krasilnikov, and A. A. Ivakin, *React. Kinet. Catal. Lett. 24*: 119 (1984).
41. V. M. Mastikhin, O. B. Lapina, and L. G. Simonova, *React. Kinet. Catal. Lett. 26*: 431 (1984).
42. V. M. Mastikhin, O. B. Lapina, and L. G. Simonova, *React. Kinet. Catal. Lett. 24*: 127 (1984).

43. J. C. Vedrine, J. M. M. Millet, and J. C. Volta, *Faraday Discuss. Chem. Soc. 87*: 207 (1989).
44. J. Li, M. E. Lashier, G. L. Schrader, and B. C. Gerstein, *Appl. Catal. 73*: 83 (1991).
45. F. D. Hardcastle, I. E. Wachs, H. Eckert, and D. A. Jefferson, *J. Solid State Chem. 90*: 194 (1991).
46. W. Zhou, *J. Solid State Chem. 76*: 290 (1988).
47. T. P. Gorshkova, R. I. Maksimovskaya, D. V. Tarasova, N. N. Chumachenko, and T. A. Nikoro, *React. Kinet. Catal. Lett. 24*: 107 (1984).
48. J. C. Edwards, R. D. Adams, and P. D. Ellis, *J. Am. Chem. Soc. 112*: 8349 (1990).
49. J. C. Edwards, J. Zubieta, S. N. Shaikh, Q. Chen, S. Bank, and P. D. Ellis, *Inorg. Chem. 29*: 3381 (1990).
50. A. T. Ashcroft, A. K. Cheetham, M. L. H. Green, C. P. Grey, and P. D. F. Vernon, *J. Chem. Soc. Chem. Commun.*: 1667 (1989).
51. C. P. Grey, C. M. Dobson, A. K. Cheetham, and R. J. B. Jakeman, *J. Am. Chem. Soc. 111*: 505 (1989).
52. C. P. Grey, M. E. Smith, A. K. Cheetham, C. M. Dobson, and R. Dupree, *J. Am. Chem. Soc. 112*: 4670 (1990).
53. F. J. Berry, *Adv. Catal. 30*: 97 (1981).
54. N. J. Clayden, C. M. Dobson, and A. Fern, *J. Chem. Soc. Dalton Trans.*: 843 (1989).
55. J. V. Nagy, A. Abou-Kais, M. Guelton, J. Harmel, and E. G. Derouane, *J. Catal. 73*: 1 (1982).
56. B. C. Lippens and J. J. Steggerda, in *Physical and Chemical Aspects of Adsorbents and Catalysts* (B. G. Linsen, ed.), Academic Press, London, 1970, p. 171.
57. H. Knözinger and P. Ratnasamy, *Catal. Rev. Sci. Eng. 17*: 31 (1978).
58. V. M. Mastikhin, O. P. Krivoruchko, B. P. Zolotovskii, and R. A. Buyanov, *React Kinet. Catal. Lett. 18*: 117 (1981).
59. C. S. John, N. C. M. Alma, and G. R. Hays, *Appl. Catal. 6*: 341 (1983).
60. T. H. Walter and E. Oldfield, *J. Phys. Chem. 93*: 6744 (1989).
61. H. D. Morris and P. D. Ellis, *J. Am. Chem. Soc. 111*: 6045 (1989).
62. W. H. Dawson, S. W. Kaiser, P. D. Ellis, and R. R. Inners, *J. Am. Chem. Soc. 103*: 6780 (1981).
63. W.H. Dawson, S. W. Kaiser, P. D. Ellis, and R. R. Inners, *J. Phys. Chem. 86*: 867 (1982).
64. P. D. Majors and P. D. Ellis, *J. Am. Chem. Soc. 109*: 1648 (1987).
65. V. M. Mastikhin, I. L. Mudrakovsky, and S. V. Filimonova, *Chem. Phys. Lett. 149*: 175 (1988).
66. S. Komarneni, R. Roy, C. A. Fyfe, G. J. Kennedy, and H. Strobl, *J. Am. Ceram. Soc. 69*: C42 (1986).
67. P. Cloos, A. J. Leonard, J. P. Moreau, A. Herbillon, and J. J. Fripiat, *Clays Clay Miner. 17*: 279 (1969).
68. L. B. Welsh, J. P. Gilson, and M. J. Gattuso, *Appl. Catal. 15*: 327 (1985).
69. J. M. Thomas, J. Klinowski, P. A. Wright, and R. Roy, *Angew. Chem. Int. Ed. Engl. 22*: 614 (1983).

70. C. Doremieux-Morin, C. Martin, J. M. Bregeault, and J. Fraissard, *Appl. Catal.* 77: 149 (1991).
71. A. P. M. Kentgens, K. F. M. G. J. Scholle, and W. S. Veeman, *J. Phys. Chem.* 87: 4357 (1983).
72. H. Hamdan and J. Klinowski, *Chem. Phys. Lett. 158*: 447 (1989).
73. E. Hallas, U. Haubenreisser, M. Hähnert, and D. Müller, *Glastechn. Ber. 56*: 63 (1983).
74. M. McMillan, J. S. Brinen, J. D. Carruthers, and G. L. Haller, *Colloids Surfaces* 38: 133 (1989).
75. M. Hunger, D. Freude, H. Pfeifer, H. Bremer, M. Jank, and K. P. Wendlandt, *Chem. Phys. Lett. 100*: 29 (1983).
76. L. B. Schreiber and R. W. Vaughan, *J. Catal. 40*: 226 (1975).
77. C. E. Bronniman, I. Ssuer-Chuang, B. L. Hawkins, and G. E. Maciel, *J. Am. Chem. Soc. 109*: 1562 (1987).
78. J. R. Schlup and R. W. Vaughan, *J. Catal. 99*: 304 (1986).
79. G. E. Maciel, J. F. Haw, I. Ssuer-Chuang, B. L. Hawkins, T. A. Early, D. R. McKay, and L. Petrakis, *J. Am Chem. Soc. 105*: 5529 (1983).
80. J. F. Haw, I. Ssuer-Chuang, L. Hawkins, and G. E. Maciel, *J. Am. Chem. Soc. 105*: 7206 (1983).
81. V. M. Mastikhin, I. L. Mudrakovskii, V. P. Shmachkova, and N. S. Kotsarenko, *Chem. Phys. Lett.*: 93 (1987).
82. N. S. Kotsarenko, V. P. Shmachkova, I. L. Mudrakovskii, and V. M. Mastikhin, *Kinet. Katal. USSR 30*: 974 (1989).
83. K. Segawa, Y. Nakajima, S. I. Nakata, S. Asaoka, and H. Takahashi, *J. Catal. 101*: 81 (1986).
84. T. T. P. Cheung, K. W. Willcox, M. P. McDaniel, M. M. Johnson, C. Bronniman, and J. Frye, *J. Catal. 102*: 10 (1986).
85. E. C. DeCanio, J. C. Edwards, T. R. Scalzo, D. A. Storm, and J. W. Bruno, *J. Catal. 132*: 498 (1991).
86. M. McMillan, J. S. Brinen, and G. L. Haller, *J. Catal. 97*: 243 (1986).
87. B. F. Chmelka and A. Pines, *Science 246*: 71 (1989).
88. H. Eckert, *Ber. Bunsenges, Phys. Chem. 94*: 1062 (1990).
89. B. F. Chmelka, K. T. Müller, A. Pines, J. F. Stebbins, Y. Wu, and J. W. Zwanziger, *Nature 339*: 42 (1989).
90. J. F. Haw, B. R. Richardson, I. S. Oshiro, N. D. Lazo, and J. A. Speed, *J. Am. Chem. Soc. 111*: 2052 (1989).
91. B. R. Richardson, N. D. Lazo, P. D. Schettler, J. L. White, and J. F. Haw, *J. Am. Chem. Soc. 112*: 2886 (1990).
92. J. C. White, N. D. Lazo, B. R. Richardson, and J. F. Haw, *J. Catal. 125*: 260 (1990).

5

NMR Characterization of Silica and Alumina Surfaces

Gary E. Maciel

Colorado State University, Fort Collins, Colorado

Paul D. Ellis*

University of South Carolina, Columbia, South Carolina

I. INTRODUCTION AND OVERVIEW

For a variety of reasons, silica and alumina structures are important frameworks for the subject of heterogeneous catalysis [1–3]. Aluminas, and perhaps to a lesser extent silicas, are employed directly as cracking catalysts or as substrates for assorted catalytic systems (see Chapter 4). The silylation of silica surfaces also provides a strategy for immobilizing a catalytic center that has been found useful in the context of homogeneous catalysis [4]. Furthermore, many heterogeneous catalytic systems based on zeolites, clays, or silica-aluminas have aluminosilicate frameworks for which silica and alumina structures serve as structural prototypes.

The structural framework of silica [5] is based on interconnected $Si(-O-)_4$ tetrahedra, with some OH terminations at the surface. Aluminas are built primarily of $Al(-O-)_4$ tetrahedra and $Al(-O-)_6$ octahedra, with some surface-OH terminations [6]. The fundamental framework structure of aluminosilicates include primarily $Si(-O-)_4$ and $Al(-O-)_6$ units plus charge compensating bridging or framework hydroxyl groups, namely structure I, which shows formal charges on oxygen and aluminum [7]. On the basis of these kinds of structures and various so-called defect structures that may include octahedral and three-coordinate aluminum sites, the investigation by NMR of a wide variety of nuclei can be anticipated. Numerous NMR studies based on ^{29}Si, ^{27}Al, ^{1}H, ^{2}H, and ^{17}O have been reported on silica, alumina, and silica-alumina systems.

Current affiliation: Battelle Pacific Northwest Laboratory, Richland, Washington.

Silica, alumina, and silica-alumina samples are often complex materials, sometimes multiphase systems, almost always having a distribution of structures at the surface. From the point of view of catalysis, structures at both the surface and in the interior (bulk properties) are important. In this chapter we focus on surfaces. Because of NMR's sensitivity to *local* structure and its forgiveness of long-range disorder, solid-state NMR is one of the methods of choice for studying the structures of these materials. NMR is, of course, also a powerful tool for studying dynamics.

This chapter does not cover the most common aspects of the solid-state NMR techniques employed in the study of heterogeneous catalysts; such techniques are described in Chapter 4. Since this chapter emphasizes the *surface* characterization of silica and alumina systems and silica aluminas by NMR methods, only those technical aspects highly relevant to surface characterization and not otherwise emphasized in this volume are explicitly discussed here. NMR studies of zeolites and clays are treated in separate chapters, and the bulk structures of silica and alumina systems are covered by Eckert. Unavoidably this chapter is also concerned with *dynamics* at the surface, although the amount of detailed work on that subject to date is limited. With the increasing availability of variable-temperature solid-state NMR equipment, however, one can expect that attention devoted to dynamics at surfaces will increase markedly during the next few years.

II. SURFACE-SELECTIVE OBSERVATION OF SUBSTRATE NUCLIDES

A. Strategies and Goals

If a particular NMR technique does not specifically favor detection of nuclei at the surface, then the resulting spectrum will be dominated by peaks due to nuclei from the bulk of a particle. There are simply many more nuclei in the main framework than at structural sites on the surface, unless the surface area is very large, say, greater than 100 m^2/g. Exceptions may occur either if observed nuclei are located largely at the surface or if the method of polarization transfer discriminates strongly in favor of surface nuclei. The former situation often applies for protons in typical silica, alumina, or silica-alumina systems, because most of the protons there exist at the surface as covalently attached -OH groups, as physisorbed H_2O, or as structures of type I. Such cases will be described in Sec. II C.

Many polarization schemes can be envisioned for preferentially polarizing surface nuclei in the presence of an overwhelmingly larger number of nuclei in analogous structural sites in the interior. Perhaps the most obvious strategy would be to use a relaxation reagent that can, at least briefly, interact with the surface and thereby relax selected nuclei. Several types of surface relaxation techniques come to mind, but surprisingly little effort seems to have been expended in this direction. The primary exception to this lack of attention is the elegant ultralow-temperature ($\sim$10 mK) NMR work of Waugh and coworkers [8], who use the effects of ^{3}He impinging on the surface for selective relaxation. Other surface-selective or preferential relaxation mechanisms would seem possible via the dipolar mechanisms, if based on large nuclear magnetic moments (say, ^{1}H or ^{19}F) of reversibly adsorbed relaxant molecules or even on the electron spin magnetic moments of paramagnetic relaxant species. In the latter case, the possibility of dynamic nuclear polarization (DNP) [9] of surface nuclei from adsorbed paramagnetic species seems reasonable, with the Overhauser mechanism operating if the adsorption process is rapidly reversible or else the solid-state mechanism if the adsorbed state is essentially static. Overhauser-type experiments can be envisioned in which an optically pumped polarization [10] of adsorbate nuclei is transferred to surface nuclei. Another possibility for a surface-selective relaxation mechanism could be based on quadrupolar relaxation of a nuclide with $I > 1/2$ (such as ^{27}Al or ^{17}O), for which rapid, reversible adsorption/desorption would cause a modulation of the local electric field gradient. However, apparently little attention has been directed along these avenues.*

B. ^{1}H-X Cross-Polarization

The most popular and generally successful surface-selective polarization strategies to date employ ^{1}H $\rightarrow$ X cross-polarization (CP) [11], where X is a nucleus present at the surface (and presumably also within the bulk) [12,13]. These strategies, which assume that essentially all the protons in the system are present at the surface, are based on the dependence of cross-polarization upon a static component of the ^{1}H-X dipolar interaction. The dipolar coupling varies as the inverse cube of the ^{1}H-X internuclear distance, and the cross-polarization rate

---

*One should note that the net effect of a surface-enhanced quadrupolar interaction may work in the opposite sense to what is implied here. That is, the surface nuclei might be rendered unobservable (vide infra).

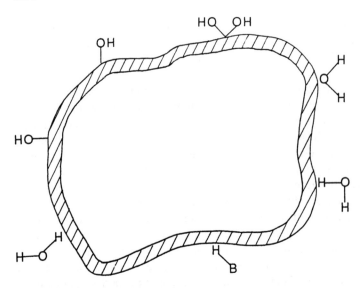

Figure 1 Cross-polarization as a surface-selective strategy, showing protons near only the surface region (///), as covalently attached hydroxyls, physisorbed water, or physisorbed acids (B-H) of some other type.

constant in turn has essentially an r^{-6} dependence. Figure 1 displays the gist of this strategy, where it is seen that only those X nuclei close enough to the surface (within, say, 5–6 Å), represented by the cross-hatched area, can be cross-polarized by surface protons. By contrast, the more remote X nuclei, namely those "interior" nuclei situated under the surface by more than one or two layers of atoms, are not cross-polarized efficiently. Hence, the dynamics of cross-polarization can be used to discriminate in favor of the surface nuclei.

1. $^1H \rightarrow ^{29}Si$ Cross-Polarization in Silicas

The first example using ^{1}H-X cross-polarization for surface-selective X NMR was the demonstration of ^{1}H-^{29}Si CP in silica gel by Sindorf and Maciel in 1980 [11,12], and since then ^{1}H$\rightarrow^{29}$Si CP has remained the most popular application of this strategy. Figure 2A shows a typical CP-MAS spectrum of a silica sample based on the ^{29}Si nuclide, a spin-1/2 nuclide with 4% natural abundance and a magnetic moment about two thirds that of ^{13}C. Three peaks are seen in the ^{29}Si CP-MAS spectrum: (1) a large peak at about -109 ppm relative to liquid tetramethylsilane (TMS), due to (-O-)$_4$Si groups with no OH groups attached to silicon (Q$_4$); (2) a large peak at about -100 ppm due to (-O-)$_3$SiOH groups with one OH group attached to each silicon (Q$_3$); and (3) a small but distinctive shoulder at about -89 ppm due to (-O-)$_2$Si(OH)$_2$ groups with two OH groups attached to each silicon (Q$_2$). Figure 3 shows the results of a variable contact-

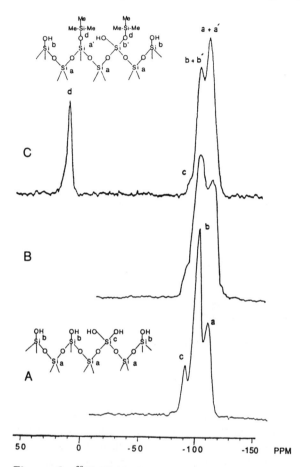

Figure 2 ^{29}Si CP-MAS spectra of silica gel samples: (A) humidified sample; (B) sample dehydrated under vacuum at 209°C; (C) sample derivatized with $(CH_3)_3SiC$.

time CP experiment in which the $^1H \rightarrow {}^{29}Si$ CP contact period is varied in order to elucidate the $^1H \rightarrow {}^{29}Si$ CP (relaxation) time constant (T_{HSi}) for each ^{29}Si peak. Analysis of the curves shows that T^{-1}_{HSi} for the -89 ppm peak is roughly twice that of the -100 ppm peak, which itself is an order of magnitude larger than that of the -109 ppm peak. This relationship is consistent with the structural assignments given above (Q_2, Q_3, and Q_4, respectively), which are also the assignments suggested by analogy with ^{29}Si chemical shift assignments of spectra obtained on silicic acid solutions.

Spectra of the type shown in Fig. 2A have served as the basis for a wide range of chemical studies involving structural changes of the silica surface, including

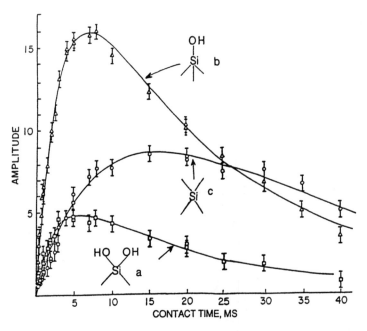

Figure 3 Variable contact time ^{29}Si CP-MAS plots for silica. For a, b, and c see Fig. 2.

dehydration and rehydration (e.g., Fig. 2B) [14] and silylation (Fig. 2C) [15–17]. In silylation, the silanol ($\geqslant$ Si-OH) groups at the surface react with a silane reagent, $>$ Si(X)R, on which there is at least one highly labile moiety (X, say -Cl or -OCH$_2$CH$_3$), to form a $\geqslant$ Si-O-SiR (siloxane) linkage. In this way a variety of chemically useful moieties such as catalytic centers, for example, can be immobilized on a substantially stable surface, thereby creating materials with a variety of technologically useful properties.

$$\text{Si-OH} + \text{X} \!\!\geqslant\!\! \text{Si-R} \rightarrow \text{Si-O—Si-R} + \text{HX} \tag{1}$$

The results of detailed studies require a partial relaxation of the hopeful assumption that all protons of underivatized silica systems are at the surface [18]. Recent NMR studies indeed show that moderately intense D$_2$O exchange conditions leave about 3% of the original proton content in a silica gel sample. This residual 3% is attributable to trapped OH groups.* On the basis of extensive

---

*In this sense the term "trapped" includes a variety of possibilities, e.g., interstitial sites, species entrapped in cavities, etc.

spin-dynamics studies, it was concluded that the trapped hydroxyls occur as single silanols (Q_3) and water, but not as geminal silanols (Q_2).

2. $^1H \rightarrow {}^{27}Al$ Cross-Polarization in Aluminas

Alumina is a material that has been studied extensively [6,19,20]. Its γ-form is characterized as a tetragonally distorted defect spinel lattice, with a unit cell composed of 32 oxygen atoms and 21⅓ aluminum atoms. There are 2⅔ vacant cation positions per unit cell. Among the 21⅓ aluminum atoms in the unit cell are a significant number of octahedral, as well as tetrahedral, sites. For the γ-form, one can consider an idealized surface to be composed of two low-index defect-spinel crystal planes, specifically the (110) and (100) planes [19]. The presence of these planes implies that there is a mixture of octahedral and tetrahedral aluminum sites exposed on the surface. Recent solid-state NMR experiments have observed these sites indirectly [21] and suggest that the surface can also be described by (110) and (111) planes. In general, however, it is best to consider the surface of γ-alumina as being composed of the (110), (100), and (111) planes.

It is the varied surface chemistry associated with the γ-form that is the most important property of γ-alumina. The termination of the various planes with hydroxyl groups and Lewis acid sites (anion vacancies) is responsible for the reactivity of γ-alumina. The reactivities of Bronsted and Lewis acid sites are dependent upon the pretreatment of the surface. The surface area of a typical γ-form ranges between 200 and 220 m^2/g. Further, the γ-form is highly reactive compared to the α-form, which is considered inert and has a typical surface area of 2 m^2/g or lower.

α-Alumina is characterized as a hexagonal close-packed array of oxygen atoms, with the aluminum atoms occupying exclusively octahedral positions within the lattice. The α-form is an important industrial support, especially in those applications that require a nonreactive support, such as the partial oxidation of ethylene to ethylene oxide [22].

To understand the varied chemistry of the aluminas, techniques need to be developed for studying the surface independent of the bulk. Structural and dynamical aspects of the surface do have their origins in the bulk, but the specific details delineating the surface will be different. Clearly it would be advantageous to apply the same surface selective CP methodology developed for the silicas [12–16] to the surface of the aluminas. Before addressing this particular point, however, we need to consider the feasibility of the experiment. Are the aluminum atoms at the surface indeed observable by NMR methods? If surface aluminum atoms are observable, we must then recognize that the spin of interest, ^{27}Al, is not a spin-1/2 nuclide (I for ^{27}Al is 5/2); hence ^{27}Al has a nonzero nuclear electric quadrupole moment. Cross-polarization from protons to a quadrupolar nucleus presents the experimenter with another layer of complication in comparison to the CP experiment between pairs of spin-1/2 nuclides. We will deal with each of the new issues in turn.

D. E. O'Reilly [23], in early ^{27}Al experiments, considered the observability of surface aluminum atoms in aluminas. As summarized in Fig. 4, O'Reilly obtained room-temperature data on a series of aluminas with varying surface areas and found that for an equal number of aluminum nuclei, each of the other aluminas had a weaker ^{27}Al NMR signal than that observed for α-alumina. Specifically, for an equal number of spins the integrated signal was inversely proportional to the surface area of the particular alumina. O'Reilly reasoned that the reduction in observed signal occurred because surface aluminum atoms

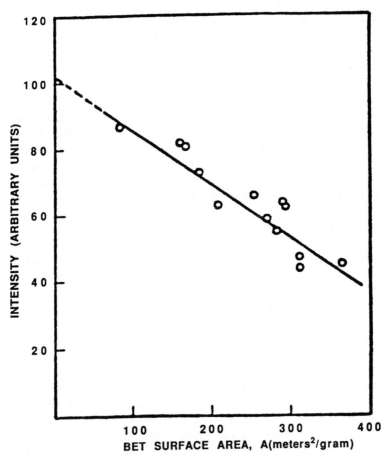

Figure 4 Integrated ^{27}Al signal intensity (for a fixed amount of alumina) as a function of BET surface area of a given alumina. (From Ref. 23.)

possess a large electric field gradient (efg). Such a large efg would make the resulting surface aluminum quadrupole coupling constant (Q_{CC}) sufficiently large as to render the signal broadened beyond detection by the NMR technique—that is, an inhomogeneous line too broad to be observed.

IR investigations [20] suggest that the structures of surface aluminum atoms are not unlike those existing in the bulk, as represented in Fig. 5. This is especially evident for those aluminum atoms associated with Bronsted acid sites. In all fairness to O'Reilly, it is difficult to predict the efg's at selected points within the structure by inspection, given that the IR and NMR data seem to be at odds. As has been demonstrated by Huggins and Ellis [24], though, the O'Reilly paradox arises because of an important feature of an alumina surface, namely its dynamical nature.

O'Reilly's [23] experimental observations are correct; the *observed* signal, for a fixed number of ^{27}Al nuclei, is inversely proportional to the surface area.

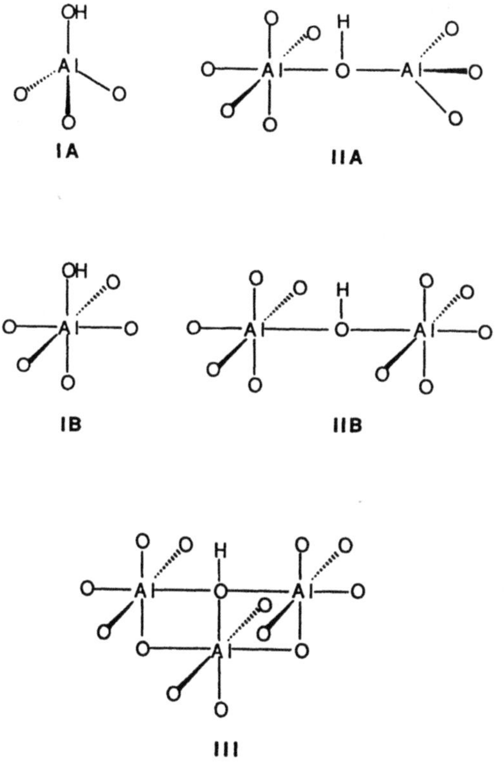

Figure 5 The five types of hydroxyls proposed to exist on the surface of γ-, η-, and higher surface area aluminas.

However, an NMR pattern can be broadened in two ways: either by an inhomogeneous mechanism (as can be inferred from O'Reilly's analysis) or by a homogeneous mechanism resulting from dynamical processes. If the dynamics are associated with a thermally activated process, these two alternatives can be separated by performing variable-temperature experiments. Possible mechanisms for the dynamical processes responsible for O'Reilly's observation have been hypothesized by Huggins and Ellis [24] and are summarized in Fig. 6. From ab initio molecular orbital calculations on aluminum clusters, selected to model the surface of an alumina, and from line shape calculations involving *chemical exchange* in the solid state, Huggins and Ellis hypothesized that it is the mobility of the Lewis acid sites that is responsible for the disappearance of *some* (not all) of the surface aluminum atoms in the ^{27}Al NMR spectrum. Furthermore, most of the so-called type IA and IB Bronsted acid sites (see Fig. 5) should still be

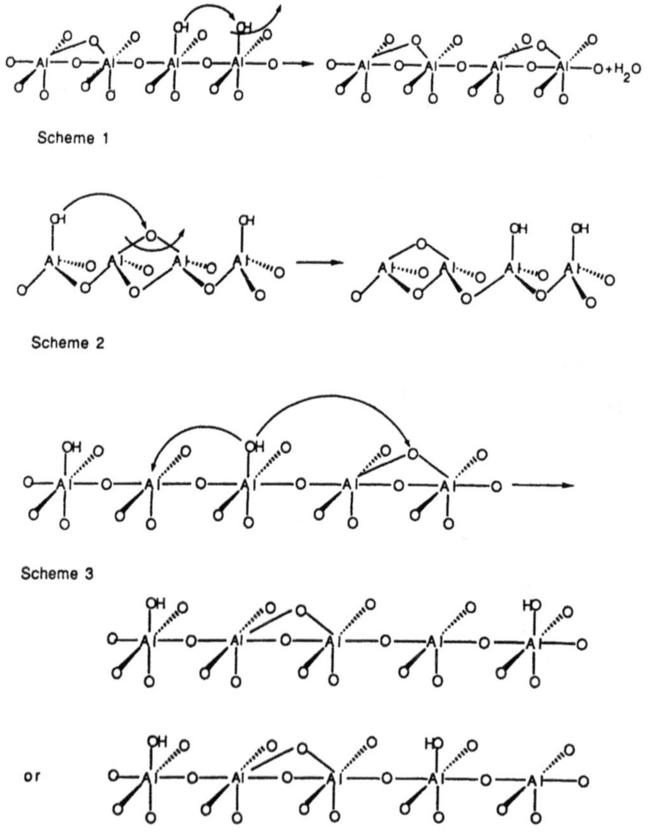

Scheme 1

Scheme 2

Scheme 3

or

Figure 6 Proposed mechanisms for Bronsted and/or Lewis acid mobility on the surface of an alumina.

observable. Detailed elaboration of the calculations is available in the original paper [24]. The main conclusion to be drawn from this work is that at room temperature one should be able to observe a subset of surface Bronsted sites. If all the sites are of interest, then further low-temperature experiments are needed. Since more than 90% of the sites should be observable at temperatures in the neighborhood of 100 K, a surface-selective CP experiment appears favorable.

As mentioned above, there are new complications associated with the CP experiment in which the magnetization is transferred from a spin-1/2 reservoir (^{1}H) to a quadrupolar, nonintegral-spin nucleus. Several groups have addressed these problems, notably those of Vega [25], Oldfield [26], Harris [27], Woessner [28], and Ellis [29–31]. There are two important points to keep in mind in the case of a quadrupolar nuclide: the Hartmann-Hahn condition is changed, and the expected gain in signal-to-noise (S/N) ratio is not given simply by the ratio of the γs for the spins of interest.

For a quadrupolar nuclide with a nonintegral spin ($I = \frac{3}{2}, \frac{5}{2}, \frac{7}{2}$, or $\frac{9}{2}$) in high magnetic field, the narrowest static-sample resonance corresponds to the $-1/2 \leftrightarrow 1/2$ (or, $\pm 1/2$) transition. This transition is broadened only in second-order by the quadrupole interaction. The line shapes from the other transitions are typically so broad that they often cannot be effectively excited. Therefore one usually is performing a selective experiment, in the sense that only a particular transition within the allowed manifold of transitions is observed. Under these circumstances the Hartmann-Hahn [11,32,33] match condition becomes

$$\alpha_S \gamma_S B_1^S = \alpha_H \gamma_H B_1^H \tag{2}$$

where γ and B_1 are the gyromagnetic ratio and the applied rf field strength, respectively, for the relevant nuclei. The quantity α denotes a scale factor for the gyromagnetic ratio of the nucleus S and is given by $\alpha = [I(I + 1) - m(m - 1)]^{1/2}$ for a transition between the m and $m - 1$ levels. I is the spin quantum number for the dilute spin (^{13}C, ^{95}Mo, etc.). In the case of a nucleus with spin $I = 1/2$, the factor α is equal to 1, whereas for a nucleus with $I = 3/2$ it takes the value 2. In the case of a spin-5/2 nucleus, such as ^{17}O, ^{27}Al, or ^{95}Mo, α is equal to 3. These conditions represent the matching of the ^{1}H rf field strength to the S rf field strength, as measured via the corresponding $\pi/2$ pulse durations.

The maximum S/N gain expected from cross-polarization relative to a solid-echo sequence is given by the ratio of the nuclear gyromagnetic ratios and the heat capacities of the two spin reservoirs [33,34]. Equation (3) describes the change of magnetization from a reference state, M_{S0} (the equilibrium magnetization obtainable in the spin-echo experiment), to a final equilibrium state, $M_{S\infty}$ (the magnetization obtainable in a spin-echo version of the cross-polarization experiment):

$$\frac{M_{S\infty}}{M_{S0}} = \frac{\gamma_I}{\gamma_S} \cdot \frac{1}{1 + \epsilon} \tag{3}$$

where ϵ is the ratio of the heat capacities of the spin reservoirs:

$$\epsilon = \frac{N_S}{N_I} \left[\frac{S(S + 1) - m(m - 1)}{I(I + 1)} \right] \tag{4}$$

and N_S = number of S spins and N_I = number of I spins. For $S = 5/2$, the term in square brackets in Eq. (4) is equal to 12.

For the surface of an alumina, the ratio of N_S/N_I is unknown but typically will be greater than 1. This is in contrast to the case for ^{13}C at natural abundance in typical organic compounds, where the figure could be on the order of 1/150. If we assume that the ratio of N_S/N_I is 1, then the *maximum* "gain" to be expected from the $^{27}Al/^{1}H$ CP experiment would be 0.29—in fact, a loss of signal in excess of a factor of 3.4. Typically, the loss is even greater owing to the failure of the simplifying assumptions used in the development of the preceding equations. Hence the experiment could be better described, in the case of ^{27}Al CP, as a filter.

Nevertheless, Morris and Ellis [29] demonstrated the feasibility of the $^{27}Al/^{1}H$ CP experiment. For a "hydrated" γ-alumina, they were able to observe O_h and T_d Bronsted acid sites on the surface. In the original work, the matching conditions favored CP to the O_h sites relative to T_d sites. Subsequent experiments, represented in Fig. 7, illustrate more balanced CP conditions for surface T_d and O_h sites. Upon partial dehydration, the population of T_d sites is reduced relative to that of O_h sites. These findings are consistent with the relative acidity of the sites. Concomitant with these observations in dehydration experiments was a significant reduction in ^{27}Al signal intensity, which is consistent with the surface ^{1}Hs being the source of the ^{27}Al magnetization.

In the case of high surface areas an important concern is the question of how much of the observed CP signal is due to interstitial ^{1}H. To address this issue, Morris and Ellis [29] prepared "completely" deuterated alumina by exhaustive exchange with D_2O. At first no measurable $^{1}H \rightarrow ^{27}Al$ CP signal was observed. To that same sample, isotopically normal pyridine was then added, and the resulting CP signal is depicted in Fig. 8. There are two important points to be learned from this experiment. First, the relative signal-to-noise ratios of a partially dehydrated alumina (PDA) sample (no deuteration) and a deuterated PDA sample with adsorbed pyridine are about the same. If interstitial sites contribute to the $^{1}H \rightarrow ^{27}Al$ CP signal, their contribution therefore is small relative to the CP signal arising from surface ^{1}Hs. Furthermore, the pyridine sample clearly gives rise to a signal distinctly different from that of the sample without pyridine. Thus the pyridine sample is reflective of Lewis acid sites, while the untreated sample is reflective of surface Bronsted acid sites.

It is important to note that the relative amounts of these T_d and O_h Bronsted sites are *not* manifested in these initial experiments. Clearly, as we have discussed above, the room temperature CP experiments are governed by a complex

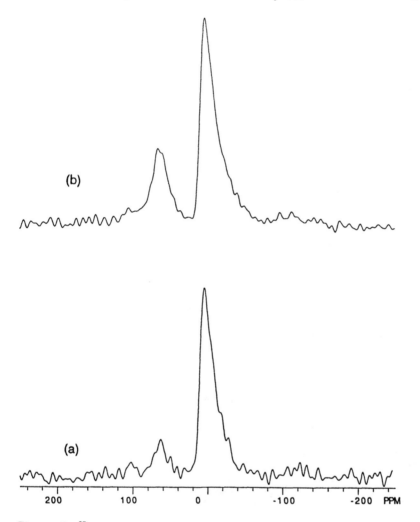

Figure 7 ^{27}Al MAS spectra of γ-alumina (out of the bottle): (a) ^{1}H$\rightarrow$^{27}Al CP-MAS (512 transients); (b) ^{1}H-decoupled Bloch decay (16 transients).

dynamic surface averaging process. More work is needed to determine these important surface parameters.

In a follow-up study, Morris et al. [30] extended the ^{1}H$\rightarrow$^{27}Al CP experiments to clays. The major rationale for this investigation was to examine the question of quantitation. That is, does the ^{1}H$\rightarrow$^{27}Al CP experiment observe all of the aluminum atoms? To investigate this point, a sample of the clay kaoline was used. The results, summarized in Fig. 9, indicate that for a static sample the

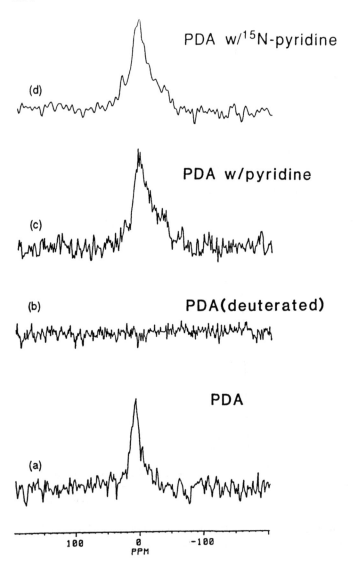

Figure 8 $^1H \to ^{27}Al$ CP-MAS spectra of (a) PDA γ-alumina, (b) deuterated PDA γ-alumina, (c) 25% of a monolayer coverage of pyridine (natural abundance) on PDA γ-alumina, (d) 25% of a monolayer coverage of [^{15}N]pyridine on PDA γ-alumina. Note the factor of 3 increase in signal-to-noise ratio between (c) and (d). The number of transients was 31168, 32000, 127, and 104, respectively. The line broadening by which the FIDs were multiplied is 100, 100, 500, and 500 Hz, respectively. The half-height linewidths for (c) and (d) are 23.8 and 25.6 ppm, respectively. (From Ref. 29.)

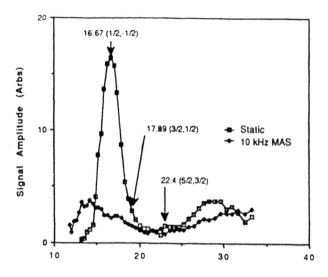

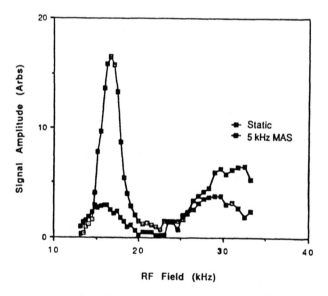

Figure 9 $^1H\rightarrow^{27}Al$ CP Hartmann-Hahn match spectra of kaoline, static, and MAS. Each point represents the signal amplitude after 64 transients have accumulated. The ^1H rf field strength remained at 50 kHz throughout these experiments. Results obtained with two different MAS speeds are illustrated. (From Ref. 30.)

$^1H\rightarrow{}^{27}Al$ CP experiment does observe all the aluminum atoms. Under conditions of MAS, however, there is a significant loss of signal intensity corresponding to a factor of between 5 and 6 relative to the static experiment. Examination of other data, for example, the ^{17}O CP results of Walter et al. [26], indicates similar losses. Recently, Vega [34,35] has shown that the losses arise because the *selective* nature of the transitions is violated: in the absence of spinning, the $\pm 1/2$ transition can be considered as an isolated fictitious-spin 1/2 system; sample spinning, however, renders the quadrupole splitting, which depends on orientation, time dependent. This time dependence has, in Vega's words, a profound effect on the spin dynamics involved in the polarization transfer portion of the CP experiment, specifically in the spin-locking condition. Under normal rf conditions, with $|Q_{CC}| >> \omega_{IS}$, this complication destroys the spin lock and thus severely interferes with the efficiency of the polarization transfer. The result is a significant loss in S/N with MAS as compared to a static sample.

The problem can be partially circumvented by using some ideas associated with the DAS or angle-flipping experiments developed by the Pines [36–38] and Maciel [39] groups. If sample spinning is the cause of a CP problem, then one might execute the polarization transfer with the rotor axis in an orientation where spinning effects are minimal, as demonstrated by Sardashti and Maciel [40] for $^1H\rightarrow{}^{13}C$ and $^1H\rightarrow{}^{113}Cd$ cross-polarization, using angles of 54.7° and 90°.

If technically feasible, CP for a quadrupolar nucleus such as ^{27}Al should be carried out with the rotor axis at 0°; then, after $^1H\rightarrow{}^{27}Al$ CP is complete, the rotor can be flipped to the complementary DAS angle of 63.4349488 . . . °. The advantage of this experiment is that a surface-selective CP spectrum devoid of second-order quadrupole effects can be obtained. For the angle-flipping approach to work, however, the T_1s for the spins of interest must be in the range of *at least* 100–200 ms, which is *not* the case for the surface aluminum atoms depicted in Figs. 7 and 8. Ellis and coworkers [41] have confronted this issue in developing a probe with a double-tuned Helmholtz coil that can be flipped as in the DAS experiment. The $^1H\rightarrow{}^{27}Al$ CP-DAS experiment, for example, is found to fail for γ-alumina at room or reduced temperatures, simply because of the T_1s at the surface sites. T_1 is estimated to be less than 3 ms for the O_h sites and less than 1 ms for the T_d site. Clearly the CP-DAS experiment has the potential to become a powerful method for probing surface sites, but the spins must have the appropriate relaxation times.

3. $^1H\rightarrow{}^{17}O$ Cross-Polarization

During the past several years the Oldfield group has contributed significantly to materials chemistry through their work in solid-state ^{17}O NMR spectroscopy [42]. Walter et al. [26], for example, have demonstrated that ^{17}O CP experiments are not only feasible, but that they can be used to "edit" ^{17}O spectra of solids by discriminating against those oxygen sites lacking directly bonded hydrogens. As an illustration of this technique, Oldfield et al. examined the ^{17}O spectrum of amorphous SiO_2, shown in Fig. 10. By comparing static and MAS spectra with and without CP, they were able to assign surface Si-O-H groups. Selective CP experiments of this sort should become increasingly important spectroscopic methods for investigating structure and dynamics of surfaces.

C. ^1H CRAMPS and SP-MAS NMR

There is an extensive literature on the application of ^1H NMR techniques to the study of silica, alumina, and silica-alumina systems [43–46]. Experimental approaches range from simply measuring the free induction decay (FID) to the

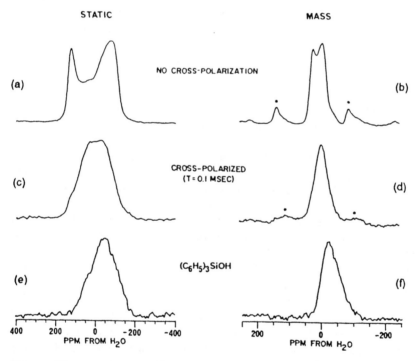

Figure 10 Static and MAS ^{17}O spectra of amorphous SiO_2 and polycrystalline $(C_6H_5)_3$ SiOH obtained at 67.8 MHz. (a) ^1H-decoupled static spectrum of SiO_2 without CP: 108 scans. (b) ^1H-decoupled MAS spectrum: 100 scans, 7.6 kHz spinning speed (* indicates spinning sidebands). (c) $^1H \rightarrow ^{17}O$ static spectrum of SiO_2: 200 scans, 0.1 ms contact time. (d) $^1H \rightarrow ^{17}O$ MAS spectrum of SiO_2, 200 scans, 0.1 ms contact time. (e) ^1H-decoupled static spectrum of $(C_6H_5)_3$ SiOH without CP: 500 scans. (f) ^1H-decoupled MAS spectrum of $(C_6H_5)_3$ SiOH: 800 scans, 4.0 kHz spinning speed. All spectra were obtained using a 2 s recycle time. (From Ref. 26.)

technically demanding CRAMPS (Combined Rotation and Multiple Pulse Spectroscopy) method [47,48]. Although much valuable information has been developed by ^{1}H NMR methods, this subject suffers substantially from the fragmentary nature of much of the work and from the *huge uncertainties associated with the effects of dynamics at surfaces*. Such effects are manifested in different ways in the data obtained by the various techniques. Issues associated with dynamics will be sorted out as more variable-temperature, especially low-temperature, experiments are reported.

The ^{1}H NMR experiment most commonly used for samples relevant to this chapter is the simple single-pulse/FID sequence with magic-angle spinning (SP-MAS). Here the implicit assumption, or hope, is that the protons are sufficiently dilute at the surface and not grouped in clusters, such that ^{1}H-^{1}H magnetic dipole-dipole interactions behave inhomogeneously under MAS. If so, then the associated dipolar broadening is efficiently averaged by MAS at modest speeds [49]. While this assumption may often be valid, especially if the MAS speed is large (say greater than 10 kHz), it has seldom been justified a priori. A useful strategy employed by Vega and coworkers is to ensure that the ^{1}H concentration is small by exchanging protons at the surface with deuterons [50]. One should note, however, that this approach might yield misleading results if there are variations in the kinetic or equilibrium reactivities of various hydrogen sites in the system.

To remove ^{1}H-^{1}H dipolar effects one can apply the CRAMPS technique, in which a multiple-pulse sequence is used to average the dipolar interaction [51,52] while MAS is employed simultaneously to average the chemical shift anisotropy (CSA) [53]. Figure 11 illustrates the CRAMPS approach.

Figure 12 compares ^{1}H SP-MAS results on a silica gel sample (obtained with MAS speeds from 0 to 11 kHz) with the corresponding CRAMPS spectrum [54]. For this particular sample the ^{1}H CRAMPS spectrum is similar, but not identical, to the associated SP-MAS spectra at modest speeds. For a more proton-rich sample, as seen for poly(methylmethacrylate) in Fig. 13, there are much more dramatic differences between SP-MAS and CRAMPS. Close inspection of Fig. 12, moreover, reveals that the SP-MAS spectra miss a substantial amount of intensity associated with a broad peak in the CRAMPS spectrum, intensity presumably due to surface OH groups that experience strong ^{1}H-^{1}H dipolar interactions (involved extensively, perhaps, in hydrogen bonding). One would expect CRAMPS to average strong dipolar interactions more efficiently than simply a modest-speed SP-MAS experiment. Nevertheless, the CRAMPS technique is not necessarily a panacea for ^{1}H NMR studies of surfaces or any other type of sample. Dynamical effects associated with motion or chemical reactions, if occurring with a time constant comparable to the cycle time (τ_C) of the multiple-pulse sequence, interfere with the multiple-pulse average of ^{1}H-^{1}H dipolar interactions [48]. An analogous problem can also arise with magic-angle

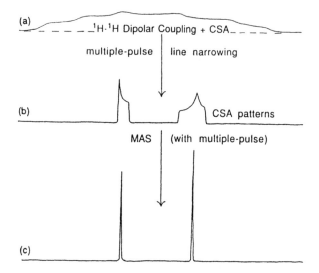

Figure 11 ^{1}H CRAMPS strategy, showing the spectra of a hypothetical solid obtained (a) with no line-narrowing applied, (b) using multiple-pulse averaging of the ^{1}H-^{1}H dipolar interaction, and (c) with a combination of MAS for averaging the chemical shift anisotropy (CSA) and multiple-pulse averaging of the dipolar interaction.

spinning, of course, for which the critical period is $\nu^{-1}{}_{MAS}$, the inverse of the MAS frequency. To quench such dynamical interference with the line-narrowing efficiencies of either MAS or the multiple-pulse sequence, one can try to alter the dynamics, perhaps by changing the sample temperature.

Another effect working against the generation of narrow ^{1}H lines in SP-MAS or CRAMPS spectra of aluminas or silica-aluminas is the nuclear electric quadrupole interactions of any ^{27}Al nuclei that may be dipolar-coupled to protons. It is well known that nuclear electric quadrupole effects interfere with the functioning of MAS in averaging heteronuclear dipolar interactions [56,57], for example, ^{1}H-^{27}Al interactions [48]. Pfeifer and coworkers have estimated the effects of ^{1}H-^{27}Al dipolar interactions on ^{1}H MAS spectra of aluminas and found them to dominate the static ^{1}H line width [45]. The interference of quadrupole effects with MAS line-narrowing, which is due to second-order perturbations, should be reduced as the static field strength (B_0) is increased, because the quantization axis of ^{27}Al is then determined more dominantly by the Zeeman interaction. Figure 14 shows an example of this effect in ^{1}H CRAMPS spectra of silica-alumina obtained at two different fields corresponding to 360 and 187 MHz [48].

The effects of dynamics and ^{27}Al quadrupole interactions, together with both the uncertainties in MAS-only efficiencies for averaging ^{1}H-^{1}H dipole-dipole

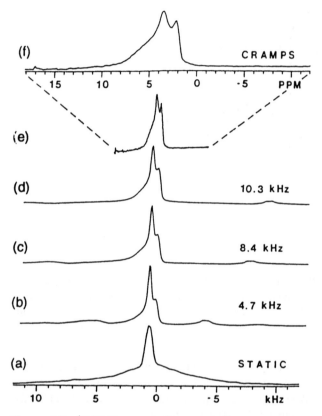

Figure 12 ^{1}H NMR spectra of untreated silica-gel: (a) static sample, single pulse; (b–d) MAS only with indicated MAS speed; (e) CRAMPS; (f) CRAMPS on an expanded scale. (From Ref. 54.)

interactions and with the typical variations in sample preparation among different laboratories, lead to substantial ambiguities in the interpretation of ^{1}H NMR results. Hence few definitive statements can be made at the present time about competing ^{1}H NMR techniques and the corresponding structural and dynamical interpretations. What is needed to remove most of these uncertainties are systematic comparisons on a common set of samples in which ^{1}H CRAMPS and ^{1}H-SP-MAS experiments are carried out as a function of B_0, ν_{MAS}, τ_{cycle} (for CRAMPS), and temperature. Furthermore, as is true for all NMR techniques described in this chapter, future work on surface systems should be carried out, as much as possible, under experimental conditions that provide reliable *absolute* intensities. In this way we will know *what fraction* of the spins are represented in a spectrum.

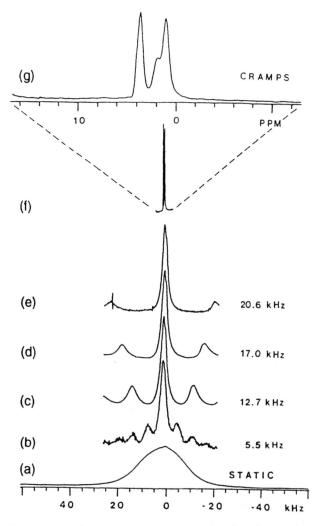

Figure 13 ^{1}H NMR spectra of a solid polymethylmethacrylate: (a) static sample, single pulse; (b–e) MAS only with indicated MAS speed, using spinner made from the solid polymer itself; (f) CRAMPS; (g) CRAMPS on an expanded scale. (From Ref. 54.)

Because of the possible complicating effects that nearby ^{27}Al nuclei have on ^{1}H NMR spectra, the ^{1}H NMR spectra of silicas are somewhat better understood than those of aluminas or silica-aluminas. A detailed ^{1}H CRAMPS study of silica gel and its dehydration was reported by Bronnimann et al. [43]. Figure 15 shows the ^{1}H CRAMPS spectra obtained from an untreated silica gel sample (Fig. 15a)

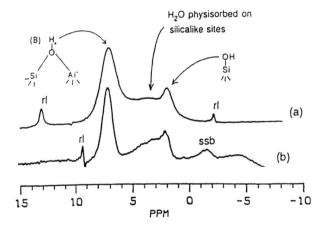

Figure 14 The (a) 187 and (b) 360 MHz ^{1}H CRAMPS spectra of amorphous silica-alumina, showing rotor lines (rl) and spinning sidebands (ssb). (From Ref. 48.)

and from samples evacuated at 25°C (Fig. 15b), 200°C (Fig. 15c), and 500°C (Fig. 15d). Computer simulation (Fig. 15a′, a″) of the untreated sample shows that the following three peaks contribute to the spectrum: (1) a sharp peak at 1.7 ppm, (2) a broader but well-defined peak centered at 3.5 ppm, and (3) a broad, irregularly shaped peak with a maximum at about 3.5 ppm. The fact that the relatively narrow 3.5 ppm peak is essentially eliminated by sample evacuation at even 25°C identifies this peak as physisorbed water.

The broad peak with a maximum near 3.0 ppm is apparently unaffected by sample evacuation at 200°C, but is eliminated by evacuation at 500°C. This behavior suggests that the signal is due to silanol (SiOH) groups clustered sufficiently close together for suitable dehydration pathways to be available. The persistence of the 1.7 ppm peak after sample evacuation even at 500°C implies that the silanol groups it represents are sufficiently isolated from each other to make dehydration difficult. If one identifies the clustering OH groups with hydrogen bonding and the mutual isolation of silanol groups with the absence of hydrogen bonding, then this interpretation is also consistent with customary views on the effects of hydrogen bonding on proton chemical shifts. For many years it has been commonly accepted that hydrogen bonding produces a shift to lower proton shielding, with the magnitude of the shift increasing along with the strength of the hydrogen bonding. The sharp, higher-shielding peak at 1.7 ppm is therefore identified with non–hydrogen-bonded SiOH groups, and the broad, lower-shielding (3.0 ppm) band with hydrogen-bonded SiOH groups. The breadth and shape of this broad peak arise from a wide distribution of hydrogen-bonding structures and strengths on the silica surface. These peak assignments

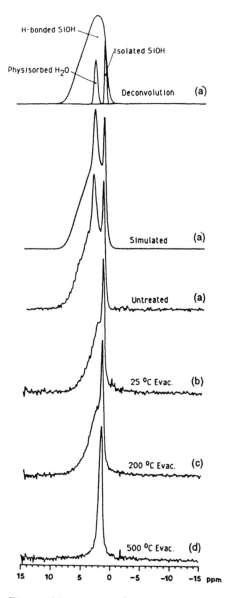

Figure 15 187 MHz ^{1}H CRAMPS spectra of Fisher S-679 silica gel: (a) untreated, (b) evacuated at 25°C, (c) evacuated at 100°C; (d) evacuated at 500°C, (a″) deconvolution of spectrum a, (a′) computer simulation based on a″. (From Ref. 43.)

based on classical chemical shift arguments are consistent with data from relaxation experiments.

Figure 16a shows a ¹H CRAMPS version [43] of the well-established "interrupted decoupling" or "dipolar-dephasing" experiment so popular in ¹³C CP-MAS NMR. In this ¹H CRAMPS version, a dephasing period 2τ is inserted between the initial π/2 pulse that generates transverse magnetization and the line-narrowing pulse train that detects the transverse magnetization. During the dephasing period, those proton magnetic moments that experience strong dipolar interactions undergo a corresponding degree of dephasing and the resulting signal detected will be attenuated accordingly. Figures 16b and c show the results of applying the dipolar-dephasing CRAMPS experiment to untreated silica gel (Fig. 16b) and to a silica gel sample evacuated at 200°C (Fig. 16c). In both cases one sees that the broad (3.0 ppm) peak is attenuated by a dephasing period (2τ) of

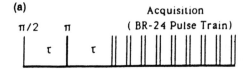

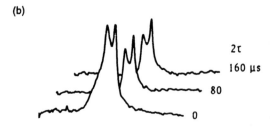

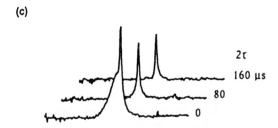

Figure 16 Dipolar-dephasing experiment: (a) pulse sequence; (b) results for untreated silica gel (showing τ values); (c) results for sample evacuated at 200°C (showing τ values). (From Ref. 43.)

160 μs, whereas the sharp (1.7 ppm) peak is largely unchanged. In view of the relationship between hydrogen bonding and proximity, and in particular the r^{-3} dependence of dipolar interactions, this behavior provides strong support for identifying the 1.7 ppm peak as non–hydrogen-bonded silanols and the broad 3.0 ppm peak as hydrogen-bonded silanols. Hence the dipolar-dephasing results lead to the same conclusion regarding proximity and hydrogen bonding that one would derive from the dehydration behavior shown in Fig. 15. That the physisorbed water peak (3.5 ppm) is only slightly attenuated by the 160-μs dephasing period (Fig. 16b) is evidence for weak resultant dipolar interactions, presumably due to a combination of mobility and proton exchange in the physisorbed water.

Other ^{1}H CRAMPS-based relaxation methods that have been applied to the study of silica surfaces are inversion-recovery (Fig. 17a) and spin-exchange (Fig. 18a) experiments. Application of the CRAMPS-detected ^{1}H inversion-recovery sequence [48] to an untreated silica gel sample (Fig. 17b) and to a silica gel sample evacuated at 25°C (Fig. 17c) shows in each case an essentially homogeneous spin-lattice relaxation behavior of all types of protons within a given sample. This implies that all the protons are in efficient spin-exchange contact with each other within the rather long time scale represented by the measured T_1^H values of these samples (0.67 s or 3.8 s for untreated or 25°C-evacuated silica gel, respectively).

Spin exchange among the protons in a sample can occur by one of two mechanisms: (1) spin-spin flip-flops

$$\uparrow \ \downarrow \ \leftarrow \rightarrow \ \downarrow \ \uparrow$$

(the basis for spin diffusion), which are brought about by strong dipolar interactions, and (2) chemical exchange. ^{1}H-^{1}H spin exchange can often be monitored directly by the CRAMPS-based procedure shown in Fig. 18a [48]. In this experiment, a 2τ dephasing period depletes magnetization of those protons experiencing strong ^{1}H-^{1}H dipolar interactions (H-bonded silanols, in the present case) before the surviving transverse magnetization is placed along the z axis (by a -π/2 pulse) for a period, t_{mix}. During t_{mix}, spin exchange can occur between the depleted component (H-bonded silanols) and the remaining component (non–H-bonded silanols and physisorbed H_2O). The rate of this exchange provides useful information on the structure and dynamics of these OH groups on the surface. From Figs. 18b and 18c, one sees that substantial spin exchange occurs among the different types of protons for mixing periods as short as 500 μs and that a very high degree of spin-polarization equilibration occurs by about 2–3 ms of mixing. Since chemical exchange at that rate would broaden the ^{1}H CRAMPS peaks more than is observed, it was concluded that spin-polarization equilibration occurs mainly by spin-spin flip-flops generated by ^{1}H-^{1}H dipolar coupling. Deconvolution of the spectra in Fig. 18b led to the conclusion that ^{1}H spin exchange between physisorbed water protons and silanol protons is almost

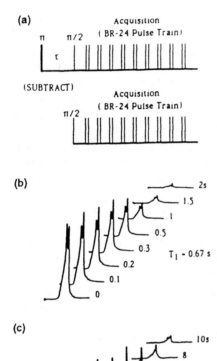

(a)

Acquisition
(BR-24 Pulse Train)

π   π/2

τ

(SUBTRACT)

Acquisition
(BR-24 Pulse Train)

π/2

(b)

2s

1.5

1

0.5

0.3 $T_1 = 0.67$ s

0.2

0.1

0

(c)

10s

8

5

3

2          $T_1 = 3.8$s

1

0.5

0

Figure 17 ^{1}H CRAMPS T_1 determination: (a) pulse sequence, showing individual
sequences employed in alternate scans (lower case subtracted from upper case in computer
memory); (b) results for untreated silica gel (showing τ values); (c) results for sample
evacuated at 25°C. (From Ref. 43.)

nonexistent on the μs-ms time scale and that spin exchange between the two
types of silanol protons occurs for both types of samples with a correlation time
in the range of 15 ms to 1s. These kinds of results are examined in greater detail
in Sec. II.D.

The interpretation of ^{1}H NMR experiments on aluminas and silica-aluminas
is perhaps less straightforward. From a combination of ^{1}H SP-MAS and ^{1}H
CRAMPS data on aluminas reported during the early to mid-1980s [45,57], one
would expect to find surface Al-OH peaks at about 1.8–2.0 ppm and a peak

(a)

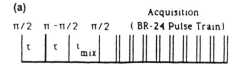

(b)

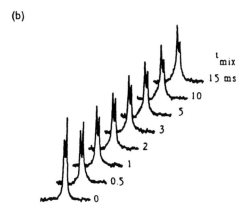

(c)

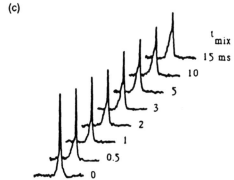

Figure 18 ¹H CRAMPS spin-exchange experiment: (a) pulse sequence; (b) results for untreated silica gel (showing τ_{mix} values); (c) results for sample evacuated at 200°C (showing τ_{mix} values). (From Ref. 43.)

centered at about 4.8 ppm identified with physisorbed water on structures of the γ-Al$_2$O$_3$ type. Both peaks are typically broad, with the detailed lineshape depending on a variety of experimental parameters (for example, SP-MAS or CRAMPS, MAS speed, magnetic field strength, observation temperature, and moisture content). For silica-aluminas the situation is more complex. All the ¹H NMR peaks described in the literature for silicas and aluminas have been reported for silica-alumina samples. The early work of Schreiber and Vaughan [58], carried out without any line-narrowing techniques, was apparently the first extensive, systematic ^{1}H NMR study of silica-aluminas ranging in composition from SiO_2 to Al_2O_3. ^{1}H NMR spectra obtained at 60 MHz by the SP-MAS approach by Hunger and coworkers [57] on a similar range of dry silica-aluminas are shown in Fig. 19. These spectra reveal peaks or shoulders at 1.8–2.0 ppm, which were identified with SiOH and AlOH protons, and at 7 ppm, which were tentatively identified with structures of type I. This last peak has been reported

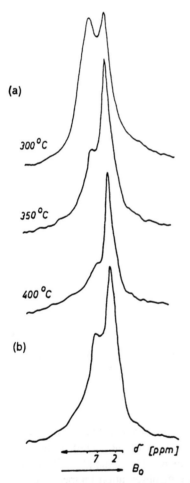

Figure 19 60 MHz SP-MAS ^{1}H NMR spectra of an amorphous silica-alumina (25 wt % Al_2O_3) for different procedures of pretreatment: (a) rehydrated and reactivated at different temperatures; (b) without rehydration. (From Ref. 57.)

only in samples containing the elements of both silica and alumina but not in a simple mixture of the those two components. The general type of structure represented by structure I is often referred to as structural hydroxyls in silica-aluminas and zeolites. Such a bridging hydroxyl species is believed to be responsible for acid catalysis in a variety of technologically important processes.

Figure 20 shows ^{1}H CRAMPS spectra of two different silica-aluminas (containing, formally, 14% Al_2O_3 and 25% Al_2O_3) over a range of sample evacuation temperatures [44]. Figure 21 provides a comparison of the spectra of silica gel, γ-alumina. and the two silica-aluminas in an untreated state and after evacuation at 155°C. These spectra show the 7 ppm peak occurring only for the two silica-alumina samples, it being largely eliminated by evacuation at 340°C. This latter observation prompted Bronnimann et al. [44] to propose that the structure (I) identified with the 7 ppm peak may have some (unspecified) degree of hydration. That reasoning is somewhat different from previous interpretations by Pfeifer and coworkers [57], based largely on SP-MAS data of the early 1980s, which viewed the 7 ppm peak in terms of a "bare" bridging hydroxyl, and from their more recent SP-MAS interpretation, which identifies the 7 ppm peak as residual NH_4^+ [45]. Figure 22 shows the ^{1}H SP-MAS spectra of silica-aluminas ranging between SiO_2 (Fig. 22a) and γ-Al_2O_3 (Fig. 22e). These spectra clearly show the 7 ppm peak or shoulder (bcd) in the silica-alumina samples.

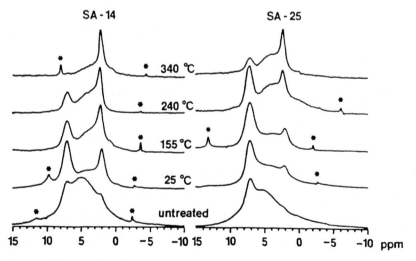

Figure 20 187 MHz ^{1}H CRAMPS spectra of silica-aluminas, SA-14 (left) and SA-25 (right), untreated (bottom) or evacuated at 10^{-2} torr at the indicated temperature. Rotor lines are indicated with asterisks. Chemical shifts are given relative to Me$_4$Si. (From Ref. 44.)

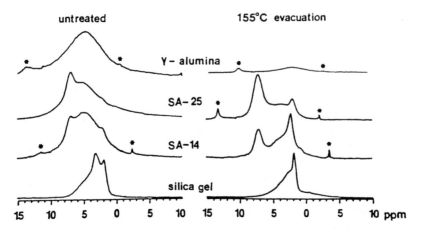

Figure 21   187 MHz ¹H CRAMPS spectra of untreated samples (left) and samples evacuated at 155°C and 10^{-2} torr (right). Rotor frequency lines are indicated with asterisks. Chemical shifts are given relative to tetramethylsilane (Me₄Si). (From Ref. 44.)

According to the NH_4^+ interpretation of the 7 ppm ¹H peak in silica-aluminas, the acidic framework OHs (structure I) display chemical shifts in the range of 3.8 to 5.8 ppm and are converted to NH_4^+ in the presence of NH_3. The disappearance of the 7 ppm peak at high temperatures is interpreted in terms of the conversion of NH_4^+ to NH_3 and desorption of the latter. It would appear that this appealing interpretation can be further tested in silica-aluminas by systematic studies of the reversibility of the postulated $\geqslant Si\text{-}O^+(H)\text{-}\overline{Al}\leqslant + NH_3 \leftrightarrows \geqslant Si\text{-}O\text{-}\overline{Al}\leqslant + NH_4^+$ process and related processes with other bases, using a combination of SP-MAS, ¹H CRAMPS, and possibly ¹H dilution by partial D_2O exchange. Investigations of this sort will be especially useful when carried out as a function of temperature (to alter the relevant chemical dynamics) and under conditions that provide *absolute* measures of ¹H peak intensities. For zeolites, such studies have already been performed to a substantial degree by Pfeifer and coworkers [46].

D.   Correlation of ¹H and X Resonances

In order to minimize the fragmentary character of NMR probes of surfaces based on just one nuclide, it is desirable whenever possible to correlate the results with NMR data obtained with other nuclides. In the case of silica, alumina, and silica-alumina surfaces, this is likely to involve ¹H-X correlations, where X is ^{29}Si or ^{27}Al. For such combinations, ¹H→^{29}Si and ¹H→^{27}Al cross-polarizations provide promising bridges on which to base the correlations, and the ¹H-^{29}Si case has received considerable attention.

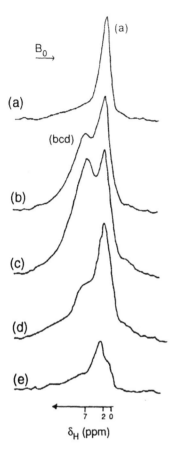

Figure 22 ^{1}H MAS NMR spectra of pretreated amorphous silica-aluminas of different composition: (a) silica, (b) 20 wt % Al_2O_3, (c) 25 wt % Al_2O_3, (d) 50 wt % Al_2O_3, (e) γ-Al_2O_3 at 60 MHz with 2.5 kHz speed of rotation. (From Ref. 45.)

A priori, one way to correlate ^{1}H and X NMR spectra is to examine the effect of some systematic chemical treatment on both types of spectra. Maciel and coworkers [59] have explored the effects of silylation with the reagent CHD_2 $(CD_3)_2SiCl$ on the ^{1}H CRAMPS and ^{29}Si CP-MAS spectra of silica gel. The spectra are shown in Fig. 23. After deconvolution and an attempt to correct the intensities of the deconvoluted peaks for differences in relaxation properties, it appeared that the ^{1}H CRAMPS peak identified with hydrogen-bonded silanols correlates *chemically* with the ^{29}Si CP-MAS peak identified with single silanols. In addition, the ^{1}H CRAMPS peak identified with isolated (non–hydrogen-

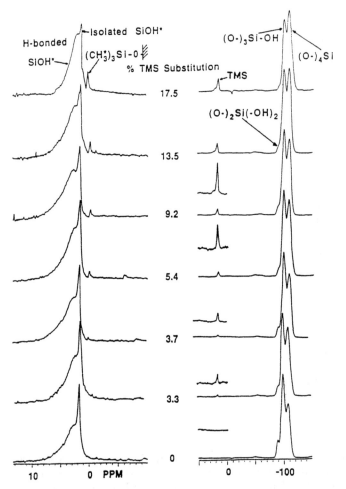

Figure 23 ¹H CRAMPS (left) and ²⁹Si CP-MAS spectra (right) of silica gel derivatized by (CH₃)₃SiCl [perdeuterated, except for 1% (C¹H₃)₃SiCl]. Higher loading levels progressing up the figure. (From Ref. 59.)

bonded) silanols correlates chemically with the ²⁹Si CP-MAS peak identified with geminal silanols, Si(OH)₂. This *apparent* correlation, which runs counter to prevailing structural models of the silica surface, is presumably the result of poorly understood CP dynamics (namely, protons that contribute very differently to ¹H CRAMPS and ¹H→²⁹Si CP-MAS spectra and that respond very differently in silylation reactivity). Other, more direct approaches lead to the opposite ¹H-²⁹Si correlations.

Chuang and coworkers [60] have used four variations of the $^1H \rightarrow {}^{29}Si$ CP-MAS experiment to study the effects of ^1H-^1H spin exchange on the ^{29}Si spectrum, thereby correlating the information available from ^1H CRAMPS experiments (e.g., which SiOHs are hydrogen bonded) with the information available from ^{29}Si spectra (how many OHs are attached to each silicon). Figure 24 shows a ^{29}Si CP-MAS experiment in which a period (2τ) of ^1H-^{29}Si dipolar-dephasing is inserted between the cross-polarization and detection periods. Figure 25 shows the ^{29}Si NMR peak intensities measured for the Q_2, Q_3, and Q_4 peaks in the ^{29}Si CP-MAS spectrum of an untreated silica gel, obtained as a function of the dipolar-dephasing time (2τ) in the experiment of Fig. 24. These results demonstrate the following two major patterns: the dephasing rates are in the order $> Si(OH)_2 >\geqslant SiOH >> Si < (Q2 > Q_3 > Q_4)$, and partial refocusing is observed in the ^{29}Si CP-MAS signals for dephasing times $2\tau = nt_{rot}$, where n is an integer and t_{rot} is the MAS rotor period (ν_{rot}^{-1}). This behavior is based on the facts that (1) the ^1H-^{29}Si dipolar interaction refocuses for $2\tau = nt_{rot}$ as long as the ^1H-^{29}Si dipolar interactions are not altered during t_{rot} by molecular motion, chemical exchange, or ^1H spin diffusion, (2) the isotropic part of the ^{29}Si chemical shift refocuses for any τ value, and (3) the ^{29}Si CSA refocuses for any $\tau = nt_{rot}$ ($2\tau = 2nt_{rot}$). The overall faster dephasing rate of the $Si(OH)_2$ peak, as reflected in its less efficient refocusing at $2\tau = 2nt_{rot}$, is consistent with the idea that ^1H-^1H spin flip-flops are more efficient among $Si(OH)_2$ protons than among SiOH protons. The same conclusion is drawn from ^{29}Si CP-MAS experiments in which ^1H decoupling is *not* applied during the detection period. Figure 26 shows a set of such spectra (lower spectrum in each pair) for six different MAS speeds, along with corresponding spectra obtained *with* ^1H decoupling (upper spectrum in each pair). These results show that MAS is least able to average the ^1H-^{29}Si

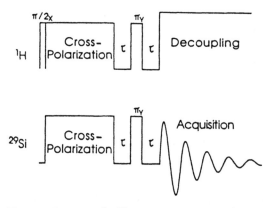

Figure 24 The ^1H-^{29}Si dipolar-dephasing ^{29}Si CP-MAS NMR experiment. (From Ref. 60.)

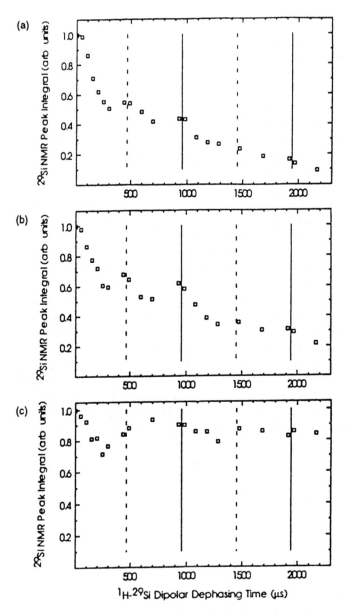

Figure 25 Plots of deconvoluted peak integrals of the 39.75 MHz ^{29}Si CP-MAS NMR spectra of Fisher S-679 silica gel vs. [1]H-[29]Si dipolar-dephasing time up to four rotor periods. CP contact time, 5 ms; magic-angle spinning speed, 2.0 kHz. Vertical dashed lines show odd numbers of rotor periods, and vertical solid lines show even numbers of rotor periods. (a) −89 ppm peak (geminal silanols); (b) −99 ppm peak (single silanols); (c) −109 ppm peak (siloxane silicons). (From Ref. 60.)

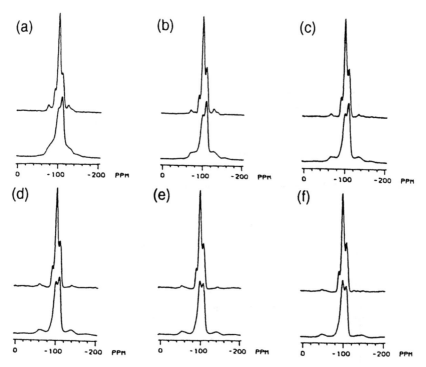

Figure 26 Proton-decoupled (top spectrum of each set) and proton-coupled (bottom spectrum of each set) 39.75-MHz ^{29}Si CP-MAS NMR spectra of Fisher S-679 silica gel at six different magic-angle spinning speeds. Cross-polarization time, 5 ms. (a) 1.0 kHz, 1960 accumulations; (b) 1.1 kHz, 3000 accumulations; (c) 1.4 kHz, 720 accumulations; (d) 1.6 kHz, 2000 accumulations; (e) 1.8 kHz, 2000 accumulations; (f) 2.0 kHz, 2000 accumulations. (From Ref. 60.)

dipolar interaction for the ^{29}Si magnetization of Si(OH)$_2$ (-89 ppm peak). The SiOH peak (at -99 ppm) is narrowed more efficiently, and the Q_4 peak (-109 ppm) is the most efficiently narrowed by MAS. This result suggests that the Si(OH)$_2$ protons are the most efficient at executing ^{1}H-^{1}H spin flip-flops—that is, they are the most extensively hydrogen bonded.

The view that the Si(OH)$_2$ protons experience the strongest ^{1}H-^{1}H dipolar interactions (and presumably the most extensive hydrogen bonding) is further supported by a ^{29}Si CP-MAS experiment in which a ^{1}H dephasing period (2τ) is inserted between the initial ^{1}H $\pi/2$ pulse and the CP contact period (Fig. 27) [60]. Results for an untreated silica gel sample are shown in Fig. 28 for four different CP contact times, each with spectra obtained with ^{1}H dipolar-dephasing times of 2 μs and 1 ms (corresponding to two rotor periods). The data for small CP contact periods of 100 μs and even 300 μs show conclusively that the ^{29}Si

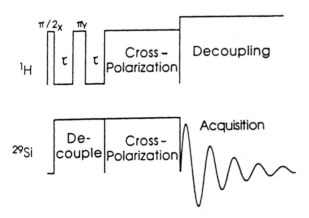

Figure 27 ^{29}Si CP-MAS NMR experiment with ^{1}H-^{1}H dipolar dephasing prior to ^{1}H→^{29}Si cross-polarization. (From Ref. 60.)

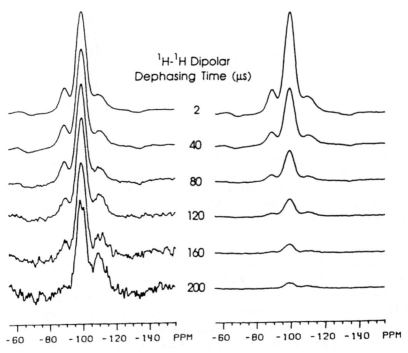

Figure 28 39.75 MHz ^{29}Si CP-MAS NMR spectra of Fisher S-679 silica gel obtained with six different ^{1}H-^{1}H dipolar-dephasing times (as indicated) prior to ^{1}H→^{29}Si cross-polarization. Cross-polarization time, 100 μs. Each spectrum is the result of 20,000 accumulations. Spectra in the left column have been scaled to the same peak height for the −99 ppm peak, and spectra in the right column are plotted on the same absolute scale. (From Ref. 60.)

266

signal whose CP proton source is most efficiently dephased during the pre-CP dephasing period is the Si(OH)$_2$ peak (-89 ppm).

All these ^{1}H-altered ^{1}H→^{29}Si CP-MAS experiments, together with a spin-exchange experiment in which ^{1}H→^{29}Si cross-polarization follows a sequence analogous to the predetection portion of the ^{1}H CRAMPS spin-exchange experiment described in Fig. 18 [60], lead to a model of the silica surface that corresponds to the 100 face of β-cristobalite for Si(OH)$_2$ silanols and to the 111 face of β-cristobalite for SiOH silanols. These faces are depicted in Fig. 29, in which one can see that the distance between OH groups on adjacent Si(OH)$_2$ groups is much smaller than between adjacent SiOH groups, or even between the

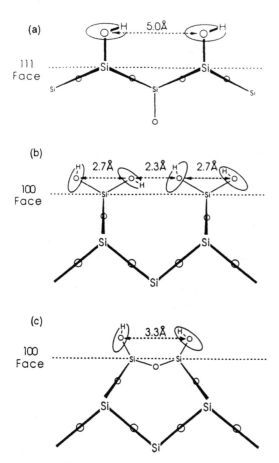

Figure 29 Side views of specific silicon planes (dashed lines representing an edge of such a plane) of β-cristobalite. Drawn approximately to scale: (a) 111 face; (b) 100 face; (c) vicinal sites from dehydration of the 100 face. (From Ref. 60.)

two OH moieties in an individual $Si(OH)_2$ group. Hence one expects that hydrogen bonding among $Si(OH)_2$ silanols should be more extensive than among SiOH silanols. Furthermore, from this figure one can also readily appreciate why dehydration of $Si(OH)_2$ groups to form structures of the type shown in Fig. 29c should be much more facile than dehydration of SiOH groups.

The most direct way of correlating ^{1}H NMR data with ^{29}Si NMR data would seem to be via a [1]H-[29]Si two-dimensional heteronuclear correlation (HetCor) strategy. Such work has been reported by Vega [61]. These interesting and important experiments were carried out using [1]H-[29]Si cross-polarization with relatively short CP contact periods (500–1000 μs) as the mixing stage in the two-dimensional strategy (a shorter CP contact period can give severe signal-to-noise problems). As reported previously [60], however, CP contact periods greater than 100 μs can result in substantial [1]H-^{1}H spin diffusion, which to some degree compromises the [1]H-[29]Si correlation one seeks; similar conclusions are drawn from ^{1}H CRAMPS spin-exchange experiments. Unfortunately, the multiple-pulse sequences required to suppress [1]H-^{1}H spin diffusion, while implementing [1]H→X coherence transfer [62,63], are technically very demanding for pairs of nuclei that are not directly bonded, such as [29]Si-O-[1]H.

III. MULTINUCLEAR STUDIES OF MODIFIED SILICAS AND ALUMINAS

A. Derivatized Silicas

The silylation of silica surfaces, as depicted in Eq. (1), has provided a strategy for the synthesis of a wide variety of useful materials, including separations materials, composites, and immobilized catalysts. Multinuclear NMR methods offer a powerful approach for characterizing such systems [15–17,64]. [29]Si, of course, plays a central role in these characterizations, as seen in Fig. 2C. Here, peak d shows the $(CH_3)_3Si-O-ⓢ$ resonance in the [29]Si CP-MAS spectrum of a silica (surface depicted by ⓢ) that was derivatized with $(CH_3)_3SiCl$. In this specific case there is only one general type of siloxane attachment to the silica surface. Silylating agents with more than one labile group can lead to more complex structures at the surface—for example, more variety in the silane-surface attachments and "horizontal" siloxane polymerization along (above) the surface. Figure 30 shows [29]Si CP-MAS spectra of a set of samples prepared by silylation of a series of silica gels, which differed in the details of dehydration/hydration, by $(CH_3CH_2O)_3SiCH_2CH_2CH_2NH_2$ (APTS), and subjected to post-treatment curing at various temperatures [65]. The positions of the siloxane peaks in the [29]Si CP-MAS spectrum range between about -49 and -66 ppm, reflecting a variety of attachments to the surface and corresponding variations in horizontal polymerization (for example, structures III and VI in Fig. 30). The

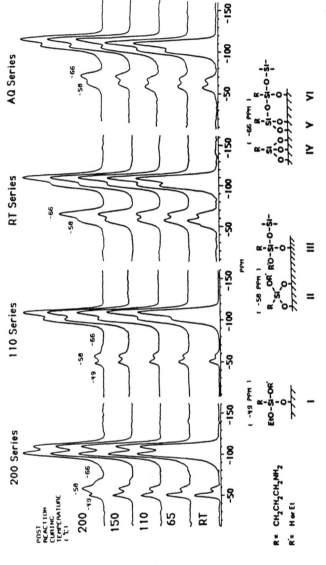

Figure 30 ²⁹Si CP-MAS spectra of APTS-modified silica gels. Each column of spectra corresponds to the temperature of silica gel drying (°C; RT = room temperature) under vacuum prior to reaction in dry toluene, or aqueous reaction conditions (AQ). Postreaction treatment (curing) temperature shown on the left. Structural assignments given at the bottom. (From Ref. 65.)

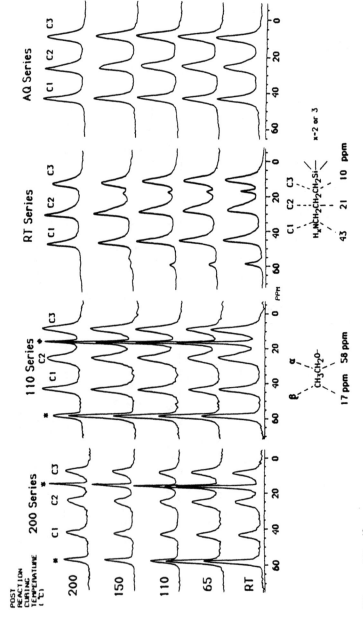

Figure 31 ^{13}C CP-MAS spectra of APTS-modified silica gels. Each column of spectra corresponds to the temperature of silica gel drying (°C; RT = room temperature) under vacuum prior to reaction in dry toluene, or aqueous reaction conditions (AQ). Postreaction treatment (curing) temperature shown on the left. Structural assignments given at the bottom. (From Ref. 65.)

associated ^{13}C CP-MAS spectra (Fig. 31) display somewhat less variety than their ^{29}Si counterparts, but do show that ^{13}C NMR can be valuable in ascertaining the extent of reaction of the labile ethoxy groups (peaks marked with asterisks).

Other nuclides have also shown promise for the study of derivatized silicas. For example, Figs. 32, 33, and 34 present the ^{1}H CRAMPS, ^{13}C CP-MAS, and ^{15}N CP-MAS (natural abundance) spectra of nonacidified and acidified samples of APTS-derivatized silica [48]. The NMR spectra of all three nuclides clearly display dramatic differences between Ⓢ-O ⩾ SiCH$_2$CH$_2$CH$_2$NH$_2$ and Ⓢ-O ⩾ SiCH$_2$CH$_2$CH$_2$NH$_3^+$ forms.

A wide range of catalyst systems are based on phosphine complexes of transition metals. Figure 35 shows ^{31}P CP-MAS spectra of samples prepared by derivatization of silica with (C$_6$H$_5$)$_2$PCH$_2$CH$_2$SiO(CH$_2$CH$_3$)$_3$ [66]. The samples for which spectra are shown correspond to the situations before and after

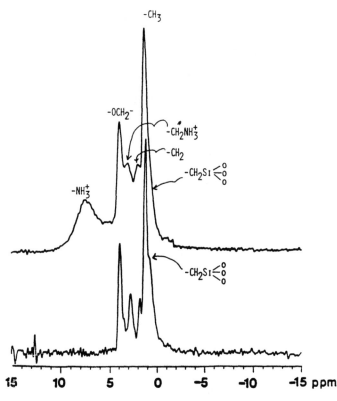

Figure 32 ^{1}H CRAMPS spectra of APTS-modified silica gel. Lower, untreated. Upper, treated with HCl. [ref. 59].

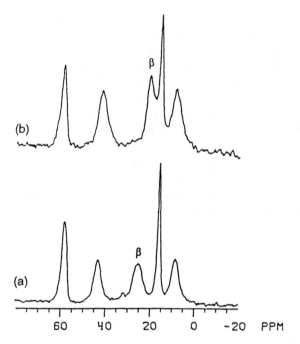

Figure 33 ^{13}C CP-MAS spectra of APTS-modified silica gel (a) and HCl-treated (b) samples. (From Ref. 59.)

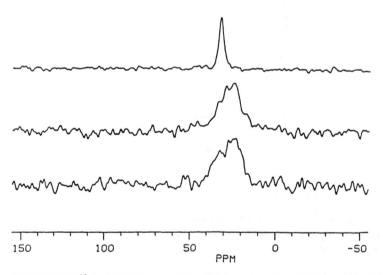

Figure 34 ^{15}N CP-MAS spectra (natural abundance) of APTS-modified silica gels: (upper) after treatment with H_2SO_4; (middle) untreated; (lower) after treatment with 0.1 M NaOH. (From Ref. 59.)

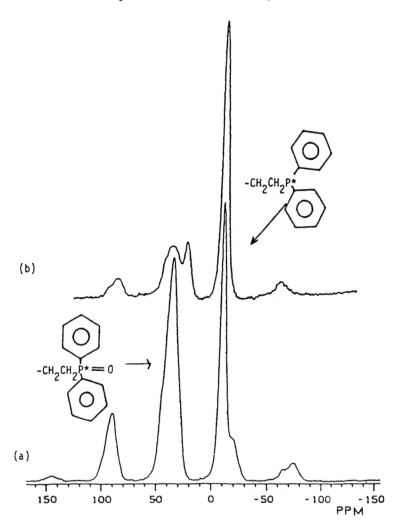

Figure 35 ^{31}P CP-MAS NMR spectra of silica gel derivatized with diphenyl-phosphinoethyltriethoxysilane, octadecyldimethylcholorosilane, and trimethylchlorosi-lane. DPP-ODS-TMS bonded phase before (a) and after (b) reduction. Extra peaks are due to spinning sidebands. (From Ref. 66.)

treatment by the reducing agent, SnCl$_2$. Clearly, the Ⓢ-O ⋛ Si-CH$_2$CH$_2$P(O) (C$_6$H$_5$)$_2$ species formed during the silylation process was largely reduced to the desired Ⓢ-O ⋛ Si-CH$_2$CH$_2$P(C$_6$H$_5$)$_2$ species by SnCl$_2$. Fyfe and coworkers and others have reported the ^{31}P MAS NMR spectra of phosphines, and their transition metal complexes, attached to silica surfaces [67].

B. Modified Aluminas

Rather than survey all of the possible modifications that can be made to an alumina surface, we will focus on a subset involved in two different types of surface-catalyzed chemical reactions, namely, the partial oxidation of ethylene to ethylene oxide (EO) and hydrodesulfurization (HDS) processes. Both of these catalytic systems have functional points in common, in that alumina serves as a support (α-alumina for the EO process and γ-alumina for the HDS process) and alkali-metal salts serve as promoters for both reactions. To illustrate this commonality, this section will be divided into three parts: (1) the adsorption of alkali-metal salts to γ-alumina, as reflected in the [87]Rb and [133]Cs solid-state NMR spectroscopy of these systems; (2) the absorption of ethylene to silver supported on aluminas in the presence and absence of cesium salts, as followed by ^{13}C NMR spectroscopy, and (3) the solid-state [95]Mo NMR of "fresh" and reduced/sulfided molybdena-alumina catalysts.

1. Adsorption of Rb$^+$ to γ-Alumina

As mentioned above, alkali-metal salts are used as promoters for various surface reactions [22]. One way to explain the enhanced efficiency in the presence of a promoter ion in, say, the ethylene oxide process is to propose that the alkali-metal cations (or solvent-separated ion pairs) absorb to "burning sites" (sites that lead ultimately to conversion to CO_2). As a result of their adsorption, the reactive site is deactivated.

To gain a better understanding of how higher alkali-metal cations interact with such reactive sites, Cheng and Ellis [68] initiated a detailed investigation of the nature of the adsorption of Rb$^+$ to γ-alumina. They found that the adsorption of the alkali-metal salts is complicated by the heterogeneity of the surface. There are two types of alkali adsorbates on the surface, which can be broadly described as "surface species" and "surface salts." By "surface species" is meant an alkali-metal ion not strongly associated with its anion, while the term "surface salt" refers to an alkali-metal ion which *is* associated with its anion. At sub-monolayer coverages, the prominent form on the surface is the surface species. Within the category of surface salt there are also at least two forms: a disordered component and one approaching that of the crystalline bulk salt. Likewise, the surface species are composed of a disordered form and one that is more ordered. The disordered species were distinguished by their NMR linewidths and by their dynamics. Further, it was demonstrated that the alkali-metal cations undergo moderately rapid motion within the disordered components on the surface and that the degree of hydration of the surface modulates the exchange rates for these processes. This motion is such that there is little, if any, exchange between the two types of adsorbates on the surface, a behavior that is consistent with the concept of adsorbate islanding.

To obtain this perspective, Cheng and Ellis [68] used a spin-echo sequence described as a "spikelet echo," $(\pi/2)_S - \tau_1 - (\pi)_S - \tau_2 - [- \tau - (\pi)_S - \tau]_m$, where τ_1 would equal τ_2 in the limit of δ-function pulses. This sequence, in which data acquisition occurs throughout each period in brackets, is analogous to that developed by Garroway [69] to establish the relative amounts of inhomogeneous (static) and homogeneous (dynamic) contributions to an observed powder pattern. In the spikelet pulse sequence the magnetization is continuously sampled from the top of the echo formed at the end of the duration τ_2. The subscript S on the $\pi/2$- and π-pulses denotes that these pulses are selective to the $\pm 1/2$ transition. Such a sequence allows the magnetization to be preselected in terms of T_2 behavior through the chosen values of τ_1 and τ_2. The spikes or lines that result from this sequence do not extend significantly beyond the static powder lineshape. Also, each spike or line in the frequency spectrum should have a normal (Lorentzian or Gaussian) shape with a width determined by T_2.

An important feature of the spikelet experiment is the scaling of the chemical shift. If one were to Fourier transform the function $f(t)$ consisting of only the points at the top of each echo, the resulting frequency spectrum would contain just a single line with a width corresponding to the natural T_2 of the spins. Furthermore, this line would appear at zero frequency. Cheng and Ellis [68] did not Fourier transform just the tops of the echos, however, but rather transformed the entire echo train. As a result, the chemical shift is scaled to a degree that depends upon the choice of τ employed. If there are two lines in the spectrum separated by a chemical shift difference $\Delta\nu$ Hz, the resulting spikelet echo spectrum may or may not contain evidence for the existence of the two lines. The chemical shift difference between the two lines will be preserved in the spikelet spectrum if the pulse repetition frequency between the π-pulses is small compared to the chemical shift separation in hertz. On the other hand, if the π-pulse repetition frequency is large, the chemical shift difference will not be evident in the resulting spikelet spectrum.

If one had two species in the sample with different chemical shifts and significantly different T_2s, the spikelet echo sequence could demonstrate the existence of both species, independent of the pulse repetition rate. The resulting spectrum would be a composite of the spikelet spectra of the two individual species with different breadths (widths or extents of the power spectra) and different linewidths for each of the spikes or lines. This capability of distinguishing species based upon their T_2s proved to be important in the identification of disordered phases or components of adsorbate material on the surface.

It was further noted that chemisorbed water has a pronounced effect on the spectra of the alkali metals. The "surface species" is highly reactive and is dramatically altered by water, whereas the surface "salt species" appears to be unaffected. This higher reactivity of the surface species is certainly related to the incomplete coordination environment expected for that species.

Huggins and Ellis [70] have made similar investigations using ^{133}Cs NMR. The Cs$^+$ results are completely analogous to those for Rb$^+$, except in the case of sensitivity to water; the surface species is exceptionally sensitive to trace amounts of water in the Cs$^+$ case. In summary, there are strong anion effects observed for either Rb$^+$ or Cs$^+$, and hence both the cation *and* anion must be considered in those systems. As will be demonstrated below, the alkali metal salts are not involved only in adsorption to various burning sites on a surface. These salts also have a dramatic impact on the adsorption of ethylene to silver.

2. Adsorption of Ethylene to Silver Supported on γ-Alumina

It is well known from theory and experiment that the interaction of ethylene with silver is nearly repulsive [71]. This weak adsorption can be investigated with solid-state ^{13}C NMR [72], and an illustration of such spectra for 1-^{13}C-C$_2$H$_4$ and 1,2-^{13}C$_2$-C$_2$H$_4$ is depicted in Fig. 36. The r_{CC} value extracted from the chemical shift/dipolar powder patterns of these spectra is 1.34 Å. This result is close to the value of 1.335–1.340 Å for ethylene in the gas phase [73], a bond distance that reflects the weak interaction between the silver and the adsorbed ethylene.

This picture is changed dramatically in the presence of the alkali metal cation, Cs$^+$. Figure 37 shows how the added Cs$^+$ leads to a dramatic increase in the r_{CC} for the adsorbed ethylene: r_{CC} goes from 1.34 Å for 0% added Cs$^+$ to 1.43 Å for

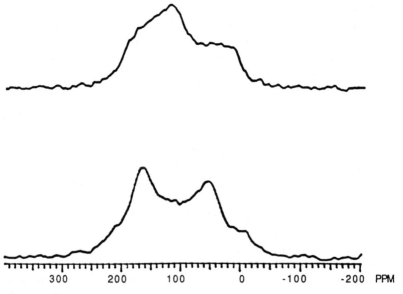

Figure 36   The static ^{1}H→^{13}C CP spectra of ethylene at 100 K: (top) 1-^{13}C ethylene; (bottom) 1,2-^{13}C ethylene.

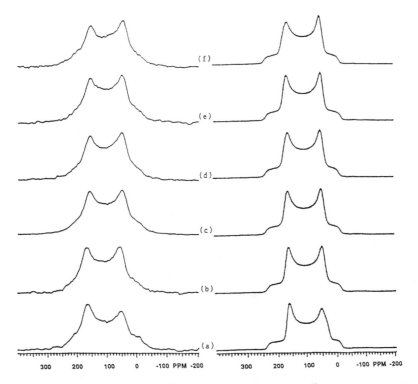

Figure 37 75.43 MHz $^1H \rightarrow ^{13}C$ CP static spectra of $[1,2-^{13}C]$ ethylene adsorbed on Ag/γ-alumina with different amounts of Cs additives. On the left side are experimental results and on the right side are the simulations. (a) No added Cs, (b) Cs is 5%, (c) Cs is 10%, (d) Cs is 14%, (e) Cs is 18%, (f) Cs is 30%. (From Ref. 72.)

30% added Cs^+. Wang and Ellis [72] have performed SEDOR[74] experiments (data not shown), which demonstrate that the Cs^+ is not directly associated with the adsorbed ethylene as suggested by Carter and Goddard [71]. In order to develop a working hypothesis as to how the alkali-metal salt can cause such a dramatic lengthening of the carbon-carbon bond, one needs to introduce some simple physical concepts concerning the nature of the work function of a metal.

It has been recognized for several years from ultra–high-vacuum investigations that alkali metals can lower the work function of metal surfaces [for review, see Ref. 75]. In its simplest form, the work function is defined as the minimum energy required to extract one electron from a metal. Although this definition is correct, it does not lend itself to description of how an adsorbate can alter the work function of a metal.

To obtain a better perspective, consider the diagram of the work function developed by Lang [76] and depicted in Fig. 38. Take a rectangular slab of some metal and assume that the positive charges in the slab can be averaged to a rectangular distribution of positive charge. That charge distribution corresponds to the region identified as "positive background" in Fig. 38. The corresponding electron density, plotted on a coordinate normal to the surface, is depicted in this figure. Note that there are regions at the surface that have a depletion of electrons; that is, just below the surface there is a net positive charge. Furthermore, there are positions above the surface where the electron distributions extend into the vacuum, creating regions of negative charge. Thus, there exists a surface dipole with the positive end below the surface, and this surface dipole corresponds, in fact, to the work function. In the process of removing an electron from the Fermi level, one has to work against the dipole.

With this picture in mind, consider now the charge distribution associated with a polarizable metal atom (for example, a cesium atom) as it approaches the surface along the normal. As the cesium atom approaches, its electron distribution nearest the surface is repulsed by the negative surface charge. The cesium's electron distribution becomes polarized in the sense opposite to that of the surface, and thus there arises an induced dipole that acts to reduce the surface dipole. As a result, the cesium atom has lowered the local work function of

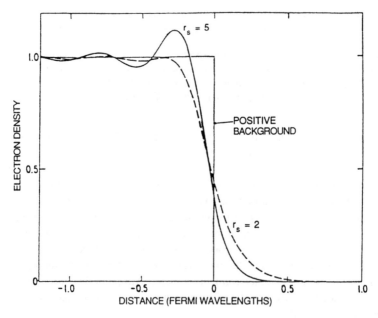

Figure 38 Electron density at a metal surface versus distance × normal to the surface, as computed by N. D. Lang [76] at two r_s values, using the planar uniform-background model.

the surface. The presence of the cesium atom on the surface makes it easier for an electron to be transferred to the appropriate orbitals of an adsorbate (say, ethylene) and, accordingly, makes possible an adsorption mechanism allowing "bonding interactions" to occur when the adsorbate is farther from the surface. The reduction of the work function induced by the alkali metal cation therefore facilitates electron transfer (back donation from the Fermi level) to antibonding π^* orbitals on ethylene. As a consequence of this back donation, the carbon-carbon bond lengthens.

The view described above is exactly what would be expected if the experiment were performed under ultra–high-vacuum conditions and if cesium atoms were sputtered onto a silver surface. It is clear, however, that these conditions do not correspond to what is depicted in Fig. 37. There are important differences between the NMR experiments represented in Fig. 37 and one based on a high-vacuum Cs-sputtered system, the most obvious being the difference between cesium atoms and cesium ions. This suggests that cesium ion pairs in the experimental system represented in Fig. 37 play the role of the cesium atom at the silver surface in the model of Fig. 38. Whether these ion pairs involve oxygen atoms at the surface or anions from the impregnation solution has not been resolved. Furthermore, from the results of SEDOR [74,77] experiments, it is clear that the cesium is not directly involved with ethylene. More research is needed on this and other systems before these results are completely understood. As can be easily seen, the alkali metal salts play several roles in ethylene oxide production; the picture is equally intriguing in HDS systems.

3. Solid-State ^{95}Mo NMR of "Fresh" and Reduced/Sulfided Molybdena-Alumina Catalysts

^{95}Mo, along with ^{87}Rb, is an example of a quadrupolar nuclide for which the lineshape of the $\pm 1/2$ transition is not dominated by the quadrupole interaction. The shielding tensor must be considered as well, and this situation complicates analysis of the resulting powder line shapes. For example, consider the dimolybdate ($Mo_2O_7^{2-}$) and heptamolybdate ($Mo_7O_{24}^{6-}$) species, the structures for which are shown in Fig. 39. The $^1H \rightarrow {}^{95}Mo$ CP spectrum [78] of $Mo_2O_7^{2-}$ is depicted in Fig. 40, and its simulation is illustrated in Fig. 41. What is clear from the simulation is that both the quadrupole and shielding tensors, as well as the three Euler angles relating the two principal axis systems, are needed to fit the observed spectrum—a total of eight parameters. The details of the lineshape theory are presented elsewhere [79].

The magnitude of the simulation problem becomes apparent when discussing the lineshape of the $Mo_7O_{24}^{6-}$ spectrum. In this modest cluster there are three distinct molybdenum sites in the ratio of 4:2:1, and each site requires eight parameters. Further, if the relative number of sites per cluster is also included in the analysis of the lineshape, the problem becomes a constrained 26-parameter fit. There is no doubt the lineshape can be fitted and the NMR parameters

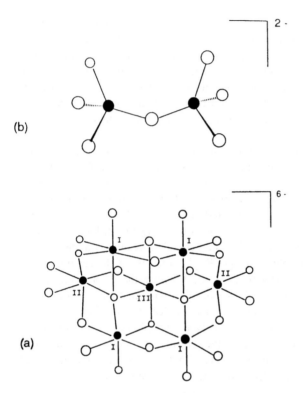

Figure 39 Structures of (a) the heptamolybdate anion and (b) the dimolybdate anion.

extracted, but what significance do they have? More work will be needed to address these points.

With these caveats in mind, Edwards and Ellis [80] have examined the solid-state ^{95}Mo NMR of a significant number of model systems for HDS catalysis. They employed solid-state static-echo, MAS-echo, and spikelet-echo ^{95}Mo NMR techniques to obtain information on reduced/sulfided molybdena-alumina catalysts, as well as the effect of addition of cobalt, cesium, and potassium to "fresh" molybdena-alumina and cobalt to reduced/sulfided molybdena-alumina. They also investigated a used catalyst exposed to thiophene for 24 hours. The results obtained for the reduced/sulfided catalysts indicate that the Mo^{VI}-O species present in the fresh catalyst undergo only partial reduction and sulfiding, with a portion of the tetrahedral Mo^{VI}-O species remaining unchanged in the reduced and sulfided catalyst. The bulk MoO$_3$ and octahedral molybdenum species, also present in the fresh catalyst, are reduced to Mo(IV) and sulfided to MoS$_2$. An unidentified species, possibly MoOS, was also found to be present. On the surface there are many other minor species present that cannot be identified, but which represent a high proportion of the molybdenum present there.

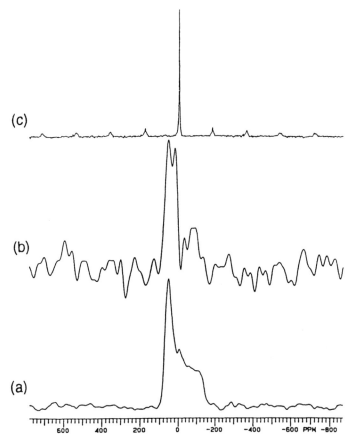

Figure 40 Static and MAS ^{95}Mo spectra of polycrystalline $(Bu_4N)_2Mo_2O_7$: (a) ^{1}H-^{95}Mo spectrum with 3477 scans, 5-s recycle delay, and Lorentzian line broadening of 500 Hz; (b) ^{1}H-decoupled Bloch decay spectrum with the same line broadening and number of transients as in (a); (c) 4.7 kHz MAS spectrum with ^{1}H decoupling, 6970 scans, 5-s recycle delay, and 50 Hz of line broadening. (From Ref. 31.)

These could consist of a whole range of Mo(VI) and Mo(IV) mixed oxygen/sulfur compounds. There is no evidence for the presence of MoO_2, except in one case. Simulation of the static powder lineshapes allowed the deconvolution of the various components, thus yielding values of quadrupole parameters as well as relative intensity data. From the spikelet experiments it was ascertained that all species are present as both static adsorbed and dynamically active phases. The ^{95}Mo spectrum of the used catalyst shows the presence of the tetrahedral molybdenum-oxo species, along with a much reduced MoS_2 resonance (relative to the fresh system), perhaps suggesting that MoS_2 is the active site in the

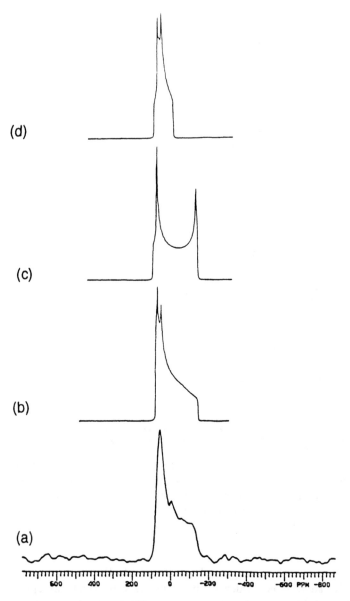

(d)

(c)

(b)

(a)

500   400   200   0   -200   -400   -600 PPM -800

Figure 41   (a) The ^{1}H-^{95}Mo spectrum of the same dimolybdate as in Fig. 40; (b) best least-square simulation of (a) with shielding and quadrupole tensors in noncoincident frames of reference; (c) with shielding and quadrupole tensors forced to have coincident frames of reference; (d) quadrupole alone. (From Ref. 31.)

thiophene HDS process and is degraded over time into a range of Mo(IV) and Mo(VI) oxo/sulfur species.

In the case of the cobalt-promoted fresh catalysts, the observed linewidths indicate the presence of a paramagnetic coupling. This broadening was found to be dependent on the method of impregnation of the promoter ion. Spikelet-echo experiments demonstrated the difference in surface chemistry between catalysts prepared by different impregnation techniques, namely, a much stronger Co-Mo interaction in the separately impregnated catalysts. Upon calcination, the usual increase in linewidth associated with polymerization of the surface molybdena was not observed for the promoted catalyst prepared by separate impregnation. The linewidth is actually smaller for these catalysts, indicating either that the paramagnetic coupling is transmitted via intervening dipole coupling among a collection of water molecules present on the uncalcined surface or that the cobalt is sequestered by the surface alumina to form $CoAl_2O_4$.

In the promoted reduced/sulfided catalysts, one observes no MoS_2 resonance in the separately impregnated promoted catalyst, while the MoS_2 is observed in the coimpregnated catalysts. This shows that the cobalt is closely associated with the MoS_2 phase in the separately impregnated catalysts. For the cesium- and potassium-promoted fresh catalysts, there is a marked increase in the ^{95}Mo NMR linewidth of the uncalcined catalysts. The large alkali-metal ions may interact with the molybdenum in such a way as to distort the molybdena species and thus produce a larger quadrupole interaction, or they may donate electron density to the molybdenum, thus creating a more shielded environment as well as perturbing the existing electric field gradient. In the calcined catalysts containing cesium or potassium, the ^{95}Mo lineshapes are considerably narrowed compared to the uncalcined catalysts. Comparison with the static and MAS spectra of Cs_2MoO_4 demonstrates the presence of predominantly Cs_2MoO_4 on the surface of high-loading catalysts. K_2MoO_4 was also found to predominate at high loadings of potassium. Spikelet-echo experiments of the uncalcined cesium promoted catalysts reveal a homogeneous broadening mechanism that increases with increasing loading.

It is clear that these model systems begin to approach real catalytic systems. They have demonstrated that solid-state ^{95}Mo NMR spectroscopy can be used effectively to extract structural information from molybdenum systems that have been difficult to investigate by other spectroscopic means.

IV. BASIC NMR PROBES OF ACID SITES ON SURFACES

The question of surface acidity and the type and relative number of acid sites on the surface are fundamental questions in the area of surface catalysis. Many experimental methods have been developed to address these questions. Space does not allow us to review all possible approaches; rather we will focus on multinuclear NMR investigations involving probe molecules. The overall strategy has been to adsorb on a surface a probe molecule that will be sensitive to Bronsted and/or Lewis acidity of the surface. This sensitivity can be manifested by chemical shift differences, dynamical differences, or the appearance of dipolar sidebands, depending on the particular experiment. Examples of each of these methods will be illustrated. However, within the context of dynamics, some clarification may first be in order.

As we will demonstrate below, several probe molecules appear to undergo rapid chemical exchange between various sites on a surface. An analysis employing typical liquid-state NMR strategies can yield both chemical shifts and populations in such instances. The other way to consider dynamics is through a partial tensor averaging experiment. That is, the products of the surface reaction, for example, protonation or Lewis acid-base complexation, can be probed by how the product moves on the surface. In this case the resulting molecular motion of the product can lead to *different* modes of tensor averaging. Hence a composite lineshape analysis will determine the relative populations of the various Bronsted and Lewis acid sites.

The greatest need for probe bases is for assessing *Lewis* acid sites, because ^{1}H NMR techniques are capable of addressing Bronsted acid sites directly, as we have seen above. In order to use a basic probe to assess the Lewis sites on a surface, however, it is often also necessary to understand the interactions between the basic probe and any Bronsted sites.

The basic probe molecules of primary interest in this chapter are pyridine, ammonia, *n*-butylamine, phosphines, and phosphine oxides. All possess several nuclei amenable to solid-state NMR experiments, and, as we shall see, this multinuclear perspective is useful in extracting chemical insights into the nature of the surface. Finally, it is important to point out the early, pioneering contributions of Ian Gay [81] to this area of NMR spectroscopy. It was those initial experiments that demonstrated the potential utility of solid-state NMR methods to address important questions in the area of surface chemistry.

A. Pyridine

Of the three amines to be discussed here, pyridine is the least basic, with $pK_b =$ 8.77. Gay and Liang [81] have recently used pyridine and substituted aromatic bases in conjunction with wideline NMR to probe reactivity differences in variously treated aluminas and mixed alumino-silicates. Difficulties were encountered with bases that bind tightly to the surface since magnetic dipolar effects then broaden the lines, causing overlap and loss of information. Pyridine gives a broad, ill-defined ^{13}C spectrum, even at elevated temperatures and relatively high surface coverages. By contrast, Ellis and coworkers [82] have found that the ambient temperature ^{13}C CP-MAS spectrum of pyridine at 0.05

BET monolayer surface coverage on γ-alumina is completely resolved, with separate resonances for each of the three types of carbon present. Within experimental error the line positions coincide with the values for liquid pyridine. The lines are noticeably broad (3–4 ppm). The use of ^{15}N enriched (90%) material results in a partial reduction (approximately 30%) in linewidth relative to the isotopically normal (>99.7% [14]N) spectrum, presumably because the ^{14}N quadrupole effect interferes with MAS averaging of [13]C-^{14}N dipolar interactions. The residual ^{13}C linewidths may be due to dipolar effects not removed at the level of ^{1}H decoupling power used, to the presence of ^{15}N scalar coupling, to heterogeneity of the surface acid sites, or to site diffusion between chemi-sites or physi-sites on the surface.

Ripmeester [83] pointed out that a better pyridine NMR probe would be [15]N. He demonstrated that for γ-alumina, in the absence of air and moisture, the ^{13}C resonances observed by Ellis and coworkers [82] were not due to protonated species on the surface. Furthermore, he found two types of Lewis acid sites on the surface. The resolution of these sites set limits on the possible exchange rates of pyridine on the surface of γ-alumina. The *upper limit* at 25°C was found to be approximately $2 \times 10^{-3} \text{ s}^{-1}$, corresponding to slow exchange. These results are in contrast to the results of Maciel and coworkers [84], who used ^{13}C and ^{15}N CP-MAS NMR spectroscopy to characterize the adsorption of pyridine on silica-alumina. They found that the combination of ^{13}C and ^{15}N spectroscopy provides a relatively detailed characterization of the pyridine/silica-alumina system. The data suggest that hydrogen bonding is the dominant interaction for samples with surface coverages of above 0.5 monolayer. The pyridine is very mobile for loading levels this high, and signals from an exchange-averaged species are observed. Rotation about the C_2 axis is one facet of a complex overall motion. [13]C spectral evidence suggests that a low-mobility Lewis acid-base complex dominates at lower loading levels. ^{15}N CP-MAS NMR of isotopically enriched samples provides evidence for two discrete forms of protonated pyridine for HCl pretreated samples [83]. Further indication that ^{15}N NMR is a more promising probe for adsorbed amines than ^{13}C NMR is the fact that some pyridine samples with much higher isotopic enrichment than Ripmeester's 30% show additional low-intensity ^{15}N peaks, including a signal at 200 ppm, which may be due to a slowly exchanging (or nonexchanging) Bronsted complex. The need for low-temperature studies is clear.

Maciel and coworkers [85] extended this work on silica-aluminas to an investigation of competitive adsorption of pyridine and *n*-butylamine. ^{15}N-enriched pyridine was used as the probe "indicator," and natural-abundance *n*-butylamine (the much stronger base, with a pK_b of 3.39) was used as a basic "titrant." From this investigation they were able to deduce that at low loadings the pyridine was essentially protonated. As they added the titrant *n*-butylamine, however, the pyridine appeared to undergo rapid exchange between hydrogen bonding and Lewis acid sites. From a preliminary analysis of these spectra, they were able to deduce site occupancies for Lewis acid, Bronsted sites, and hydrogen-bonding sites for both *n*-butylamine and pyridine. This study was the first to apply the combination of site competition and solid-state NMR to surface chemistry. More research is clearly needed for this type of investigation, especially as a function of temperature. Furthermore, some of the assumptions made regarding the absorption of *n*-butylamine could be reinforced by a complementary ^{15}N study examining the *n*-butylamine directly as well.

The observation by Ripmeester [83] of two different Lewis acid sites stimulated Majors and Ellis [86] to look at the pyridine/alumina system in more detail. They examined the ^{15}N spectrum of pyridine on γ-alumina with modest coverages corresponding to 25% of a monolayer. They also compared the ^{15}N spectra for three different pretreatments of the alumina surface; the corresponding ^{15}N spectra are shown in Fig. 42. Three individual methods for vacuum drying the γ-alumina samples were employed. Partially dehydrated alumina (PDA) was prepared by evacuating the sample for 8 hours at 350°C to a final pressure of 10^{-5} torr. An intermediary dehydrated alumina (IDA) sample was prepared by first evacuating the starting material at 510–550°C, then calcining the sample under O_2 for 30 minutes (3- to 10-minute intervals with intermittent evacuation), and finally evacuating the surface for 3 hours at 610°C. An extensively dehydrated alumina (EDA) sample was obtained by calcining the starting material under O_2 for 1 hour at 815°C and then evacuating for 5 hour at the same temperature to an ultimate pressure of 10^{-6} torr. From this work, Majors and Ellis [86] hypothesized that the two resonances arise from Lewis acid sites corresponding to positions associated with O_h and T_d sites on the surface. They then asked why the ^{15}N NMR spectrum shows only two distinct Lewis acid peaks rather than several, or else simply a continuous distribution of sites, which would yield a single broad peak. Using IR data critically compiled by Knozinger and Ratnasamy [20], Majors and Ellis [86] were able to demonstrate that the ^{15}N NMR data are consistent with the notion of a γ-alumina surface composed of equal mixtures of (111) and (110) planes. The calculated 64% tetrahedral anion vacancy for the 1:1 surface compares closely with the 61% tetrahedral site pyridine population that is observed for the PDA sample. Both experimental results (EDA sample) and model considerations indicate an approximate 50% tetrahedral site distribution at elevated surface-preparation temperatures. This result conflicts with previous studies, which suggest that the combination of (110) and (100) planes gives a better description of γ-alumina surfaces [20].

The (110) and (100) combination predicts a majority of O_h anion vacancies, in disagreement with the experimental NMR results. Since the interpretation of the experimental results depends upon the tentative assignments of peaks due to the Lewis acid-base complex, one might suspect an error in these assignments (perhaps a reversed assignment of the T_d and O_h resonances). However, the calculated distribution for the combination of (110) and (100) planes fails to

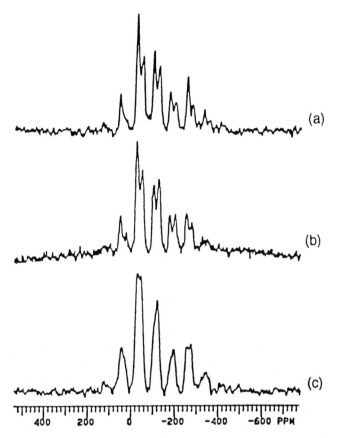

Figure 42 30.4 MHz $^1H \rightarrow ^{15}N$ CP-MAS spectra of pyridine-^{15}N on γ-alumina (25% monolayer coverage): (a) PDA sample, 12160 1-s repetitions, 3-ms contact time, $\nu_r = 2350$ Hz; (b) IDA sample, 25025 repetitions, 3-ms contact time, $\nu_r = 2330$; (c) EDA sample, 35072 repetitions, 3-ms contact time, $\nu_r = 2300$ Hz. (From Ref. 86.)

match the dependence of the experimental distribution on surface preparation temperature. Specifically, it displays no dependence on the surface preparation temperature at higher temperatures, and it deviates significantly from the approximately 50:50 population ratio for $T_d:O_h$ experimentally observed at higher temperatures.

Majors and Ellis [86], applying solid-state ^2H NMR, concluded that the motion of the pyridine ring was a mixture of ring flips and multisite diffusion. The idea of using ^2H to investigate dynamics on a surface suggests that motion can also be exploited to examine the nature of the species on the surface. As we shall see, ammonia is an ideal probe to study this type of motion.

B. Ammonia

Ammonia, with a pK_b of 4.75, is a probe molecule that potentially could reflect both the Lewis and Bronsted acidity of a given surface. The most obvious parameter to be examined is the ^{15}N chemical shift. Using γ-alumina, however, Majors et al. [87] demonstrated that even with MAS methods the ^{15}N resonance is too broad to be used as such a probe. Nevertheless, they did conclude that ^2H would be an outstanding probe for both sites. The reason for this choice is clear: the motion details of a tetrahedral NH_4^+ ion are quite distinct from those of an ammonia molecule tethered at a Lewis acid site.

The motion of the ND_3H^+ ion on the surface is quite complicated. This species virtually skates around the surface with three of its protons directed toward the surface and the other proton lying along the threefold (or tunneling) axis. The ND_3H^+ ion tumbles when it encounters a defect on the surface, causing an exchange of roles and environments between one of the protons directed toward the surface and the proton lying along the rotation axis. This averaging results in a nearly isotropic line for the ND_3H^+ species. However, the motion of the Lewis acid-base surface complex is distinctly different, being simply rotation about the threefold Ⓢ-N axis. As a result, the ^2H spectrum of the mixture is, in principle, easily separable.

The basic idea is simple. One generates a surface with a very low concentration of Bronsted sites and obtains the ^2H spectrum, e.g., Fig. 43a. Such a powder pattern is reflective of rapid motion about the threefold axis of the Lewis acid-base surface complex, and the integrated area of the spectrum can be easily quantitated. It is well known that protons from the surface can exchange with ammonia protons. Therefore, a second reference spectrum can be obtained on a partially deuterated sample of the alumina. With these two serving as reference spectra, the surface exchange spectrum can be subtracted from the spectrum of interest (see Fig. 43d). What remains is a spectrum that contains the partially averaged ^2H spectrum of the ND_3H^+ ion and surface Lewis acid-base complex of ND_3. The latter can be subtracted from the previous difference spectrum, yielding the ^2H spectrum of the ND_3H^+ ion. Hence one can obtain the relative amounts of Bronsted and Lewis acid sites for a given surface. Ellis and co-workers examined only the γ-alumina surface by this method. It is clear, however, that the Lewis:Bronsted ratio of peak areas, normalized to a fixed amount of ammonia, would lead to a relative acidity scale for alumina surfaces.

C. Alkylamines

The strongest amine discussed so far is *n*-butylamine, with a K_b of 3.39. Ellis and coworkers [88] have used NMR techniques to investigate the adsorption of *n*-butylamine on γ-alumina. The solid-state ^{13}C spectrum of *n*-butylamine on γ-alumina is depicted in Fig. 44a. There are *six* prominent resonances, two of

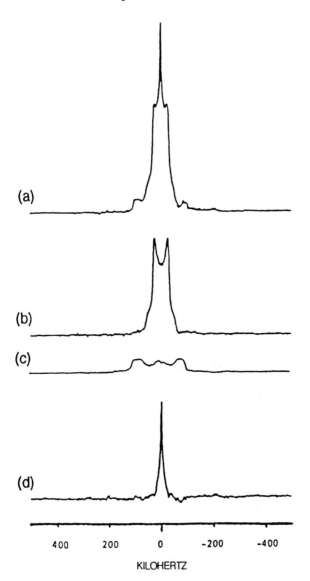

(a)

(b)

(c)

(d)

400   200   0   -200   -400

KILOHERTZ

Figure 43 The subtraction of identified ^{2}H lineshape components from the spectrum of ND$_3$/PDA-alumina: (a) ND$_3$/PDA-alumina, (b) ND$_3$/EDA-alumina, (c) reacted D$_2$O/ PDA-alumina, (d) optimized difference spectrum displaying the approximate lineshape of the Bronsted ammonium species. (From Ref. 87.)

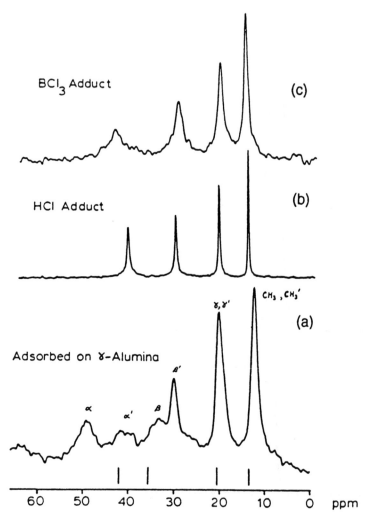

Figure 44 ^{1}H→^{13}C CP-MAS spectra of *n*-butylamine (a) adsorbed to the surface of γ-alumina (38000 repetitions). The vertical bars indicate carbon chemical shifts for liquid-phase *n*-butylamine; (b) solid HCl adduct (46 accumulations); (c) solid BCl$_3$ adduct (212 accumulations). (From Ref. 88.)

which are readily assigned to the γ-methylene and methyl carbons by comparison with the liquid-phase spectrum. These resonances are shifted only slightly upon adsorption. The remaining four resonances arise from the two α- and β-methylene carbons, from which we conclude that at least two types of chemically different butylamine species are present on the surface. The existence of well-resolved resonances demonstrates that, if surface diffusion occurs, its rate is not comparable to the spinning rate (~3.2 kHz). Furthermore, if site exchange is occurring, then its rate must be slow compared to the chemical-shift difference between the two species on the surface; otherwise two distinct peaks would not be observed. This spectrum, shown in Fig. 44a, is therefore consistent with a picture in which the nitrogen of the *n*-butylamine is firmly anchored to the surface. In all probability the system executes rapid, albeit limited angular diffusion about the bond axis connecting the surface to the nitrogen, with the motion of the alkyl chains increasing as one moves away from the surface.

The appearance of four peaks in the region expected for the two α and β carbons of the alkyl group is consistent with there being two quite different sites available to the amine. Two candidates are the classic Lewis and Bronsted sites. In order to check this possibility, Ellis and coworkers [88] obtained the CP-MAS ^{13}C spectra of two solid adducts of *n*-butylamine (Fig. 44b,c). The ^{13}C peaks of the solid HCl adduct match closely with four of the resonances of the surface-adsorbed species. The resonances corresponding to the α and β carbons of a solid BCl_3 adduct are at lower shielding than for the corresponding carbons within the HCl adduct, but are at higher shielding than for the analogous carbons of the adsorbed amine. Hence it was concluded that the least shielded set of carbons correspond to the *n*-butylamine that is attached to the surface via Lewis acid-base bonds to an aluminum atom. The fact that the chemical shifts of such species are at lower shielding than those of the BCl_3 adduct was attributed to the acid site of this surface being a stronger Lewis acid than BCl_3.

From the three amine systems described here, it is readily seen that significant information concerning relative acidity and dynamics can be obtained from the multinuclear NMR investigation of surfaces with amine-based probe molecules.

D. Phosphines and Phosphine Oxides

Other useful classes of basic probe molecules used to examine silica, alumina, and silica-alumina surfaces (as well as zeolite systems) include small organic phosphines and phosphine oxides, which rely on the highly convenient ^{31}P nuclide ($I = 1/2$, 100% natural abundance). As Lunsford and coworkers demonstrated for zeolites [89], the ^{31}P NMR signal of trimethylphosphine is a useful probe for Bronsted acid sites on surfaces. The basis for this approach is the formation of R_3P^+-H BⓈ sites at surface Bronsted acid sites, H-BⓈ.

Very different intensities of the ^{31}P NMR signals from $(CH_3)_3$P-derived chemical species on surfaces can be observed by using different NMR pulse modes for generating transverse ^{31}P magnetization [90]. Figure 45 shows the CP and single-pulse (FT) MAS spectra for two surface concentrations of $(CH_3)_3$P adsorbed on γ-alumina. The CP-MAS spectrum reveals the presence of Bronsted (-4 ppm) and Lewis (-48 ppm) complexes in addition to physisorbed phosphine (-58 ppm). At small (<1 ms) CP contact time the cross-polarization

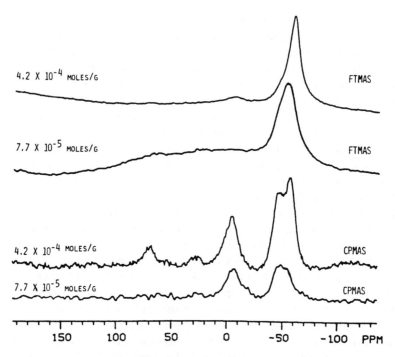

4.2 X 10⁻⁴ MOLES/G   FTMAS

7.7 X 10⁻⁵ MOLES/G   FTMAS

4.2 X 10⁻⁴ MOLES/G   CPMAS

7.7 X 10⁻⁵ MOLES/G   CPMAS

150   100   50   0   -50   -100   PPM

Figure 45 ^{31}P MAS single-pulse (FT-MAS) and cross-polarization (CP-MAS) spectra (60.7 MHz) of trimethylphosphine on γ-alumina at different surface coverages (as indicated). (From Ref. 90.)

experiment discriminates strongly against the physisorbed species relative to the more tightly bound acid-base complexes. The physisorbed phosphine peak grows relative to the other peaks at longer contact times (up to 10 ms) but does not achieve the full relative intensity seen in the single-pulse spectra. Clearly the spectra obtained from single-pulse experiments should reflect more straightforwardly the true populations of the different species on the surface if proper attention is paid to spin-lattice relaxation considerations. It is usually rather convenient to satisfy this requirement.

Figure 46 shows SP-MAS ^{31}P NMR spectra of $(CH_3CH_2)_3P$ adsorbed on a silica-alumina over a range of surface loading levels. As in the $(CH_3)_3P$ case outlined above, the ^{31}P chemical shift of the Bronsted-bound R_3P^+H $\overline{B}\text{Ⓢ}$ species was determined by analogy to model systems, for example, R_3P adsorbed on a sulfonic acid resin (where the dipolar pattern of ^1H-^{31}P pairs can be observed and where J_{PH} splittings of about 490 Hz are seen in the spectra obtained without ^1H decoupling). At the lowest $(CH_3CH_2)_3P$ surface coverage studied, only the Bronsted complex is observed, with a peak at 20 ppm relative to

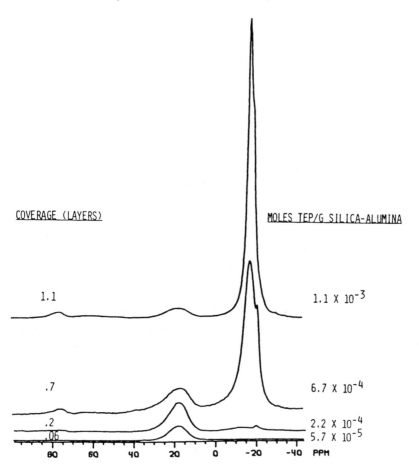

Figure 46 ^{31}P single-pulse MAS NMR spectra (60.7 MHz) of triethylphosphine on silica-alumina at different surface coverages (as indicated). (From Ref. 90.)

85% phosphoric acid. At higher $(CH_3CH_2)_3P$ concentrations, two additional peaks are observed, which differ from each other in chemical shift by only a few ppm. The lower-shielding member of these two peaks is assigned to Lewis acid-base complexed phosphines, since its chemical shift corresponds most closely to the Lewis acid-base phosphine complex observed in studies of model compounds. The higher-shielding peak is assigned to physisorbed phosphines. The chemical shift of this peak agrees with the chemical shift reported for solution studies of $(CH_3CH_2)_3P$. In experiments on $(CH_3CH_2)_3P$ adsorbed on silica gel, it was found that only one resonance is observed, with the same chemical shift reported for $(CH_3CH_2)_3P$ in solution. This observation confirms the expectation that phosphines adsorbed on silica gel exist essentially as physisorbed species. Additional support for the assignment of the $(CH_3CH_2)_3P$/Lewis-acid assignment was obtained from the dependence of linewidth on B_0, based on the effects of the ^{27}Al quadrupole interaction on MAS averaging of the ^{31}P-^{27}Al dipole-dipole interaction.

From experiments of the type represented in Fig. 46, one can determine the populations of Bronsted-bound sites, Lewis-bound sites, and physisorbed phosphines (or a combination of the last two) as a function of surface R_3P-loading level. With the assumption that the R_3P species present at the surface are in a state of equilibrium, plots of this dependency can be considered as "titration" curves, with the R_3P reagent as the titrant. Figure 47 shows such a ^{31}P NMR titration curve for $(CH_3)_3P$ adsorbed on silica-alumina, obtained by using populations derived from relative peak areas and total phosphorous content determined by elemental analysis [90]. Since there was no loading level at which peaks arising from only Bronsted and Lewis complexes were observed, the data allow a straightforward assay of only Bronsted acid sites accessible to each probe molecule. The "endpoints" occur at 0.23, 0.19, and 0.15 mmol H-B(S) per g of silica-alumina for such experiments carried out with $(CH_3)_3P$, $(CH_3CH_2)_3P$, and $(CH_3CH_2CH_2CH_2)_3P$, respectively. These findings may be compared with ^{15}N NMR results from a study of pyridine/*n*-butylamine competition for Bronsted sites on the same silica-alumina, namely 0.17 mmol H-B(S)/g. Some combina-

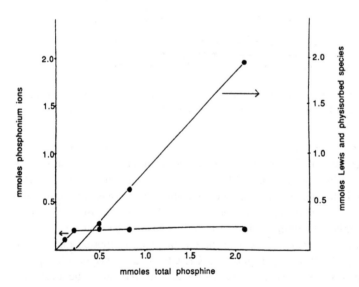

Figure 47 Experimental phosphine titration curve of trimethylphosphine on silica-alumina. (From Ref. 90.)

tion of several mechanisms could account for the observed variation. It is possible that some of the Bronsted sites may be situated at the bottom of crevices and are accessible only to small probe molecules. Topographically determined differences of this sort could also contribute to the chemical shift heterogeneity in both Lewis and Bronsted complexes, demonstrated by T_2 measurements. Alternatively, the Bronsted sites may be clustered on flat surfaces in such a way that larger probe molecules interfere with one another, preventing a complete assay of sites. A third possibility is that the three probes differ enough in basicity toward *surface* Bronsted sites that the "apparent" Bronsted titration endpoints, namely, the appearance of peaks due to Lewis complex and physisorbed species, are not at the true equivalence points and thus represent only a lower limit to the concentrations of Bronsted sites. The data cannot distinguish at present among these possibilities, and since they are not mutually exclusive it may be impossible to isolate such effects for an amorphous material. In any case, the apparent discrepancy in assaying Bronsted sites highlights the need for using more than one type of probe molecule to characterize surfaces with specific binding sites. Systematic studies with varied temperature should also prove useful in sorting out these kinds of issues.

The ^{31}P resonances of trialkylphosphine oxides have also shown great promise as NMR probes of acidic sites on surfaces [91]. Figures 48 and 49 show CP-MAS and SP-MAS ^{31}P spectra of $(CH_3)_3PO$ and $(CH_3CH_2)_3 PO$, respectively, on silica-alumina. Especially in the case of CP-MAS spectra of $(CH_3CH_2)_3PO$, one can see three separate peaks in the spectrum: the peak due to the Bronsted complex involving $(CH_3CH_2)_3POH^+$ at 76 ppm, the Lewis acid/$(CH_3CH_2)_3PO$ complex at 63 ppm, and the peak due to physisorbed and excess crystalline $(CH_3CH_2)_3PO$ at 49.5 ppm. Also of interest is the fact that the spinning sidebands, marked by asterisks in Figs. 49 and 50, manifest T_1^P values the same as those of the corresponding pure crystalline R_3PO samples (much larger than the T_1^P values of the more rapidly relaxing physisorbed components of the large high-shielding peaks).

The different characters of the ^{31}P NMR behaviors of $(CH_3)_3PO$ and $(CH_3CH_2)_3PO$ shown in Figs. 48 and 49 suggest that variation of R in the R_3PO series might yield even better ^{31}P probes of surface acidic sites. Figure 50 shows preliminary ^{31}P CP-MAS results obtained on $(n\text{-octyl})_3PO$ adsorbed on silica-alumina. The tentative peak assignments show separate peaks for physisorbed, crystalline, Lewis-complexed, and Bronsted-complexed R_3PO in this system. Studies as a function of the surface loading level should be illuminating. Figure 51 shows the effect of the size of R on the behavior of R_3PO probes on a HY zeolite system. The ^{31}P CP-MAS spectrum of $(CH_3)_3PO$ has a broad peak due to a distribution of Bronsted sites and a peak due to Lewis and/or physisorption sites. When the much larger $(n\text{-octyl})_3PO$ is used with the same zeolite, apparently it is unable to penetrate the channel/cage structure, and only signals identified as representing crystalline R_3PO are observed.

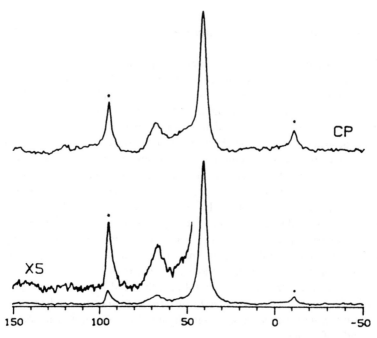

Figure 48 ^{31}P MAS spectra of (CH$_3$)$_3$PO on silica-alumina (2.8 × 10^{-3} mol/g of silica-alumina): (top) cross-polarization; (bottom) single pulse. Asterisks designate spinning sidebands. (From Ref. 91.)

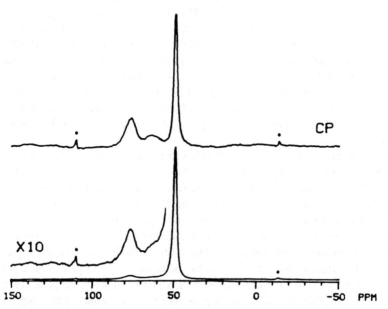

Figure 49 ^{31}P MAS spectra of (CH$_3$CH$_2$)$_3$PO on silica-alumina (3.0 × 10^{-3} mol/g of silica-alumina): (top) cross-polarization; (bottom) single pulse. Asterisks designate spinning sidebands. (From Ref. 91.)

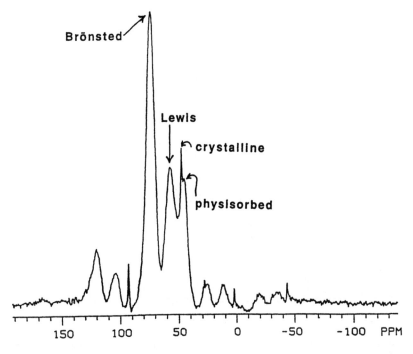

Figure 50 ^{31}P CP-MAS spectrum of (n-octyl)$_3$PO adsorbed on silica-alumina, with tentative assignments.

E. Other Basic Surface Probes

Preliminary results of Mastikhin and coworkers [92] on dilabeled ^{15}N$_2$O have shown promise for probing *Lewis* acid sites on the surfaces of silicas, aluminas, and zeolites. Figure 52 shows that the peak due to the terminal nitrogen (-235 ppm in gaseous N$_2$O) is sensitive to the details of surface adsorption at Lewis acid sites, the concentrations of which in Al$_2$O$_3$ (Fig. 52 b–d) increase at higher alumina pretreatment temperatures (c,d). As seen in this figure, the terminal nitrogen chemical shift is insensitive to the presence of *Bronsted* acid sites on the surface (Fig. 52a). Hence it has been concluded that the ^{15}N$_2$O probe is primarily sensitive to Lewis acid sites.

Other basic NMR probes that show promise for elucidating Lewis acid sites on surfaces are CH$_3$^{13}CN and ^{13}CO [47,93–95]. All three basic probes mentioned in this section apparently show a sensitivity and selectivity for Lewis sites. They also share the important feature of having compact, essentially linear shapes (if one views the CH$_3$ group as a cylinder) and are thus capable of accessing Lewis sites that might not be accessible to more bulky probes. From promising prelimi-

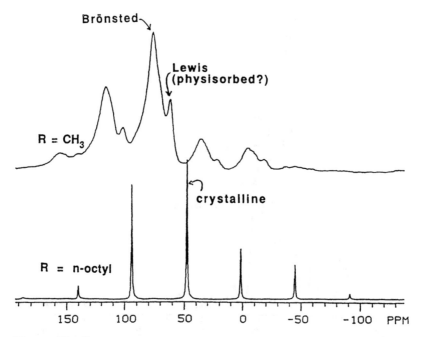

Figure 51 ^{31}P CP-MAS spectra of R_3PO adsorbed on HY zeolite: (top) R = CH_3; (bottom) R = *n*-octyl.

nary results, it would appear that more systematic studies of these kinds of systems are warranted and needed.

F. Analogous Studies of Basic Sites via Acidic Probes

By analogy to the use of small bases (amines, phosphines, phosphine oxides, N_2O) as probes of acidic sites on surfaces, one might expect that small acidic molecules could serve as probes of basic sites on surfaces. For this purpose BR_3 compounds show some promise, based on the [11]B nuclide (I = 3/2, 80.4% natural abundance) (P. Marchetti and G. E. Maciel, unpublished results).

V. MULTINUCLEAR NMR STUDIES OF SUPPORTED DISPERSE TRANSITION METALS

In Section III.2 we discussed ethylene adsorption to a dispersed silver supported on alumina. This section is separated from the earlier discussion for the principal reason that the spectroscopic methods utilized in the silver investigations are generally not applicable to transition metals with partially filled d-orbitals. The most relevant metals (primarily platinum, rhodium, ruthenium, and palladium)

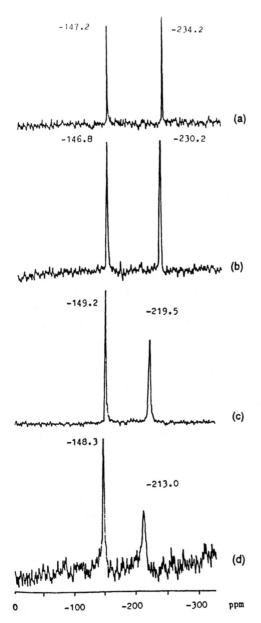

Figure 52 ^{15}N NMR spectra of N_2O adsorbed on 10% H_3PO_4/SiO_2 catalyst and on γ-Al_2O_3. (a) 80 μmol/g N_2O adsorbed on 10% H_3PO_4/SiO_2; $T = 293$ K, NS = 1000. (B) 120 μmol/g N_2O adsorbed on γ-Al_2O_3 pretreated at 473 K for 4 h in vacuum; $T = 293$ K, NS = 1300. (C) 120 μmol/g N_2O adsorbed on γ-Al_2O_3 pretreated at 773 K for 8 h in vacuum; $T = 293$ K, NS = 1200. (D) 120 μmol/g N_2O adsorbed on γ-Al_2O_3 pretreated at 873 K for 8 h in vacuum; $T = 293$ K, NS = 1000. (From Ref. 92.)

are weakly ferromagnetic. As a result, there are large anisotropic effects on the magnetic susceptibility at the surface. This results in significant shifts of the resonance frequency of each isochromat of a surface residue. One consequence of this shifting is a large offset of the ^1H resonance frequency which, in turn, will lead to a dramatic loss in efficiency of a cross-polarization experiment. To further complicate matters, the same kind of shifts of the surface resonances will lead to significant broadening of the ^{13}C resonances as well. Hence, the ability to observe the resonances via cross-polarization techniques for those molecules associated with the surface becomes improbable at best. The magnitude of these surface shifts depends upon the particular metal and how it was prepared, etc. Shifts on the order of a few gauss would not be unexpected. One gauss corresponds to approximately 4 kHz and 1 kHz, for ^1H and ^{13}C, respectively. Therefore, it is easy to understand why the cross-polarization experiment will not be an effective means to observe ^{13}C with these surfaces. Slichter and his coworkers [96–99] recognized this problem from the beginning and developed a novel solution. The application of these methods to metals is described in detail by Slichter and coworkers and is reviewed elsewhere [96].

Their solution, was to observe the rare spin directly via a spin echo. This experiment exploits the fact that at the top of the echo the inhomogeneous broadening caused by the surface has refocused. Furthermore, due to the small gyromagnetic ratio of the rare spin, direct observation of the rare spin reduces any errors caused by not being at the exact echo maximum. However, given the possible line broadening resulting from chemical shift anisotropy and dispersion, that does not appear to be a significant loss. The genius of Slichter's experiment was his choice to examine the dipolar interaction. By performing the experiment at low temperatures (to increase sensitivity and reduce motional averaging) and utilizing isotopically enriched adsorbates, his group could select structural information (CC and CH bond distances) from an analysis of echo amplitudes as a function of time. This has lead to the first unambiguous structure data to be obtained for surface adsorbed species [74]. In this section we will summarize Slichter's spin echo experiment and its application to the adsorption of small molecules to metal surfaces. Some novel results have been reported by Duncan and Zilm [100–102] concerning the adsorption of CO to ruthenium, rhodium, and palladium.

Slichter's spin echo approach is a variation on the spin echo double resonance (SEDOR) experiment originally proposed by Hahn [103]. To see how this experiment works, consider the spin echo sequences shown in Fig 53. The first case, Fig 53a, describes Hahn's original echo experiment [103]. In a heteronuclear case (always observing the rare spin) any heteronuclear dipolar interactions (along with the chemical shift and inhomogeneous line broadening processes) will be refocused as a result of this sequence. The second example, Fig 53b, is Hahn's SEDOR experiment [74]. Here, the heteronuclear dipole-dipole interaction is *not* refocused because of the application of the second π pulse to

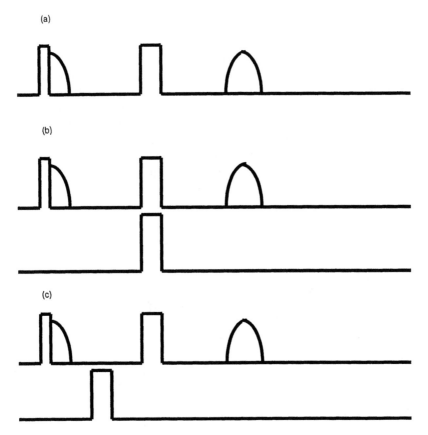

Figure 53 (a) Hahn's original spin echo experiment. The pulse spacing is τ. (b) The SEDOR experiment as proposed by Kaplin and Hahn. Again the pulse spacing in τ. (c) The SEDOR experiment as proposed by Slichter, et al. [74]. The abundant spin π pulse here is at t_1. All times begin at the end of the first pulse.

the abundant spins. The signal at the top of the echo is devoid of chemical shift information and the influence of inhomogeneous line broadening effects. The modern equivalent of the Hahn's SEDOR experiment was proposed by Slichter and coworkers [74]; it is shown in Fig 53c. Here, just the top of the echo is measured. That point is incremented in time by changing the position, t_1, of the π pulse applied to the abundant spins. The resulting signal can be described by the following equation:

$$S(t_1) = \int_0^\pi (\cos (Dt_1(3\cos^2\theta - 1)))\sin\theta d\theta$$

where D is given as

$$D = k \ \frac{\gamma_I \gamma_S}{r^3} \ h,$$

and k is either 1 or 3/2 depending on whether the interaction is heteronuclear or homonuclear, respectively. The remaining constants have their usual meaning.

An example of this approach is the determination of r_{CO} in CO adsorbed to Pd particles [98,104]. Here the $\pm 1/2$ transition of the ^{17}O resonance was observed. The SEDOR results are depicted in Fig. 54. The top figure shows the experimental data with a fit corresponding to 1.2Å; the bottom portion of the figure illustrates attempts to fit the data with 1.15Å and 1.25Å. Clearly, the fit corresponding to 1.2Å is much better, and one can expect that this technique will provide accurate determinations of bond distances.

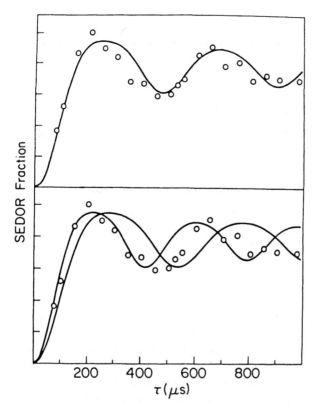

Figure 54 Bond length of CO chemisorbed on Pd measured from the time dependence of the SEDOR fraction. Top; fit with a bond distance of 1.20Å. Bottom; prediction for bond lengths of 1.15Å and 1.25Å. (From Ref. 98.)

Perhaps the solid state NMR investigation of surface adsorbed species with the highest impact was that of Wang, Slichter, and Sinefelt [74]. In this classic work they deduced the structure of an ethylidyne species ($\equiv$ C-CH$_3$) with a C-C bond length of 1.49 $\pm$ 0.02Å. They determined that the methyl group (-CH$_3$) of the ethylidyne freely rotates about the C-C direction at 77 K, and that carbon-carbon bond rupture occurs above 390 K, with the decomposition complete at about 480 K. Further, they demonstrated that the ethylidyne was not present on a supported Pt surface when acetylene was the percussor. However, it did form when ethylene was used. This ended several years of controversy with regard to the formation of ethylidynes on supported Pt surfaces. The collaboration of Slichter and Sinfelt represents a *tour de force* of expertly controlled surface chemistry and magnetic resonance spectroscopy at its best.

Another way to demonstrate the same chemistry and identify the surface species is to utilize the specific motional averaging of a methyl group as a signature of the ethylidyne. For this example Slichter and coworkers [105] chose solid-state ^{2}H NMR. For a static deuteron, the resonance frequency of a given isochromat, relative to the Zeeman frequency is given by the expression

$$\omega(\alpha,\beta) = \pm 1/2\omega_Q[(3\cos^2\alpha - 1) - \eta\sin^2\alpha \cos 2\beta]$$

Here α and β refer to powder angles, which relate the orientation of a given isochromat with respect to the applied magnetic field. The factor ω_Q is proportional to the quadrupole coupling constant for the deuteron. The remaining terms have their usual meaning. Consider now the possibility that the deuteron (assume for simplicity that $\eta = 0$) jumps between N sites about an axis which makes Euler angles θ and ϕ with respect to the principal axis system (PAS) of the deuteron. The details of the transformations are involved and are given elsewhere [106]. To be specific, for the case of a methyl group the angle θ is 70.528779°, i.e. 180° $- \theta_{tetra}$, and ϕ can take on values 0°, 120°, and 240°, i.e. ϕ is the jump angle in this example. In the fast motion limit, the final expression becomes

$$\omega(\alpha,\beta,70.528 \ldots) = \pm 1/6\omega_Q[3\cos^2\alpha - 1]$$

Hence, the motionally averaged line shape is the same as in the static case (when $\eta = 0$), except that it has been scaled by a factor of 1/3. Klug, Slichter, and Sinfelt [105] utilized this information in the investigation of the conversion of vinyldene to ethylidyne. Again, because of the relatively low gyromagnetic ratio of the deuteron the line broadening caused by the surface was small compared to the motionally averaged quadrupole coupling constant of the deuteron, i.e., $\sim$42 kHz. Similar results have been found by Chin and Ellis [107].

Another classic work in the area of the solid-state NMR of dispersed metal systems is the work of Duncan and Zilm [100]. In this work CO was examined on samples of rhodium supported on a silica. CO adsorbs on oxide-supported transition metals in three forms linear, bridge-bonded, and multicarbonyls. As

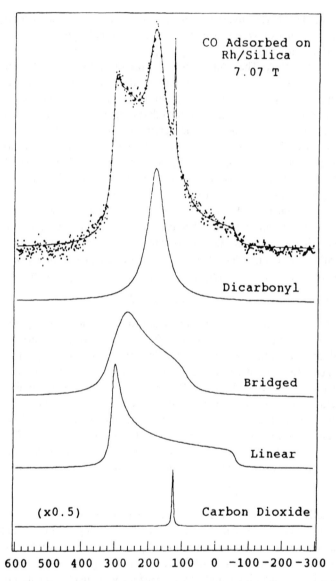

Figure 55 ^{13}C NMR spectrum of CO adsorbed on silica-supported Rh measured at 7.07T with a stationary sample. The solid line through the data is the sum of the four components that are plotted separately. (From Ref. 100.)

```
   O              O
   |||            |||         O              O
   C              C           \\\           ///
   |             /  \           C           C
   M  ─── ───M ──── M ───       \          /
                                  \       /
                                     M
 Linear        Bridging         Multicarbonyl
```

depicted in the static samples' ^{13}C NMR spectra shown in Fig. 55, Duncan, Zilm and coworkers [100] were able to quantify each of these species on the supported rhodium surface. Further, by careful comparison of static powder spectra and spectra obtained by magic angle spinning, they concluded that the resonance of the bridge-bonded CO is not sufficiently narrowed by MAS. They demonstrated that the relative populations in high resolution (MAS) spectra are inconsistent with independently calibrated broad line spectra. The main rationale for the discrepancies is postulated to arise for two reasons. First, they argue that the bridge-bonded species is not a well defined structure, i.e., it is disordered on at the surface. Secondly, Zilm and Duncan propose two possible sources of further broadening, i.e., irregularities in the metal surface and particle size distribution. The bridging sites should be more sensitive to differences in geometries; hence, the incomplete averaging of the sites by MAS methods.

ACKNOWLEDGMENTS

The authors gratefully acknowledge partial support by National Science Foundation Grants CHE-9021003 and CHE-8921632 of the Colorado State University and University of South Carolina research described in this article.

REFERENCES

1. J. F. Bradzil, in Characterization of Catalytic Materials (I.E. Wachs, ed.), Butterworth-Heinemann, Stoneham, MA, 1992.
2. A. W. Sleight in *Solid State Chemistry*, Vol. 2 (A. K. Cheetham, ed.), Oxford, Univ. Press, Oxford, 1992, p. 166.
3. R. K. Grasselli and J. F. Bradzil, eds., *Solid State Chemistry in Catalysis*, American Chemical Society, Washington, DC, 1985.
4. J. Evans in *Silanes, Surfaces and Interfaces* (D. E. Leyden, ed.), Gordon and Breach, New York, 1986, p. 203.
5. R. K. Iler, *The Chemistry of Silica. Solubility, Polymerization, Colloid and Surface Properties, and Biochemistry*, Wiley-Interscience, New York, 1979.
6. J. W. Newsome, H. W. Heiser, A. S. Russell, and H. C. Stumpf, Alumina

Properties, Technical Paper No. 10, Second Revision, Alumina Company of America, 1960.

7. G. Engelhardt and D. Michel, *High Resolution Solid-State NMR of Silicates and Zeolites*, John Wiley, New York, 1987.

8. P. C. Hammel, P. L. Kuhns, O. Gonen, and J. S. Waugh, *Phys. Rev. B 34*: 6543 (1986).

9. R. A. Wind, M. J. Duijvestijn, C. van der Lugt, A. Manenschijn, and J. Vriend, *Progr. NMR Spectrosc. 17*: 33 (1985).

10. D. Raftery, H. Long, T. Meersmann, P. J. Grandinetti, R. Leven, and A. Pines, *Phys. Rev. Lett. 66*: 584 (1991).

11. A. Pines, W. G. Gibby, and J. S. Waugh, *J. Chem. Phys. 59*: 569 (1973).

12. G. E. Maciel and D. W. Sindorf, *J. Am. Chem. Soc. 102*: 7606 (1980).

13. D. W. Sindorf and G. E. Maciel, *J. Phys. Chem. 86*: 5208 (1982).

14. D. W. Sindorf and G. E. Maciel, *J. Am. Chem. Soc. 105*: 1487 (1983).

15. G. E. Maciel, D. W. Sindorf, and V. J. Bartuska, *J. Chromatogr. 205*: 438 (1981).

16. D. W. Sindorf and G. E. Maciel, *J. Am. Chem. Soc. 103*: 4263 (1981).

17. D. W. Sindorf and G. E. Maciel, *J. Am. Chem. Soc. 105*: 3767 (1983).

18. I-S. Chuang, D. R. Kinney, and G. E. Maciel, *J. Am. Chem. Soc. 115*: 8695 (1993).

19. B. C. Gates, J. R. Katzer, and G. C. A. Schuit, *Chemistry of Catalytic Processes*, McGraw-Hill, New York, 1979, pp. 249–260.

20. H. Knozinger and P. Ratnasamy, *Catal. Rev. Sci. Eng. 17*(1): 31 (1978).

21. P. D. Majors and P. D. Ellis, *J. Am. Chem. Soc. 109*: 1648 (1987).

22. A. S. Brengle, E. Sobrante, and H. R. Stewart: U.S. Patent 2,709,173; D. K. Sacken: U.S. Patent 2,671,764; R. P. Nielson: U.S. Patent 3,702,259; R. P. Nielson and J. H. LaRochelle: U.S. Patent 3,962,136; 4,010,116.

23. D. E. O'Reilly, *Adv. Catal. 12*: 31 (1960).

24. B. A. Huggins and P. D. Ellis, *J. Am. Chem. Soc. 114*: 2098 (1992).

25. S. Vega, *Phys. Rev. A 23*: 3152 (1981).

26. T. H. Walter, G. L. Turner, and E. Oldfield, *J. Magn. Reson. 76*: 106 (1988).

27. R. K. Harris and G. J. Nesbit, *J. Magn. Reson. 78*: 245 (1988).

28. D. E. Woessner, *Z. Phys. Chem.* (Munich) *152*: 51 (1987).

29. H. D. Morris and P. D. Ellis, *J. Am. Chem. Soc. 111*: 6045 (1989).

30. H. D. Morris, S. Bank, and P. D. Ellis, *J. Phys. Chem. 94*: 3121 (1990).

31. J. C. Edwards and P. D. Ellis, *Magn. Reson. Chem. 28*: S59 (1990).

32. S. R. Hartmann and E. L. Hahn, *Phys. Rev. 128*: 2042 (1962).

33. M. Mehring, *High Resolution NMR in Solids*, Springer, Berlin, 1983, p. 135.

34. A. J. Vega, *J. Magn. Reson. 96*: 50 (1992).

35. A. J. Vega, *Solid State Nuclear Magn. Reson. 1*: 33 (1992).

36. K. T. Muller, B. Q. Sun, G. C. Chingas, J. W. Zwanziger, T. Terao, and A. Pines, *J. Magn. Reson. 86*: 470 (1990).

37. K. T. Muller, G. C. Chingas, and A. Pines, *Rev. Sci. Instrum. 62*: 1445 (1991).

38. K. T. Muller, E. W. Wooten, and A. Pines, *J. Magn. Reson. 92*: 620 (1991).

39. A. Bax, N. M. Szeverenyi, and G. E. Maciel, *J. Magn. Reson. 52*: 147 (1983).

40. M. Sardashti and G. E. Maciel, *J. Magn. Reson.* 72: 467 (1987).
41. P. D. Ellis, F. D. Doty, A. S. Lipton, and M. Whitlow, in preparation.
42. E. Oldfield and R. J. Kirkpatrick, *Science* 227: 1537 (1985).
43. C. E. Bronnimann, R. C. Zeigler, and G. E. Maciel, *J. Am. Chem. Soc. 110*: 2023 (1988).
44. C. E. Bronnimann, I-S. Chuang, B. L. Hawkins, and G. E. Maciel, *J. Am. Chem. Soc. 109*: 1562 (1987).
45. H. Pfeifer, D. Freude, and M. Hunger, *Zeolites 5*: 274 (1985).
46. H. Pfeifer, *Colloids and Surfaces 36*: 169 (1989).
47. C. E. Bronnimann, B. L. Hawkins, M. Zhang, and G. E. Maciel, *Anal. Chem. 60*: 1743 (1988).
48. G. E. Maciel, C. E. Bronnimann, and B. L. Hawkins, in *Advances in Magnetic Resonance: The Waugh Symposium*, Vol. 14 (W. S. Warren, ed.), Academic Press, San Diego, CA, 1990, pp. 125–150.
49. R. Eckman, *J. Chem. Phys.* 76: 2767 (1982).
50. Z. Luz and A. J. Vega, *J. Phys. Chem. 91*: 374 (1987).
51. J. S. Waugh, L. M. Huber, and V. Haeberlen, *Phys. Rev. Lett.* 20: 180 (1986).
52. D. P. Burum and W. K. Rhim, *J. Chem. Phys.* 71: 944 (1979).
53. E. R. Andrew, *Progr. NMR Spectrosc.* 8: 1 (1971).
54. S. F. Dec, C. E. Bronnimann, R. A. Wind, and G. E. Maciel, *J. Magn. Reson.* 82: 454 (1989).
55. J. Bohm, D. Finkze, and H. Pfeifer, *J. Magn. Reson.* 5: 197 (1983).
56. J. G. Hexem, M. H. Frey, and S. J. Opella, *J. Chem. Phys.* 77: 3847 (1982).
57. M. Hunger, D. Freude, H. Pfeifer, H. Bremer, M. Jank, and K.-P. Wendlandt, *Chem. Phys. Lett. 100*: 29 (1983).
58. L. B. Schreiber and R. W. Vaughan, *J. Catal.* 40: 226 (1975).
59. G. E. Maciel, C. E. Bronnimann, R. C. Zeigler, I-S. Chuang, D. R. Kinney, and E. A. Keiter, in *The Colloid Chemistry of Silica, Advances in Chemistry 234* (H. Bergna, ed.), American Chemical Society (in press).
60. I-S. Chuang, D. R. Kinney, C. E. Bronnimann, R. C. Zeigler, and G. E. Maciel, *J. Phys. Chem. 96*: 4027 (1992).
61. A. J. Vega, *J. Am. Chem. Soc. 110*: 1049 (1988).
62. D. P. Burum and A. Bielecki, *J. Magn. Reson.* 94: 645 (1991).
63. C. E. Bronnimann, C. F. Ridenour, D. R. Kinney, and G. E. Maciel, *J. Magn. Reson.* 97: 522 (1992).
64. D. W. Sindorf and G. E. Maciel, *J. Phys. Chem.* 87: 5516 (1983).
65. G. S. Caravajal, D. E. Leyden, G. R. Quinting, and G. E. Maciel, *Anal. Chem. 60*: 1776 (1988).
66. W. E. Rudzinski, T. L. Montgomery, J. E. Frye, B. L. Hawkins, and G. E. Maciel, *J. Catalysis 98*: 444 (1986).
67. L. Bemi, H. C. Clark, J. A. Davis, C. A. Fyfe, and R. E. Wasylishen, *J. Am. Chem. Soc. 104*: 438 (1982).
68. J. T. Cheng and P. D. Ellis, *J. Phys. Chem.* 93: 2549 (1989).
69. A. N. Garroway, *J. Magn. Reson.* 28: 365 (1977).
70. B. A. Huggins and P. D. Ellis, in preparation.

71. E. A. Carter and W. A. Goddard, III, *Surf. Sci. 209*: 243 (1989).
72. J. Wang and P. D. Ellis, *J. Am. Chem. Soc. 113*: 9675 (1991).
73. K. Kuchitsu, *J. Chem. Phys. 44*: 906 (1966).
74. D. E. Kaplin and E. L. Hahn, *J. Phys. Radium 19*: 821 (1958); P. K. Wang, C. P. Slichter, and J. H. Sinfelt, *Phys. Rev. Lett. 53*: 82 (1984); S. E. Shore, J-Ph. Ansermet, C. P. Slichter, and J. H. Sinfelt, *Phys. Rev. Lett. 58*: 953 (1987).
75. H. P. Bonzel, *Surface Science Rep. 8*: 43 (1987).
76. N. D. Lang, in *Solid State Physics* (H. Ehrenreich, F. Seitz, and D. Turnbull, eds.), Academic Press, New York, 1973, pp. 225–300.
77. J. Wang and P. D. Ellis, in preparation.
78. J. C. Edwards and P. D. Ellis, *Magn. Reson. Chem. 28*: S59 (1990).
79. J. T. Cheng, J. C. Edwards, and P. D. Ellis, *J. Phys. Chem. 94*: 553 (1990); W. P. Power, R. E. Wasylishen, S. Mooibroek, B. A. Pettitt, and W. Danchura, *J. Phys. Chem. 94*: 591 (1990).
80. J. C. Edwards, R. D. Adams, and P. D. Ellis, *J. Am. Chem. Soc. 112*: 8349 (1990); J. C. Edwards and P. D. Ellis, *Langmuir 7*: 2117 (1990).
81. I. D. Gay and S. H. Liang, *J. Catal. 44*: 306 (1976); I. D. Gay, *J. Catal. 48*: 430 (1977); S. H. Liang and I. D. Gay, *J. Catal. 66*: 294 (1980).
82. W. H. Dawson, S. W. Kaiser, P. D. Ellis, and R. R. Inners, *J. Phys. Chem. 86*: 867 (1982).
83. J. A. Ripmeester, *J. Am. Chem. Soc. 105*: 2295 (1983).
84. G. E. Maciel, J. F. Haw, I-S. Chuang, B. L. Hawkins, T. A. Early, D. R. McKay, and L. Petrakis, *J. Am. Chem. Soc. 105*: 5529 (1983).
85. J. F. Haw, I-S. Chuang, B. L. Hawkins, and G. E. Maciel, *J. Am. Chem. Soc. 105*: 7206 (1983).
86. P. D. Majors and P. D. Ellis, *J. Am. Chem. Soc. 109*: 1648 (1987).
87. P. D. Majors, T. E. Raidy, and P. D. Ellis, *J. Am. Chem. Soc. 108*: 8123 (1986).
88. W. H. Dawson, S. W. Kaiser, P. D. Ellis, and R. R. Inners, *J. Am. Chem. Soc. 103*: 6780 (1981).
89. J. H. Lunsford, W. P. Rothwell, and W. Shen, *J. Am. Chem. Soc. 107*: 1540 (1985).
90. L. Baltusis, J. S. Frye, and G. E. Maciel, *J. Am. Chem. Soc. 109*: 40 (1987).
91. L. Baltusis, J. S. Frye, and G. E. Maciel, *J. Am. Chem. Soc. 108*: 7119 (1986).
92. V. M. Mastikhin, I. L. Mudrakovsky, and S. V. Filimonova, *Chem. Phys. Lett. 149*: 175 (1988).
93. D. Michel, A. Germanus, and H. Pfeifer, *J. Chem. Soc. Far. Soc. 78*: 237 (1982).
94. L. Juenger, W. Meiler, and H. Pfeifer, *Zeolites 2*: 310 (1982).
95. A. Michael, W. Meiler, D. Michel, and H. Pfeifer, *J. Chem. Soc. Far. Trans. 1 82*: 3053 (1986).
96. C. P. Slichter, *Sur. Sci. 106*: 382–396 (1981); C. P. Slichter, *Ann. Rev. Phys. Chem. 37*: 25–51 (1986).
97. Z. Wang, J-Ph. Ansermet, C. P. Slichter, and J. H. Sinfelt, *J. Chem. Soc. Faraday Trans. 84*: 3785–3802 (1988).
98. J-Ph. Ansermet, C. P. Slichter, and J. H. Sinfelt, *Prog. in NMR Spectroscopy 22*: 401–421 (1990).

99. C. A. Klug, C. P. Slichter, and J. H. Sinfelt, ACS Symp. Series, "Surface Science of Catalysis," No. 482, 219–229 (1991).
100. T. M. Duncan, K. W. Zilm, D. M. Hamilton, and T. W. Root, *J. Phys. Chem. 93*: 2583–2590 (1989).
101. K. W. Zilm, L. Bonneviot, G. L. Haller, O. H. Han, and M. Kermarec, *J. Phys. Chem. 94*: 8495–8498 (1990).
102. K. W. Zilm, L. Bonneviot, D. M. Hamilton, G. G. Webb, and G. L. Haller, *J. Phys. Chem.* 1463–1472 (1990).
103. E. L. Hahn, *Phys. Rev. 80*: 580–594 (1950).
104. L. R. Becerra, C. P. Slichter, and J. H. Sinfelt, *J. Phys. Chem. 97*: 10–12 (1993).
105. C. A. Klug, C. P. Slichter, and J. H. Sinfelt, *J. Phys. Chem. 95*: 7033–7037 (1991).
106. M. S. Greenfield, A. D. Ronemus, R. L. Vold, R. R. Vold, P. D. Ellis, T. E. Raidy, *J. Magn. Reson. 72*: 89–107 (1987).
107. Y. H. Chin and P. D. Ellis, *J. Amer. Chem. Soc. 115*: 204–211 (1993).

6

NMR of Layered Materials for Heterogeneous Catalysis

Grant W. Haddix and Mysore Narayana

Shell Development Company, Houston, Texas

I. INTRODUCTION: LAYERED SOLIDS AS HETEROGENEOUS CATALYSTS

Layered solids have been of catalytic interest for quite some time. Materials such as clays, for example, were used over 50 years ago for the cracking of heavy petroleum feedstocks to gasoline-range products. The advent of zeolites led to the demise of these early systems, but then the U.S. oil crisis of the 1970s rekindled research since it was perceived that clays could, in principle, be used as essentially "very large-pore zeolites" for catalytic cracking. This revival of interest has brought several important findings, among which is the method of including inorganic "pillars" between clay layers via ion exchange. The technique was developed in the mid- to late 1970s, and NMR has played a significant role in characterizing the clay lattice (including the pillars), in the charge-compensating cations, and various compounds intercalated or adsorbed in the clay. Structure, orientation, and dynamics of adsorbates on clays have been studied in addition.

In the course of this chapter we will treat a number of other novel materials as well. Solids such as zirconium phosphates also possess layered structures and represent some of the newest materials being explored as heterogeneous catalysts. Although the zirconium phosphates have not yet been as widely studied by NMR as the clays, we review here some of the salient work done so far. This subject promises to be a thriving area of research in the near future.

Continuing the theme of "very large-pore zeolite emulators," we also review recent work on aluminum phosphates. While not necessarily layered materials, these systems attempt to fulfill similar requirements as the clays and zirconium phosphates. Important NMR measurements have recently been made on the aluminum phosphates, and this is a current area of interest for many catalytic chemists.

MoS_2 is a layered material present in most hydrotreating catalysts used today. Unfortunately, owing to the paramagnetic nature of Mo in MoS_2, NMR of the Mo or any adsorbates is virtually impossible, although ESR has been applied with success to this system. Other metal sulfides more amenable to analysis by NMR may also be promising from a catalytic standpoint. Some metal sulfides have been investigated using solid-state NMR methods, and we attempt to include those studies relevant to heterogeneous catalyis.

Graphite is not tremendously interesting as a catalyst itself, but there have been a few studies in which graphite is intercalated with an appropriate active site and used as a catalyst. There is a large body of research mainly from the surface physics community pertaining to thermodynamics of adsorbed layers on graphite. Some of this work involves NMR. We do not comprehensively review all papers in this area, but we do point out some that are perhaps of catalytic interest.

II. STUDIES OF CLAYS

A. Clay Structure

Before delving into NMR measurements on clay systems, it is worthwhile to review clay structure very briefly and point out some of the reference material on this subject. An excellent review was published in 1988 that focuses mainly on the catalytic properties and structure of pillared clays [1].

Clays are made up of layers of oxide sheets stacked much like a deck of cards. One of the major building blocks is a sheet of silicon oxide tetrahedra arranged in a hexagonal pattern of Si atoms. Another major building block is a sheet of octahedrally coordinated aluminum oxide. The aluminum- and silicon-containing sheets can bond to each other, forming what is known as a 1:1 clay. These types of clay structures, typically called kaolins, have little or no cation exchange capacity and are not able to intercalate water. Such systems are interesting from an intercalation/adsorption standpoint, since many organic molecules can be successfully intercalated in between the oxide layers. Most clays of catalytic interest are "2:1" clays, which involve, in the most general sense, a layer of octahedral aluminum oxide sandwiched between tetrahedral silicon oxide layers (Fig. 1). This composition leads to $(Si_8)_{tet}(Al_4)_{oct}O_{20}(OH)_4$ as the most general formula. Trioctahedral 2:1 clays have all possible positions in the octahedral layer occupied, while dioctahedral 2:1 clays have only two thirds of the octahedral sites filled. The stoichiometry of oxide layers in 2:1 clays typically

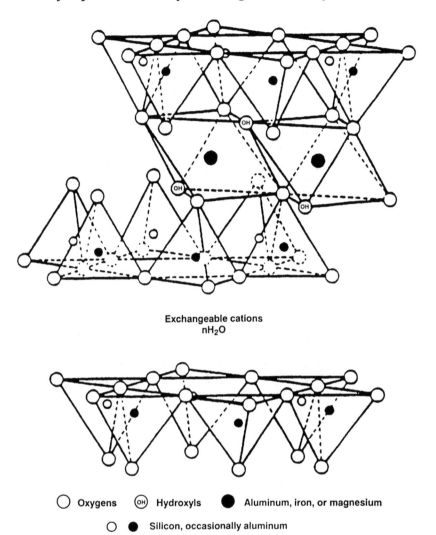

Exchangeable cations
nH₂O

O Oxygens   (OH) Hydroxyls   ● Aluminum, iron, or magnesium

O ● Silicon, occasionally aluminum

Figure 1 Parent structure of many 2:1 clays. (Courtesy Shell Development Company, Houston, TX)

requires that some interlamellar cations be present. Almost any monovalent, divalent, or trivalent cation can be used as a charge compensator. The nomenclature of clays is given in Table 1.

The smectite and vermiculite clays are particularly interesting from a catalytic standpoint since their lamellae are typically held together by hydrated cations that can be exchanged for other cations. The idea of exchanging cations in clays

Table 1 Nomenclature of Clay Types (Phyllosilicates)

Group name	Layer type	Charge (x)	Interlayer type	Octahedral occupancy	Subgroup	Species
Serpentine-kaolin	1:1	v. low or zero ($x \approx 0$)	—	tri- / di-	serpentines / kaolins	chrysotile / kaolinite, dickite
Talc-pyrophyllite	2:1	zero	—	tri- / di-	talcs / pyrophyllites	talc / pyrophyllite
Smectite	2:1	low ($0.6 < x < 1.2$)	hydrated cation (may be monovalent)	tri- / di-	saponites / montmorillonites	saponite, hectorite / montmorillonite, beidellite
Vermiculite	2:1	medium ($1.2 < x < 1.8$)	hydrated cation (usually Mg)	tri- / di-	trioctahedral vermiculites / dioctahedral vermiculites	trioctahedral vermiculite / dioctahedral vermiculite
Mica	2:1	high ($x \approx 2$)	unhydrated cation (usually K)	tri- / di-	trioctahedral / dioctahedral micas	phlogopite, biotite / muscovite, illite
Brittle mica	2:1	very high ($x \approx 4$)	unhydrated cation (divalent)	tri- / di-	trioctahedral brittle micas / dioctahedral brittle micas	clintonite / margarite
Chlorite	2:1	variable	hydroxy-octahedral layer	tri- / di-	trioctahedral chlorites / dioctahedral chlorites	chlinochlore / donbassite
Sepiolite-palygorskite (inverted ribbons)	2:1	variable	any	tri- / di-	sepiolites / palygorstites	sepiolite / palygorskite

Source: Ref. 1.

to prop open the interlamellar region was demonstrated in the 1950s [2], but the idea of using large inorganic cations as "pillars" was not developed until the mid-1970s [3]. This effort was the result of rising interest in using clays as cracking catalysts. The advent of solid-state NMR techniques for studying ^{29}Si and ^{27}Al [4,5] directly followed the inorganic pillar development, and thus it was only a short time before the first structural studies of pillared clays were made using these spectroscopic methods. Of course, there are many other structural studies of clays present in the literature for which the focus is not exclusively catalysis. We have made an effort to include reviews of those papers that may be of interest to the catalytic chemist.

It should be noted that, at this point, it is unlikely that pillared clays will replace zeolites for fluidized catalytic cracking, the reason being the hydrothermal instability of the clays at the conditions typically used in a modern riser cracker. Nevertheless, there is ongoing interest in clays and pillared clays as shape-selective catalysts for other, more specific reactions or separations.

B. Structural Studies of Clays and Pillared Clays Using ^{29}Si and ^{27}Al NMR

Structural investigations of clays pillared by the $(Al_{13})^{7+}$ cation using ^{29}Si and ^{27}Al NMR were reviewed in 1988 [6]. Results included in that review will be briefly covered here.

For the most part, structural questions that arise in clays and pillared clays are essentially the same as those that arise in solid-state NMR of silicates and zeolites. The pillaring process, especially, has been of great interest, and structural issues related to pillaring (at least for $(Al_{13})^{7+}$ as pillars) are nicely outlined by Fripiat [6].

It appears that the first solid-state NMR study using ^{29}Si NMR on a clay was performed in 1980 by Lippmaa and coworkers [7]. This work showed that the now commonly observed $Si(OSi)_{4-n}$ peaks in zeolitic aluminosilicates are clearly resolved in clays (talc) also.

Watanabe and coworkers [8] explored the effect of the inclusion of Fe^{3+} on ^{29}Si spectra of clays. ESR was used to measure the concentration of various paramagnetic species in samples of several 1:1 and 2:1 clays. At that early stage, it was presumed that clays high in Fe content would be very difficult to analyze using ^{29}Si MAS NMR, since virtually all the Si is found in the tetrahedral silica sheets (except in beidellites, for which Al substitutes for Si, much as in zeolites). It was feared that Q(OAl) in any clay would look virtually the same if much line broadening resulted from paramagnetic centers. It is certainly true that inclusion of Fe^{3+} in the clay sheets produces shorter relaxation times, but what was generally found was that the linewidths did not suffer to the point of making shift measurements useless. In fact, the paramagnetic centers make the clays easier to study owing to shortened T_1s for Si in the lattice. Normally (in quartz or a zeolite), lattice vibrations are the only effective means of ^{29}Si longitudinal relaxation, but clays containing Fe^{3+} are able to relax much more quickly. On the other hand, at some concentration, chemical information is lost because of broadening by the paramagnetic centers, and thus some clays cannot be studied successfully with NMR. Other early investigations [9,10] were concerned with the presence of paramagnetic centers in imogolite and kaolin samples and their effect on the ^{29}Si MAS spectra.

One of the first measurements using ^{27}Al MAS for clays was performed on talcs and micas [11]. Tetrahedral (Al_t) and octahedral (Al_o) aluminum were readily resolved, although poor agreement was observed for the expected ratio of Al_t to Al_o. The lack of agreement was attributed to second-order quadrupolar effects, a problem that has plagued such measurements severely and is routinely cited as the source of quantitation difficulties. Recent developments such as double angle rotation [15] undoubtedly will resolve this issue. Sanz and Serratosa [11] also used ^{29}Si MAS NMR on the talc and mica samples and concluded that the short-range order in clays could be readily studied using this approach also.

Another study by Sanz and Serratosa that same year [12] reached similar conclusions for some other 2:1 clays, but in addition they found that second nearest neighbors to the Al_o (namely the metals in adjacent octahedra) seemed to have a measurable effect on the Al chemical shifts. The clays studied were low in Fe ($<1\%$) and were purposefully chosen to represent three extremes: pyrophillite should have only Al_o, muscovite should contain both Al_o and Al_t, while the phlogopite should contain only Al_t. Despite the obvious difficulties of quantitation, it was found that for the trioctahedral clays (with few Al in the octahedral positions, mainly filled by Mg), the line for the octahedral Al shifted upfield to around 5–6 ppm, whereas for the dioctahedral clays (with only Al in the octahedra), a shift around 1–2 ppm was common. Differences were also observed for the Al_t depending on the composition of the octahedral layer. Thus it was clear that ^{27}Al NMR could be used for more than simply trying to determine the Al_t/Al_o ratio. There was no discussion in this brief paper concerning whether the observed shifts were indeed chemical shifts or second-order quadrupolar shifts.

Also in 1984 Thompson [13] studied several low-Fe clays. Again, broadening due to Fe^{3+} is recognized as a problem, as is the quantitation of Al types. This study also covered the effects of the degree of hydration, inorganic cation, and alkylammonium cations on the ^{29}Si spectra. One of the most interesting results is that a small chemical shift is observed upon intercalation of the organic cation, which is attributed to change in the hydrogen bonding of an oxygen bonded to the Si under study.

Goodman and Stucki [14] also investigated the quantitation issue for ^{27}Al. They found reasonable agreement between chemically derived Al speciation and ^{27}Al NMR.

Other early MAS-NMR studies avoided the paramagnetic issue by synthesizing clays hydrothermally [16–18]. Their conclusions were similar to previous conclusions: ^{29}Si is a powerful means of studying short-range order around Si in the lattice, while ^{27}Al NMR is somewhat useful, but plagued by broadening and thus quantitation problems. Samples studied were synthetic beidellite [16], synthetic mica montmorillonites [17], and a series of synthetic 2:1 trioctahedral clays [18].

^{29}Si solid-state NMR of 1:1 clays shows some interesting features. Barron et al. [10] attributed different Si signals in kaolinites to inequivalence induced by distortion of the tetrahedral layer. Later, Thompson [13] posited that the inequivalence is due to interlayer hydrogen bonding. This question was addressed again in 1987 [19], and it was concluded that the inequivalence of Si in the tetrahedral sheet is most likely attributable to hydrogen bonding at the surface of the silicate layer. Introduction of stacking faults between the tetrahedral and octahedral sheets using a hydrazine treatment resulted in no change in the ^{29}Si CP/MAS spectrum of kaolinite, and also the -OH stretch region of the infrared spectrum was essentially unchanged. The fact that hydrazine treatment induced stacking faults but the ^{29}Si spectrum did not change suggests that the resolution of two Si resonances is due to hydrogen bonding in the interlayer.

Another structural study of 1:1 clays was carried out by Komarneni et al. [20], the focus of which was order and disorder in the clay layers. XRD was used as a way to measure long-range order in some kaolins, while ^{27}Al and ^{29}Si NMR were used to study short-range order. Presumably, if all Al or Si sites have virtually the same surroundings, then one expects a single chemical shift or quadrupolar coupling (that is, no chemical shift or quadrupolar coupling dispersion) from many very similar sites. As noted by the authors, however, linewidths can be a dangerous means of quantifying short-range order in such systems.

A study using ^{29}Si and ^{27}Al MAS NMR on a variety of clay samples was performed by Kinsey et al. [21]. ESR was used on all of the samples to check for the presence of unpaired electrons due to Fe^{3+}. Since motional correlation times for the Fe^{3+} and for framework nuclei of interest are rather long, and since the electron relaxation is relatively very short (treat the electron as an average dipole creating a field at the nucleus, rather than as a spin-½ particle), the problem can be posed as basically one of magnetic susceptibility. That is, shifting and powder broadening can arise from the "external" field of the surrounding unpaired electrons. This model explains the large sideband envelope observed in many cases. It should also be noted that the authors saw little shifting as they changed the field, and thus the isotropic part of this interaction is believed to be small. We should note, that, in general, MAS does not necessarily "spin out" this interaction, because terms from the second-rank dipolar tensor (electron-nucleus dipolar) and g tensor (spin-orbit coupling of the electron) form products that are not averaged to zero when rendered time-dependent under MAS. In any case, some interesting trends have come out of this comprehensive study:

1. ^{29}Si resonances shift upfield with increasing Al in the tetrahedral layer. The ^{27}Al tetrahedral resonances show the same trend.
2. Al_t/Al_o ratios determined by NMR agree reasonably well with chemical analysis.
3. ^{29}Si spectra of clays with no Al_t or with ordered Al_t arrangements are generally the most narrow.
4. Paramagnetic impurities give rise to a wide spinning sideband envelope for most of the clays considered.

Weiss et al. [22] carried out another study of several clays using ^{29}Si and ^{27}Al MAS-NMR. They interpret the ^{29}Si shifts observed upon changing the octahedral layer composition as being at least partially due to the rotation of the SiO_4 tetrahedra that must take place so that the oxygens can be shared with the octahedral layer. The rotation can be measured using XRD methods, and some convincing correlations between ^{29}Si chemical shift and the rotation angle of the Si-containing tetrahedra result. There are other correlations drawn to parameters such as bond lengths in the tetrahedral layer, amount of Al in the octahedral layer, and total layer charge. From the data it is clear the ^{29}Si shifts are very much affected by the amount of structural distortion in the tetrahedral sheet and by the composition of the octahedral sheet. Basically, shear strain on the tetrahedral and octahedral sheets exists because the sheets are not completely commensurate as far as position of the shared oxygen atoms. These correlations can, in principle, be used to predict a variety of properties of the clays.

Continuing the theme of comprehensive studies, Woessner [23] recently carried out another major investigation of clay structure. The main improvement here is the use of high-speed MAS. Some of the spectra were collected at 9 kHz, which was significantly faster than for most spectra of clays published to that point. Several of the conclusions reached are in agreement with previous structural studies. One additional parameter that was extracted was the SOQE (second-order quadrupole effect), which is simply the isotropic part of the second-order quadrupolar interaction. The QCC (quadrupolar coupling constant) and η (asymmetry parameter) are defined in this paper as follows

$$SOQE = QCC \, (1 + \eta^2/3)^{1/2}$$

The second-order quadrupolar isotropic shift can be found by collecting MAS spectra at two different fields. The first moment of the central transition was measured, and Woessner shows that the determined SOQE values correlate well with parameters related to distortion in the tetrahedral sheet. In general the SOQE tends to increase with increasing Al content in the tetrahedral layer. Also, it is noted that quantitation attempts seem to improve greatly with the higher-speed spinning. Al_t/Al_o ratios determined from MAS-NMR at 9 kHz generally agree very well with those based on the structural formulae.

Attempts by de la Caillerie and Fripiat to alter the structure of a vermiculite through dealumination using $(NH_4)_2SiF_6$ [24] resulted in obvious structural changes. These authors also measured the cation exchange capacity (CEC) before and after dealumination and then calculated the expected change in CEC based on the structural changes observed by elemental analysis, XRD, and ^{29}Si and ^{27}Al NMR. A resonance assigned to pentacoordinated Al was observed in some cases at 30 ppm. It was also reported that dealumination does not cause large Al-containing cations to perform in situ pillaring. Some Al-containing cations are apparently formed that result in somewhat close basal spacing between sheets.

There have been several studies pertaining to the pillaring process. Solution NMR measurements have been used to study solution species that serve as pillaring agents, with most of the focus on identification of the $[Al_{13}O_4(OH)_{24}(H_2O)_{12}]^{7+}$ Keggenlike cation (Al_{13}) (Fig. 2) [25–28]. An excellent review of solution NMR of Al-containing species appeared recently in the literature [29], containing, in particular, a good section of Al hydrolysis. An especially interesting point concerning the Al_{13} ion is that typically only the tetrahedral Al is observed, even in solution NMR. The octahedral Al sites apparently experience a rather large quadrupolar coupling, and the time-dependent quadrupolar interaction due to tumbling of the ion presumably results in severe lifetime broadening of the octahedral resonance (Fig. 3).

The main reason for interest in the Al_{13} cation was that it is clearly a large cation. If it is possible to leave the species intact between clay layers, it will "prop" open the sheets (Fig. 4). We should recognize that the main thrust of developing pillared clays was for catalytic cracking applications. The Al_{13} cation

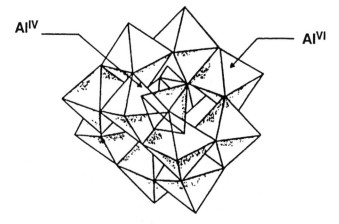

Figure 2 The Al_{13} Kegginlike cation. (From Ref. 6.)

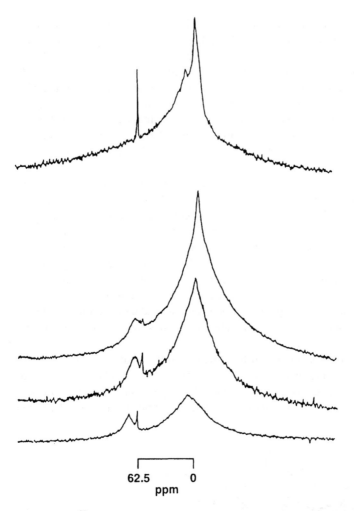

Figure 3 ^{27}Al solution spectra of almost fully hydrolyzed aluminum chloride solutions. The sharp peak at 62.5 ppm is due to the central AlO$_4$ unit of the Al$_{13}$ ion. (From Ref. 25.)

is roughly 8.4 Å in diameter, and thus separation between layers of this order were conceivable. Also, it was thought that perhaps the pillar itself may serve as a catalytic site owing to both Bronsted and Lewis acidity associated with it. 8.4 Å pores were considered quite large for a zeolitic material in the late 1970s, so it is clear why there was so much interest in these materials for cracking very heavy streams. Today, pore diameters of this size can be achieved in synthetic zeolites

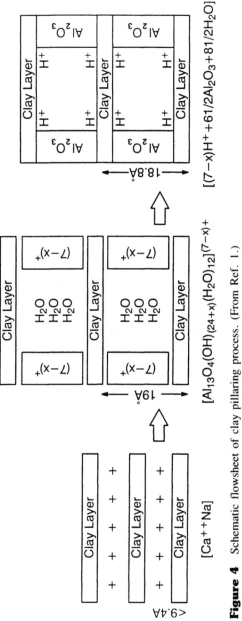

Figure 4   Schematic flowsheet of clay pillaring process. (From Ref. 1.)

or aluminophosphates, and thus the original interest in pillared clays for catalytic cracking has diminished.

The work of Fripiat and coworkers, and Pinnavaia and coworkers in the mid-1980s makes up most of the NMR literature pertaining to the pillaring process [30–32]. Plee et al. [30] were the first to apply ^{27}Al and ^{29}Si NMR to clays pillared with the Al_{13} cation. Up to that point, it had been assumed that the Al_{13} cation was indeed the pillaring species, but this assumption was based on indirect evidence such as solution NMR of the pillaring species and basal layer spacings from XRD. The study focused on pillared 2:1 clays. Hectorite and laponite with fully occupied silicate sheets and no Al in the octahedral layer were studied. A beidellite clay with some tetrahedral Al substituted for Si in the tetrahedral layers was also examined. The authors conclude from ^{27}Al MAS spectra that the pillaring species is indeed the Al_{13} cation, since octahedral Al is observed in the calcined product. Not surprisingly, the ratio of tetrahedral to octahedral Al measured by NMR is higher than expected, and the authors attribute this discrepancy to incomplete detection of some highly distorted octahedral Al. In the clays with no Al in the tetrahedral layer, it appears there is little reaction of the Al_{13} ion with the sheets. For the beidellite clay, however, it is clear that upon calcination the pillar is bonding to the sheets, most likely at tetrahedral Al sites (Fig. 5).

Pinnavaia [31] showed similarly that fully occupied silicate layers do not eliminate the possibility of reaction of the Al_{13} cation with the clay layers. A synthetic fluorohectorite is found to be pillared by Al_{13}, and, more importantly, it is apparent from the ^{29}Si and ^{27}Al spectra that structural rearrangements do occur for a clay that has Si fully occupying the silicate layer. Thus Si–O–Al linkages are formed between the cation and the silicate sheet (Fig. 6).

Later studies also showed ^{27}Al and ^{29}Si MAS NMR to detect covalent bonding between pillars and the clay sheets [32,33]. The work by Tennakoon et al. [33] is particularly interesting since it gives a good idea of the structure of the pillar after calcination. The conclusion based on spectra collected for a dried and calcined pillaring solution and for a calcined pillared hectorite (little Al in original clay) was that the pillar is basically alumina. Interestingly, they also determined that the pillar is actually covalently bonded to Al and Mg atoms in the clay octahedral layer. In addition, this study notes a curious effect on the observability of the octahedral Al in the Al_{13} ion. The octahedral peak is severely broadened in solution, but upon exchange into the hectorite the resonance sharpens considerably. Thus it is clear that the exchanged Al_{13} is either in a more symmetric environment between the clay sheets than it is in solution, or else it has a reduced tumbling rate so that rapid spin-lattice relaxation via time-dependent quadrupolar interaction is stopped. The first scenario seems more likely since the correlation time for isotropic tumbling of Al_{13} between the layers is undoubtedly long, the motion not being in the extreme narrowing limit.

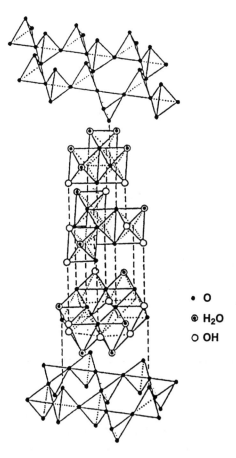

Figure 5 Expanded view of calcined pillared beidellite. Note the modified pillar structure that shares three oxygen atoms with inverted AlO_4 tetrahedra of the tetrahedral layers. (From Ref. 30.)

Though much of the interest in pillared clays as cracking catalysts seems to have waned by the early 1990s research is still going on to improve the stability of the materials. One parameter that has certainly been of great interest in catalyst preparation in the past is distribution of metals on a support pellet. How the metal is distributed strongly affects how the catalyst pellet responds under different kinetic and transport regimes. One of the major problems with pillared clays is their instability at high temperatures. It has been postulated that perhaps the instability is related to the distribution of pillars throughout clay aggregates. A large Al_{13} cation presumably could have a rather low diffusivity when confined to two-dimensional pores effectively smaller than its ionic radius. A report

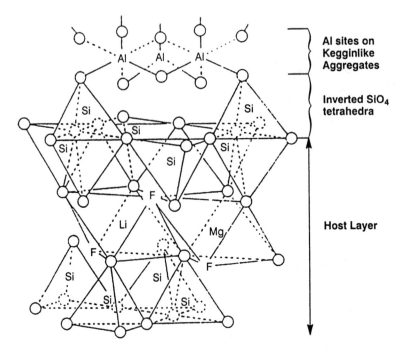

Al sites on
Kegginlike
Aggregates

Inverted SiO$_4$
tetrahedra

Host Layer

Figure 6   Structure of pillared fluorohectorite showing inverted SiO$_4$ tetrahedra. (From Ref. 31.)

by Figueras et al. [34] studied this aspect by intercalating the Al$_{13}$ cation in competition with NH$_4{}^+$ to give a more uniform pillar distribution in pillared montmorillonite. Solid-state NMR was not used to characterize the final materials, but solution NMR was employed to study the pillaring solutions. The net result was that competitive ion exchange appears to result in higher-stability, higher-acidity, and perhaps higher-surface-area pillared clays. Clearly there is room for further NMR measurements to investigate this improvement.

Another recent study shows the applicability of ^{1}H-^{27}Al cross-polarization (CP) MAS for examination of clay structure as well as drawing some interesting conclusions pertaining to the distribution of Fe^{3+} [35] in kaolin and montmorillonite clays. In some of the low-Fe samples considered, little or no tetrahedral Al signal was observed. This deficiency suggests the tetrahedral Al are close to Fe^{3+} centers. Also, for one of the montmorillonite samples studies, one would not expect to see octahedral aluminum because of the presence of Fe^{3+} in the octahedral layer, but nevertheless it is clearly observed. These results strongly point to a nonhomogeneous distribution of Fe^{3+} in the materials. It is also noted that clays provide a good system for testing quantitative aspects of ^{1}H-^{27}Al CP experiments, which are generally confounded by high levels of Fe^{3+}.

In some cases, ^1H $T_{1\rho}$ values are shortened to the point that CP is ineffective. The authors concluded that there are three types of octahedral Al in the low-Fe clays used in their experiments: nearby Fe, cross-polarizable internal, and surface Bronsted sites.

The question of the reaction site between pillar and the clay layers has been revisited recently, but for the case where the pillar is Si-based rather than Al-based [36]. Details of the Si-containing pillaring species are not given. The MAS-NMR and IR data suggest that there is little reaction as the pillars are exchanged for Na^+ cations, but there are structural rearrangements upon calcination. The authors suggest a mechanism similar to that proposed by Pinnavaia et al. [31] in which some SiO_4 tetrahedra are inverted (Fig. 6) and extend into the interlayer where they can react with the pillaring species. This study also claims that there are different tetrahedral Al sites in a rectorite sample, arising perhaps from some Fe in the tetrahedral layer.

Several other studies pertaining mainly to ^{29}Si and ^{27}Al MAS NMR of clays and other layered aluminosilicates are found in the literature, but the focus here is far afield from catalysis [37,38]. Many papers by soil and mineral researchers also report using NMR methodology, and there is, of course, an obvious connection between clay and soil science. A number of particularly interesting aluminosilicate structures that arise naturally have indeed been studied using MAS-NMR, some of which may be of catalytic interest. Several reports describe solid-state NMR on samples of imogolite [9,39–41], which is believed to have a characteristic tubular morphology (Fig. 7). It seems such a structure could be useful at least as a catalyst support for shape-selective reactions. An interesting result is that pentacoordinated Al may be formed as the imogolite structure is dehydroxylated. Another example would be the allophanes, which are naturally occurring aluminosilicates that seem to be made up of defective kaolinlike (curved layers) structures and also imogolitelike tubular structures. A model of the structure based on crystallographic and NMR data is presented in the literature [42]. Such morphologies certainly ought to be interesting from the standpoint of shape-selective catalysis.

C. Structural Studies of Clays and Pillared Clays Using Other Nuclei

There appear to be very few structural studies of clays involving nuclei other than ^{29}Si and ^{27}Al, perhaps for obvious reasons. Many clays, however, do include fluorine in place of some structural hydroxyls, and a study of ^{19}F second moments in sepiolite appeared recently [43]. The moment analysis indicates F mainly replaces internal structural OH between the octahedral and tetrahedral layers.

Evidently there are virtually no structural studies of Mg, Li, or other nuclei typically included in the octahedral layer framework. Also few studies seem to

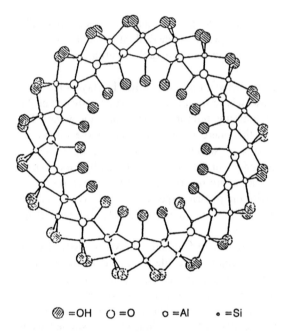

⊘ =OH   ◯ =O   o =Al   • =Si

Figure 7 Imogolite structure viewed down tube axis. (From Ref. 39.)

have been performed using ¹H MAS-NMR to probe Bronsted acidity as on zeolites [44]. In the next section we describe some ¹H NMR experiments, not necessarily dealing with acidity, and offer our apologies if any relevant work has been overlooked.

D. Studies of Cations in Clays

Cations in clays can be broken down into inorganic and organic classes. Having already covered NMR pertaining to large pillaring cations, here we will discuss work involving small exchangeable inorganic cations. The soil research community is interested in intercalation of organic cations in clays, and we will review some recent papers pertinent to the study of intercalated or adsorbed organic molecules.

Starting with the exchangeable inorganic cations, one of the earliest investigations of cations in a clay was carried out by Conard [45]. ¹H and ⁷Li static NMR measurements were made on hectorite and montmorillonite clays, and the observed spectra were attributed to trihydrated Li in the interlayer region. Details of the motion of the trihydrate species may be derived from the Li and H spectra as well. The EFG at the Li nucleus arising from coordinated water apparently changes drastically on cooling (Fig. 8).

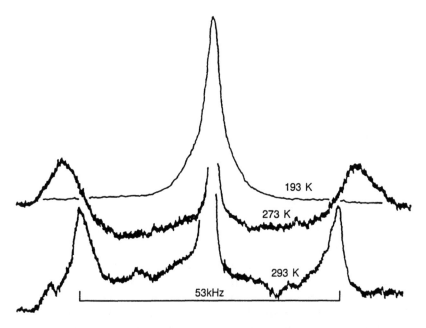

Figure 8 ^{7}Li spectra of hydrated Li$^+$-hectorite. The loss of the outer transitions upon cooling is attributed to loss of some motional averaging. The central transition broadens because of dipolar coupling of ^{1}H in water. (From Ref. 45.)

Ellis and coworkers [46], in a study of ^{113}Cd adsorbed on montmorillonite, point out that there are several places one would expect a metal ion to coordinate within the clay. Three possibilities include the interlayer region, within the sheets, and at edge positions (the two-dimensional sheets are obviously not infinite in extent). Their work then deals with Cd exchange between edge sites and the interlayer region. Based on experiments aimed at refocusing interactions such as chemical shift, the authors conclude that the ^{113}Cd linewidths observed are mainly due to the presence of Fe^{3+}. ESR is used to verify the electron relaxation and thus confirm that the ^{113}Cd resonance is broadened by Fe^{3+} in the octahedral layer. It is concluded that a broad component of the ^{113}Cd spectrum arises from cations in the interlayer region that are highly mobile and that interact with Fe^{3+} in the octahedral layer. The narrow component originates from ^{113}Cd coordinated at edge sites, where it is more strongly held in a manner similar to a defect in a solid rather than a hydrated cation.

A study by Fripiat and coworkers [47] covers ^{23}Na, ^{111}Cd, and ^{133}Cs in the interlayer region of a vermiculite sample. The main purpose was to examine chemical shifts of the cations under different states of hydration, with the hope that identification of different adsorption sites could help in understanding exchange selectivities. The results of the study are indeed quite useful. From XRD measurements to acquire basal spacings under different degrees of hydration and from corresponding static and MAS-NMR spectra of the cations, it is clear that the chemical shift of the cation is related to the degree of hydration of the clay. In general, a downfield shift of the resonances is observed as the degree of hydration increases. Other shift effects are accounted for when necessary (as, for example, second-order quadrupolar shift for ^{23}Na and ^{133}Cs). An argument is made that attributes this downfield shift basically to diamagnetic susceptibility (external shielding) arising from coordinated water molecules. This study also clearly points out the discreteness of water layers in the clay interlayer. It should be noted that the trends depicted are in agreement with some of the NMR results for hydrated cations on zeolites [48].

Weiss et al. [49] carried out another study of Cs$^+$ cations in clays using ^{133}Cs NMR on a variety of Cs-exchanged materials. The focus here was to measure the Cs chemical shift as a function of different treatments in order to determine what sites the Cs occupies. In 2:1 clays these are external basal surface sites, interlayer sites, and edge-interlayer sites. A variety of 2:1 clays with a Cs^{1+} cation under different states of hydration were studied. Samples were examined as a slurry in CsCl solution, at 100% relative humidity, at ~25% relative humidity, after partial dehydration at 100°C, and after drying at 450°C. The clay slurries all seem to show two Cs peaks, which are assigned to solution Cs and to Cs adsorbed on the clay. Nutation experiments were used in an earlier experiment [50] to determine that the quadrupolar interaction is fully averaged for the solution Cs, while only the central transition is observed for the adsorbed species. There are significantly different shifts observed for the adsorbed Cs from clay to clay. Motional averaging is clear for Cs down to a temperature of about -20°C. Also, there seems to be evidence for essentially two populations of Cs atoms in the interlayer: a tightly bound species adsorbed in the framework, and a hydrated species in the middle of the interlayer. The separation of the peaks sets upper limits on the exchange rate between them. Upon drying at 450°C, in the 2:1 clays there still appear to be two peaks, and the authors attribute these to different coordination sites for the Cs. Differences in the shifts for the dried samples show definite correlations with parameters such as Al$_t$, total layer charge, and tetrahedral rotation angle, much as the structural studies involving ^{29}Si and ^{27}Al [22,23].

There are several investigations of organic cations in clays in the literature, most of which originate from the soil research area. Many studies involving NMR were done by Mortland and coworkers [51]. A recent report involving alkylammonium cations in vermiculite and montmorillonite further depicts the kinds of interactions to expect when examining interlayer species [52]. ^{13}C MAS spectra show that tetramethylammonium cations are able to reorient essentially isotropically in the clays, whereas the larger hexadecyltrimethylammonium is restricted. The T_1 of the ^{13}C is apparently dominated by fixed paramagnetic impurities (Fe^{3+}) in the clays. The estimated electron T_1 is about 10^{-9} s, while τ_r, the tumbling correlation time of the TMA, is still considerably shorter. This situation produces an angular dependent paramagnetic shift, and the net spectroscopic result is many sidebands in the MAS spectra (beyond what one would expect for chemical shift anisotropy, for example, as noted in Ref. 21).

E. Studies of Water in Clays

Most studies of water in clays are not necessarily related directly to heterogeneous catalysis, but the state of the water in the interlayer is clearly important in the pillaring process. Woessner at Mobil pioneered this area 20 years ago. It was discovered that interlayer water in clays was undergoing anisotropic motion, the result being a partially motionally averaged Pake doublet spectrum [53–55]. Under typical hydration conditions, many clays appear to maintain the equivalent of one or two monolayers of water between the clay sheets, so it is not surprising that the motion is restricted.

Woessner [56] also considered the effects of proton exchange on the dipolar oscillations arising from oriented water in chabazite and vermiculite. This paper involves a nice application of the density matrix for calculation of the effect of exchange on the FID. Comparisons are made between the proton exchange rate measured for interlayer water and pure water, and it was found that the rate was very dependent on the clay structure and cation involved. Fripiat reviewed the area of proton exchange on acid catalysts in 1976 [59]. The findings of Woessner [56] are not in complete agreement with conclusions reached in this review, which assert that dissociation of water is more pronounced in layered materials. Woessner's results seem to indicate that this is not necessarily the case.

In a subsequent work, Woessner [57] used deuterium T_1s of D_2O adsorbed at different saturation levels on a hectorite sample to monitor water mobility in the vicinity of a clay surface. It is clear from the data that the first two monolayers of water on the clay sheets experience restricted motion.

A comprehensive review concerning NMR of water in clays published in 1982 by Stone [58] also covers much of the earlier work on the hydration state of various cations in the clays. It appears that most of the research on the state of water in clays using NMR was performed in the 1970s and early 1980s, but not much has been done in this field recently. There is undoubtedly some literature pertaining to water diffusion in clays, but we will not attempt to cover that area here.

F. Studies of Adsorbates in Clays

Studies of organic adsorbates in clays from the point of view of reactant molecules seem to be few, although the soil and mineralogy literature makes many such references. One phenomenon of general interest to soil researchers is the conversion of various types of humic molecules on or in the clays. In this sense, clays are possibly the most abundant heterogeneous catalysts on earth. We review here some of these papers in addition to more traditional applications of heterogeneous catalysts.

Resing and coworkers performed ^{13}C NMR measurements on a benzene/hectorite system [60]. A "single-crystal" study was performed in essence, since the clay platelets were preferentially oriented and then the clay/adsorbate system was rotated in the magnetic field. The benzene molecules were found basically to stand on edge in the interlayer space and undergo rotation about their hexad axes. They also seem to rotate about an axis normal to the clay layers, but this motion is frozen out at 77 K.

Fyfe et al. [61] studied *p*-xylene and γ-butyrolactone in hectorite in 1981 using ^{1}H and ^{13}C NMR with no MAS. The relatively good resolution of these spectra indicates a high degree of mobility of the adsorbates in all cases. The ^{13}C chemical shifts are nearly identical to those in solution spectra of intercalants, suggesting that magnetic susceptibility effects are not large, at least in this clay.

Clayden and Waugh used ^{31}P NMR to study the chemical shift tensor of a phosphate-containing nucleotide and showed that coordination of the nucleotide to Zn^{2+} in bentonite is via a single oxygen bridge [62].

Tennakoon and coworkers [63] used ^{13}C, ^{1}H, and ^{29}Si NMR to study *n*-pentanol dehydration and isobutene oligomerization on Gelwhite-L (montmorillonite), bentonite, and laponite. XRD was used to determine interacalation of the reactants, and the changes in the d_{001} spacings as a function of temperature give an indication of reaction in the interlayer region. Unfortunately, many of the interesting catalytic results are exhibited by the montmorillonite sample, which has a somewhat high Fe content, and thus the NMR of intercalated species is not very revealing. On the laponite sample, it is clear that addition of isobutene to a hydrated clay results in formation of *t*-butanol. Also, intercalation of methanol followed by addition of isobutene leads to formation of methyl-*t*-butyl ether. Clearly, the reactants and products must be relatively mobile since reasonable resolution can be achieved without MAS in the ^{13}C spectra.

Tennakoon and coworkers [64] also did a later study of some low-Fe hectorite and montmorillonite samples. These experiments included ^{29}Si and ^{27}Al MAS-NMR for examination of the clay sheets, ^{13}C NMR of adsorbates on the clays, and also ^{1}H NMR for examination of structural hydroxyls. Neither the ^{1}H nor the ^{13}C measurements of the adsorbate was made under MAS. Adsorption of 1-hexene on the Al^{3+}-exchanged hectorite sample followed by "high-resolution" ^{13}C NMR revealed no olefinic ^{13}C resonances. Inspection of the [27]Al MAS spectrum indicates an apparent change in the ratio of tetrahedral Al to octahedral Al. ^{1}H NMR of a deuterated sample (with D_2O replacing water of hydration of the clay) reveals that the olefin is indeed intact and therefore the olefinic end must be bonding to some site in the clay (Fig. 9). The main evidence for this would seem

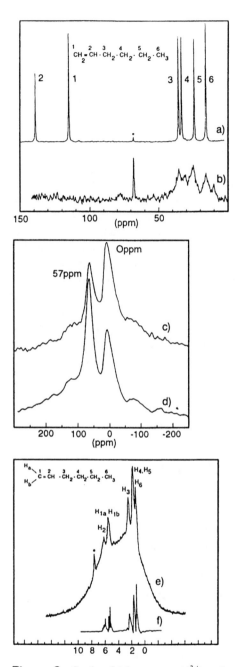

Figure 9 Study of 1-hexene on Al^{3+} exchanged Laponite: (a) ^{13}C spectrum of liquid 1-hexene; (b) ^{13}C spectrum of 1-hexene adsorbed on Al^{3+}-Laponite; (c) ^{27}Al MAS spectrum of Al^{3+}-Laponite prior to 1-hexene exposure; (d) ^{27}Al MAS spectrum of Al^{3+}-Laponite after 1-hexene exposure; (e) ^1H spectrum of 1-hexene on D_2O treated Al^{3+} Laponite; (f) ^1H spectrum of liquid 1-hexene. (From Ref. 64.)

to be the changes in the ^{27}Al spectra, since the behavior observed for the ^{13}C spectra can be explained by chemical shift anisotropy differences between aliphatic and olefinic ^{13}C's (the anisotropy being much larger for olefinic). This study is also one of the few in which ^{1}H MAS-NMR has been applied to structural hydroxyls of the clay after dehydration. Basically, all that is seen is a peak assigned to hydroxyls at 1.2 ppm.

Botto and coworkers [65] have investigated oxygen-methyl carbon bond cleavage in pillared bentonite, montmorillonite, and fluorhectorite by ^{13}C CP/ MAS and Bloch decay MAS. Reactant molecules studied were *m*-methyl anisole, 4-hydroxy-3-methoxy toluene, 4-phenoxy-3-methoxy toluene, and guaiacol, all of which are meant to model lignin. First, clay samples were intercalated with the reactant and spectra collected. The samples were then heated to 150°C for several hours, and the ^{13}C spectra were recorded after reaction. Figure 10 shows an example of a ^{13}C spectrum of 4-hydroxy-3-methoxy toluene before and after reaction. *O*-methyl cleavage clearly is occurring, and it is also clear that the mobility of the products is less than the reactants as evidenced by the increased linewidth and the failure of the reactants to cross-polarize. Rapid motion of the reactant molecules eliminates dipolar interactions and also the chemical shift dispersion that apparently exists for the products.

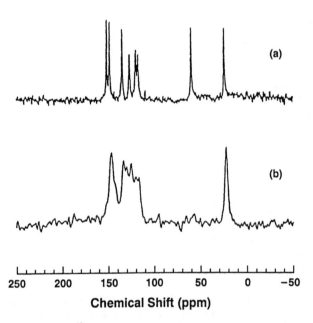

Figure 10 ^{13}C MAS spectra of 4-hydroxy-3-methoxy toluene on pillared bentonite: (a) prior to heating (no cross-polarization); (b) after heating to 150°C for 5 days (cross-polarization used). (From Ref. 65.)

O'Brien and coworkers [66], in a recent study of the binding of triethyl phosphate (TEP) in montmorillonite, have quantitatively measured the adsorption of the TEP using ^{1}H MAS. ^{31}P cross-polarization and Bloch decay spectra of the adsorbed TEP were also obtained. Two chemically shifted peaks are observed, one of which originates from bound TEP, while the other originates from isotropically tumbling TEP. Principal components of the chemical shift tensor as determined from the sideband intensities are consistent with a single oxygen bridging to the Mg^{2+} cation. This result is similar to the previous study of phosphates in clay [62]. ^{13}C CP and Bloch decay spectra indicate that the situation is more complex, however, than simply a mobile and immobile species. The methylene carbon resonance in the cross-polarization spectrum shows asymmetry at essentially all TEP loadings, which may provide evidence of another immobile species.

Several studies have been made of various intercalants of adsorbates on kaolinite [67–70]. Since most of these applications are somewhat far afield from catalysis, we have opted not to review them here.

III. STUDIES OF ZIRCONIUM PHOSPHATE

A. Structure of Zirconium Phosphates

Tetravalent metals form acid salts of the type $M(HXO_4)_2 \cdot nH_2O$ where X can be either phosphorus or arsenic, the metal atom being Zr, Ti, etc. These salts are known to exist in amorphous as well as crystalline forms, with the latter of more interest in heterogeneous catalysis because of their layered nature [71]. It is known from the early work of Clearfield and coworkers [72–75] with zirconia gels that the conditions of synthesis play a crucial role in the crystallinity as well as the nature of the final phosphate material. Amorphous gels with controlled amounts of crystallinity have been synthesized. Since catalytic applications depend on surface area, which in turn depends on the degree of crystallinity, the ability to control the latter becomes extremely important.

Several types of crystalline zirconium phosphates (henceforth referred to as ZrP) have been identified. The α phase, $Zr(HPO_4)_2 \cdot H_2O$, has a layer structure consisting of metal atoms in a plane bounded by the bridging phosphate groups above and below the plane of the metal atoms, with P-OH groups directed inwards between the layers (Fig. 11). The structure for the dihydrate, known as the γ phase [72b] is still being debated, even though it also exhibits ion-exchange capacity and possibly has a layered structure. Evidence for layering is based on an increase in the lattice spacing upon ion exchange [76,77]. The anhydrous-form ϵ phase also is supposed to have a layered structure. The resemblance of these zirconium phosphates to 2:1 clays arises from the sandwiching of the metal atoms between the phosphate groups. The metal atom is coordinated to six

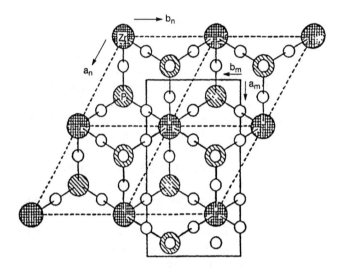

Figure 11 Idealized layer in α-zirconium phosphate, showing the relationship of the hexagonal cell (dashed lines) to the monoclinic cell (solid lines). (From Ref. 72b.)

different phosphate groups, an arrangement that requires the bridging of three metal atoms by each phosphate group since the fourth oxygen carries most of the charge and is associated with a hydrogen atom. The OH is pointed toward the interlayer space. For comparison with typical smectites containing octahedral metal ions and tetrahedral silicon units, the molecular formula for the α phase of ZrP can be written as $(Zr_4)_{oct}(P_8)_{tet}O_{24}(OH)_8$. The similarity ends here, however, since there is no substitution in either the octahedral or tetrahedral positions; the phosphate tetrahedra are inverted, and no swelling takes place in water.

B. Structural Studies of Zirconium Phosphate Using NMR

Segawa et al. [78,79] have examined the catalytic activity of several zirconium phosphates for 1-butene isomerization. They concluded that the ε phase remains active even after evacuation at 773 K, while the zirconium phosphate gel and the α form show significant decreases in activity. The pyrophosphate has the highest activity. All three forms dehydrate upon evacuation at temperatures above 1000 K and eventually become the pyrophosphate ZrP_2O_7 above 1300 K as the P-OH groups condense to form P-O-P linkages simultaneous with the total loss of catalytic activity for butene isomerization. ^{31}P MAS-NMR [79] of zirconium phosphate gel, the α form, and the ε form of ZrP (Fig. 12) show that the α form has its phosphorus resonance at -16.6 ppm, while in the ε form it occurs at -21.7 ppm (with reference to H_3PO_4 at 0 ppm). The gel shows peaks at -11.8,

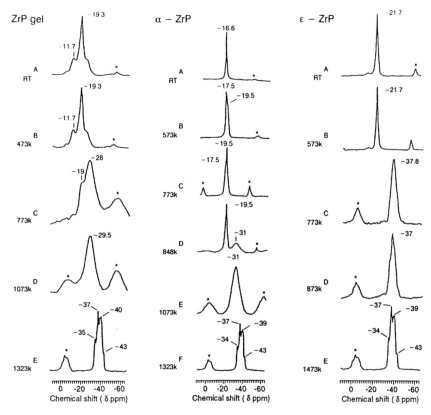

Figure 12 ^{31}P MAS NMR of Zirconium phosphate gel, the α- and the ε-morphologies. (From Ref. 78.)

-19.7, and -25.5 ppm, the middle one being the major resonance. The chemical shifts clearly indicate that the phosphorus environment in each of the three phases of ZrP is distinctly different, with multiple environments being present in the gel phase. From phosphorus NMR of samples evacuated at various temperatures, these authors conclude that: (1) ^{31}P chemical shifts move upfield as the interlayer separation decreases, (2) catalysts with the highest activity show the farthest upfield shift at -38 ppm, and (3) remnant P-OH groups after evacuation at 773 K enhance the protonic characteristics of the surface sites, since the excess electron density moves from the surface phosphorus atoms to those located between the Zr atom planes. Multiple resonances observed for the samples evacuated above 1300 K correlate well with the multiple phosphorus environments proposed by Corbridge [80] based on *cis* and *trans* forms of linear or bent P-O-P linkages.

As pointed out earlier, the ZrPs in all the crystalline forms exhibit ion-exchange capacity, with the cation taking the place of the proton in the P-OH groups between the layers. Thus each layer can be represented as a planar macro-anion $[Zr_n(PO_4)_{2n}]^{2n-}$, where most of the negative charge is on one oxygen and is balanced by an equivalent amount of protons or other cations residing between these layers. In addition to the ready exchange of cations (similar to other systems such as clays and zeolites) it is known that some of the phosphate groups in the γ form are easily exchanged with phosphate and phosphonate esters (Fig. 13) [81–83]:

$$Zr(HPO_4)_2 \cdot 2H_2O + xROPO_3^{2-} \leftrightarrow Zr(ROPO_3)_x(HPO_4)_{2-x} + xHPO_4^{2-}$$

Dines et al. [83] explored the possibility of pillaring in such a process by exchanging the phosphate groups from γ ZrP as well as synthesizing a mixed component phase followed by hydrolysis to generate the P-OH groups. They chose the system $Zr\{O_3PO(CH_2)_6OPO_3\}_{1/2}\{O_3P(CH_2)_8PO_3\}_{1/2}$ phosphate ester/phosphonate mixed component, which exhibits a layer spacing of 14.1 Å. Phosphorus NMR of this mixed-phase precursor clearly showed two envelopes of multiple resonances, these being attributed to disordered placement of the two groups between the layers. Even though thermogravimetric data showed two distinct decomposition points at 320 and 590°C (presumably due to ester and phosphonate decomposition, respectively), extended thermolysis at 350°C did not yield clean product. They were able to generate relatively clean pillared phosphate by hydrolyzing the mixed precursor in concentrated HBr. The [31]P

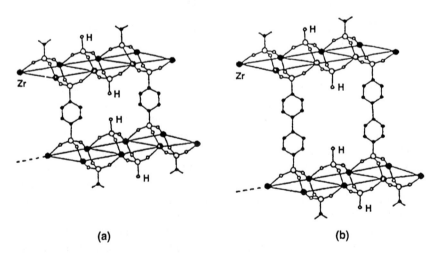

(a)   (b)

Figure 13 Idealized schematic representation of mixed component pillared zirconium phosphate derivatives: (a) *p*-phenylene-diphosphonate/monohydrogen phosphate; (b) 4,4'-biphenylphosphonate/HPO$_4$. (From Ref. 75b.)

NMR of such hydrolysis product showed two relatively sharp resonances at 10.5 and -17 ppm, and elemental analysis indicated it to be $Zr\{(O_3POH)\}\{(O_3P(CH_2)_8PO_3\}_{1/2}$. The observed resonances were assigned to the phosphonate (10.5 ppm) and disordered orthophosphate (-17 ppm), respectively. The dried product had a surface area of 209 m^2/g versus the precursor's 49 m^2/g, which implies a substantial increase the void space within the particles. Attempts to prepare such pillared products by anionic substitution of γ ZrP with a bis-phosphonic acid of *p,p*-biphenyl resulted in what appeared to be a bilayered product with a d-spacing of 26.4 Å and a pendant-free phosphonic acid group on each layer. This assignment was based on observation of three ^{31}P signals, attributed to phosphate and free and sheet-bound phosphonate, respectively. The desired pillared product can indeed be made by employing a more flexible bisphosphonic acid that had alkyl linkages, and such a species does not exhibit a free phosphonate resonance. These authors [83] concluded that the steric accessibility of two exchange sites to a bisphosphonic acid determines whether the product has a pillared or bilayered structure. It should be noted that the α ZrP is not amenable to this kind of anionic substitution and that the phosphate ester/phosphonate structures can only be introduced at the synthesis step itself. The difference may be attributable to the larger d-spacing of 12.2 Å in the γ form versus 7.6 Å in the α form, since the narrower spacing presumably imposes diffusional constraints on the relatively larger phosphate and phosphonate esters. However, Barret et al. [84] have shown that the layer phosphate groups of α ZrP can also be exchanged with HPO_4^{2-} in dilute solutions.

Clayden [85] studied by ^{31}P NMR the reaction between the γ ZrP and phenyl phosphonic acid $\{C_6H_5PO_3H_2\}$ and phenyl dihydrogen phosphate $\{C_6H_5OPO_3H_2\}$. His data clearly show (Fig. 14) that γ ZrP has *NMR inequivalent phosphorus sites*, while α ZrP does not. Despite the presence of two crytallographically inequivalent phosphorus atoms, a single resonance at -18.7 ppm is seen for the α form, while the γ form shows two resonances at -9.4 and -27.4 ppm, respectively. In other phosphates [86], crystallographically inequivalent phosphorus sites have generally been resolved in phosphorus NMR. Thus the microenvironments of the phosphorus atoms in α ZrP must be sufficiently similar to be indistinguishable in solid-state NMR.

From the low sideband intensities and the static powder pattern, Clayden concluded that the chemical shift anisotropy (CSA) is not very large in this phase. In contrast, the two widely separated resonances observed in γ ZrP indicate that there are significant chemical differences in the environment of phosphate groups. Clayden was able to estimate the CSA of the two sites by exploiting the differences in the cross-polarization efficiencies. One site requires only a short contact time on the order of 100 μs, whereas the other requires a long contact time, about 7000 μs. The sites appear to have significantly different CSA, with the upfield resonance showing very small anisotropy and thereby implying that the local environment is relatively more symmetric. The chemical

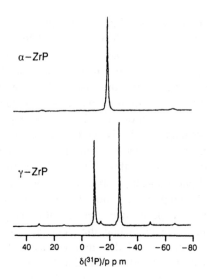

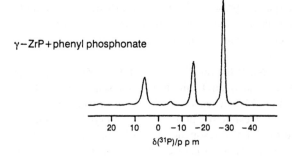

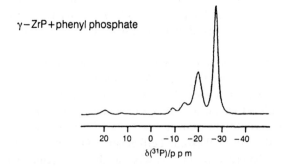

Figure 14 ^{31}P MAS NMR of zirconium phosphates. (From Ref. 85.)

shift of this resonance at -27.4 ppm is very similar to that observed in phosphates such as $HZr_2(PO_4)_3$ [87], which further implies that all four oxygens of this phosphate are shared with zirconium atoms in a $P(OZr)_4$ configuration. The second resonance at -9.4 ppm, on the other hand, indicates that some of the zirconium atoms in the second shell must have been replaced by protons, such as in $P(OZr)_2(OH)_2$. Clayden found the cross-relaxation characteristics of the two resonances in γ ZrP to be significantly different, which he interpreted as supporting to his assignment of the downfield resonance to a dihydrogen phosphate group. This was further substantiated by the ease of ion exchange of the phosphorus species responsible for the -9.4 ppm resonance with either phenyl phosphate or phenyl phosphonate in aqueous dispersions of γ ZrP with either of the esters. The phosphorus species responsible for the -27.4 ppm resonance does not change at all. Yamanaka et al. [81,82] also noted that only about half the phosphate groups could be exchanged with the esters, consistent with Clayden's hypothesis that half the phosphates in the structure are involved in the framework and the other half are the "free" or dihydrogen phosphate type.

The groups RPO_3 and $ROPO_3$ appear to behave quite differently both in the degree of exchange and in their influence on neighboring phosphate groups. Introduction of the $ROPO_3$ between the layers causes a larger d-spacing, as expected. Exchange with phenyl phosphate appears to move the remnant dihydrogen phosphate resonance from -9.4 to -14.4 ppm, the shift being attributed to the influence of nearby aryl phosphate groups. In samples exchanged with phosphonate ester, Clayden assigned the peak at $+5.6$ ppm to the phosphonate group, which conflicts with the assignment of Dines et al. [83]. From the ^{13}C data of the phosphate-exchanged ZrP, Clayden concluded that the phenyl rings are rigid at high levels of exchange, presumably owing to steric reasons.

Ortiz-Avila and Clearfield [88] demonstrated that even the α ZrP can be forced to undergo exchange with phosphate esters if the reaction is carried out either by refluxing in neat alkyl phosphate solutions or by enlarging the interlayer spacing first by intercalating with an amine. Upon exchange of the γ ZrP with butyl phosphate, these authors found phosphorus resonances at -27.7, -22.1, and -14.4 ppm, which were assigned to the framework phosphate, di-butyl phosphate, and mono-butyl phosphate, respectively. They argued that the peak at -14.4 ppm cannot be the remnant dihydrogen phosphate species as proposed by Clayden, since the bulk of organic phosphorus cannot be accounted for by such an assignment. They also found that upon exchanging with the phosphate ester of glyceric acid (2-hydroxy propionic acid), the γ form undergoes a phase transition and becomes the α form.

Using ^{31}P solid-state NMR, MacLachlan and Morgan [89,90] studied the intercalation of α ZrP with a series of amines (Fig. 15). They examined the reasons for the appearance of a single phosphorus resonance in α ZrP even though there are two crystallographically inequivalent sites. The conclusion on

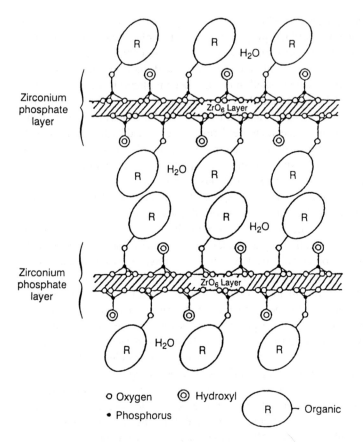

Figure 15 A schematic structural model of the organic derivatives of γ-zirconium phosphate. (From Ref. 82c.)

the basis of simulations was that the resonance is actually an envelope of two overlapping Lorentzians with different linewidths. The lack of resolution in the observed spectra of α ZrP was attributed to residual dipolar broadening and to fortuitous cancellation of donor and acceptor type hydrogen-bonding effects. Upon intercalation, protonation occurred for all the monoamines and diamines as confirmed by ^{13}C NMR. The intensity and isotropic chemical shifts of phosphorus were found to depend on the type of amine present in the layered material (Fig. 16), and chemical shifts changed significantly by as much as 10 ppm on deprotonation. Upon intercalation, linear alkyl chain primary amines produced downfield shifts, while the secondary and tertiary amines produced upfield shifts. MacLachlan and Morgan proposed that, at low amine contents, the amine replaces the interlayer water and positions itself parallel to the phosphate layers.

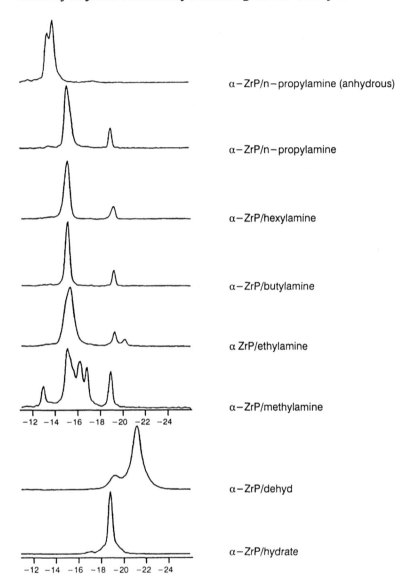

α−ZrP/n−propylamine (anhydrous)

α−ZrP/n−propylamine

α−ZrP/hexylamine

α−ZrP/butylamine

α ZrP/ethylamine

α−ZrP/methylamine

α−ZrP/dehyd

α−ZrP/hydrate

Figure 16 ^{31}P MAS NMR of intercalates of α-Zr(HPO$_4$)$_2$ with different amines. (From Ref. 89.)

With increase in the amine loading, the alkyl chain adopts a more upright position, thus significantly increasing the interlayer spacing. The transition from a parallel to an inclined orientation produces interlayer disorder for the *n*-propylamine intercalates (Fig. 17) but not for the *n*-butylamine intercalates. The changes in the phosphorus NMR spectra upon addition of water were interpreted in terms of rapid exchange within the interlayer region.

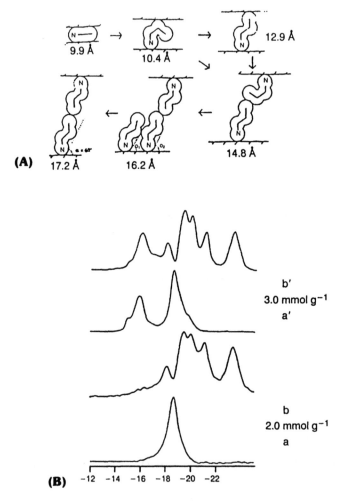

Figure 17 (A) Proposed amine orientations for the phases formed on titration on *n*-propylamine with α-Zr(HPO$_4$)$_2$-H$_2$O. (B) Subspectra generated for sample with different loadings of n-butylamine: (a, a′) slow relaxing component; (b, b′) fast relaxing component. (From Ref. 89.)

Li et al. [91] studied by multinuclear NMR the intercalation of α ZrP with 3-[(triethoxy)silyl]-1-propylamine, $NH_2(CH_2)_3Si(OC_2H_5)_3$. From the ^{13}C, ^{29}Si, and ^{31}P NMR and infrared data, they concluded that the organosilicon compound undergoes polymerization, with the dimers and trimers intercalating the interlayer spacing of the ZrP. Pillaring by a silicalike species was found to occur upon heating to 400°C. One group of phosphates interacts more strongly with the pillar than the other, as indicated by the appearance of multiple resonances in phosphorus NMR. In addition, a number of silanol (-Si-OH) species were observed via cross-polarization experiments in the calcined material, and *n*-hexane adsorption indicated that the material remains porous.

IV. STRUCTURAL STUDIES OF ALUMINOPHOS-PHATES AND THEIR DERIVATIVES USING NMR

It is well known that dense phases of aluminum phosphate are isostructural to crystalline silica phases such as tridymite and cristobalite. Moffat [92] described the use of aluminophosphates in catalytic systems. Wilson et al. [93] reported a new class of microporous aluminophosphates encompassing several zeolitelike three-dimensional porous structures and some two-dimensional layered structures. This discovery spawned a significant amount of characterization efforts involving magnetic resonance techniques. Of these new structures, $AlPO_4$-5 received the most attention since the three-dimensional network could be described by alternating Al and P atoms, each surrounded by four oxygens to form $Al(O-P)_4$ or $P(O-Al)_4$ units, respectively. Blackwell and Patton [94] compared the ^{27}Al and ^{31}P MAS-NMR data of several of these new AlPO structures to the known phases of phosphates. Typically the chemical shifts of phosphorus in dense phases were downfield of the values observed for the microporous phases. By carrying out the experiments at different field strengths and employing ^1H-^{27}Al cross-polarization, they attributed some of the unusual aluminum chemical shifts to secondary interactions of tetrahedral framework aluminum with occluded template or water within the micropores. In particular for $AlPO_4$-17, adsorption of water causes an apparent chemical shift normally observed for octahedral aluminum complexes even though the framework aluminum is still nominally tetrahedral. Muller et al. [95] studied the high-field NMR of $AlPO_4$-5 and its precursor gel. Using an empirical correlation between chemical shift and mean Al-O-P bond angle that had been determined for the dense AlPO phases, these authors arrived at a mean angle value of 150° for the $AlPO_4$-5 structure. Fedotov et al. [96] showed that the phosphorus chemical shifts vary additively with the number of aluminum atoms in the second shell, while the aluminum chemical shifts vary additively upon substitution of water by phosphate ions. Martens et al. [97] examined the structure of $AlPO_4$-H_3 by solid-state aluminum NMR and found two types of aluminum: AlO_4 tetrahedra with a chemical shift of

41.1 ppm and $AlO_4(H_2O)_2$ octahedra at -16.8 ppm, alternating at the framework sites. Meinhold and Tapp [98] reported NMR studies of $AlPO_4$-5. They also observed readily reversible reactions of the framework tetrahedral aluminum sites with water to form octahedral Al. Retention of cations by the supposedly charge-neutral framework was ascribed to the presence of defect sites. Increasing the pH of the solution increased the degree of retention but also destroyed the crystallinity of the phosphate.

Since the stoichiometric aluminophosphate is charge neutral, several researchers have attempted to synthesize and characterize materials with other tetrahedral metal atoms of differing valences so as to mimic the ion-exchange and acid site capabilities of zeolitic structures. Of these, the SAPOs (aluminophosphates) with some silicon in the framework) have received considerable attention. Flanigen et al. [99a], Pyke et al. [99b], and Kikhtyanin et al. [99c] showed that these materials exhibit unique conversion capabilities for a broad spectrum of hydrocarbon reactions. Freude et al. [100] observed two types of bridging hydroxyl groups (Fig. 18) in the ^1H MAS-NMR studies of SAPO-5. Resonances at 3.9 and 4.9 ppm were found to be acidic Si-O(H)-Al species by their ability to protonate pyridine, and the resonance at 1.5 ppm was ascribed to nonacidic P-OH groups. These authors concluded that silicon substitutes mostly for phosphorus, and silicon NMR indicated the presence of crystalline silica occluded in the structure. Based on silicon NMR data, Appleyard et al. [101] also concluded that attempts to increase silicon content in the aluminophosphate framework results in formation of significant amounts of amorphous silica. Wang et al. [102] observed that the amount of silicon incorporation into the framework depended on the source of silicon and on the template used in the synthesis of SAPOs (Fig. 19). Hasha et al. [103] studied the carbon, phosphorus, silicon, and aluminum NMR of a series of SAPOs with sodalite structure. They concluded that Si exclusively occupies phosphorus sites in the framework and thus avoids Si-O-P linkages. Consequently the aluminum atoms were found in two distinct environments, one phosphorus rich and the other silicon rich.

Davis and coworkers [104] studied ^{129}Xe NMR of xenon adsorbed in several SAPOs, ALPOs, and Y zeolites. From a comparison of the xenon chemical shift extrapolated to zero pressure, these authors concluded that Xe atoms feel significantly smaller electrostatic fields and field gradients in the aluminophosphates compared to aluminosilicates. The extrapolated chemical shift decreased from 97 ppm in erionite to 60 ppm in Y zeolite and to 27 ppm in $AlPO_4$-5, with the values for SAPOs being intermediate to Y zeolites and AlPOs as would be expected from the acidity trends. They concluded as well that SAPO-37 does not contain separate aluminophosphate and aluminosilicate islands. Dumont et al. [105] also carried out xenon NMR experiments in SAPO-37. From xenon sorption capacity and the decrease in the chemical shift, their conclusion was that the framework of calcined SAPO-37 is unstable when exposed to moist air.

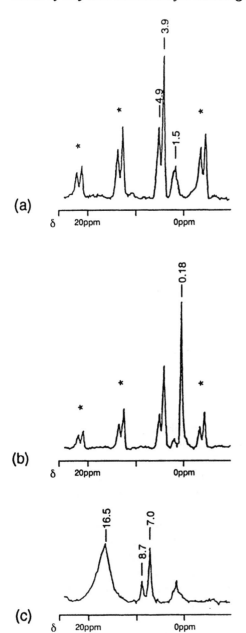

Figure 18 ¹H MAS NMR spectra of activated SAPO-5: (a) without sorbate; (b) with sorbed methane; (c) with sorbed NC_5D_5. (From Ref. 100.)

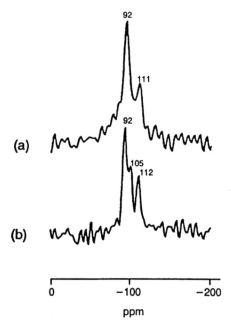

Figure 19 ^{29}Si MAS-NMR of (a) No. 6 SAPO-5, and (b) Si(OEt)$_4$ treated No. 6 SAPO-5. (From Ref. 102.)

Anderson et al. [106], using gas chromatography as well as ^{13}C and ^{1}H NMR, monitored the shape-selective catalytic conversion of methanol to low molecular weight olefins and aliphatics by SAPO-34. This phosphate has the framework topology of the naturally occurring zeolite chabazite. The primary limiting factor determining the length of the hydrocarbon chain appears to be the size of the eight-membered window. They claimed that the main species present in the intracrystalline space are branched C4 and C5 aliphatics, which owing to their size are trapped and impose additional constraints on the diffusion of the linear species. Furthermore, they hypothesized that by preparing the catalyst either with partial occlusion or with very large cations, formation of branched hydrocarbons could be prevented and thus the diffusional constraints on C2 and C3 hydrocarbons could be reduced. As a result, selectivity for propylene would improve.

Kuhl and Schmitt [107] discussed the similarities between the phosphorus-containing zeolites ZK-21 and ZK-22 and the silico-aluminophosphate SAPO-42. The phosphorous chemical shifts in all three microporous systems clearly indicate that phosphorus is in framework tetrahedral sites in each and that some phosphate is occluded into the sodalite cages. Montes et al. [108] described the substitution of cobalt and vanadium at the framework sites. Strong interaction with paramagnetic cobalt could be seen in both phosphorus and aluminum NMR spectra, indicating cobalt to be present in its divalent state.

Barrie and Klinowski [109] discussed ordering in the magnesium aluminophosphate network on the basis of aluminum and phosphorus NMR. MgAPO-20 crystallizes in a structure similar to naturally occurring mineral sodalite, with Mg, P, and Al occupying the tetrahedral sites. Barrie and Klinowski postulated that the alternation of the Mg and Al in the tetrahedral sites adjacent to phosphorus creates a situation similar to that in zeolitic aluminosilicates and that the phosphorus spectrum can be interpreted in terms of $P(O-Al)_n$ where n ranges from 0 through 4 (Fig. 20).

Molecular sieve topologies with 18- or 24-membered rings were hypothesized many years ago, but only recently was an aluminophosphate possessing such large windows reported by Davis and coworkers [110]. This material was designated VPI-5. Crystallization of VPI-5 from the gel could be followed by phosphorous NMR [110c], and this phosphate structure is unique (Fig. 21) in that three resonances of almost equal intensity are observed in ^{31}P NMR. This pattern is in contrast to the single resonance observed for most other phosphates of dense as well as microporous structure. The aluminum NMR also shows two resonances, one tetrahedral at about 40 ppm and the other at -19 ppm. This second peak was initially thought to be either an impurity or a precursor aluminum species. Derouane and coworkers [111], however, reported the synthesis of a VPI-5 analog with some silicon in the framework. Grobet et al. [112] then argued that the resonance is sensitive to hydration-dehydration cycles, being very similar to that observed in $AlPO_4$-H_3, and hence belongs to an octahedral framework aluminum species. They proposed that the three resonances in phosphorus NMR could be assigned to (1) phosphorus at the center of a double four-ring, coordinated to $3(Al)_{tet}$ and $1(Al)_{oct}$ at -33.2 ppm, (2) phosphorus in a six-ring coordinated to $3(Al)_{tet}$ and $_1(Al)_{oct}$ at -27.5 ppm, and (3) phosphorus in a six-ring coordinated to $2(Al)_{tet}$ and $2(Al)_{oct}$ at -23.7 ppm, respectively. The octahedral configuration for a nominally tetrahedral framework aluminum arises from coordination of two water molecules. The resulting distorted arrangement causes a significant second-order quadrupole contribution to the lineshape, which explains the broad asymmetric line at -19 ppm in the aluminum NMR of VPI-5 (Fig. 22). More recent NMR experiments on VPI-5 with double angle rotation (DOR) clearly showed [113,114] that the aluminum spectrum actually has three resonances (corresponding well with the phosphorus spectrum), and that the distortion of the octahedral aluminum resonance is indeed due to second-order quadrupolar effects.

Van Braam Houckgeest et al. [115] examined the phosphorus NMR of VPI-5 in a sealed rotor as a function of temperature. Here the three-line spectrum of

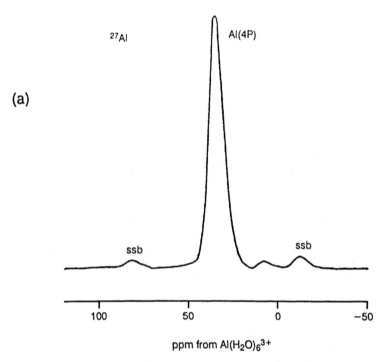

Figure 20 MAS NMR spectra of MgAPO-20 at 4.5 kHz spinning speed: (a) ^{27}Al spectrum with 0.2-s recycle delay; (b) ^{31}P spectrum with a 5-s recycle delay. (From Ref. 109.)

1:1:1 intensity is replaced by a two-line spectrum of 2:1 intensity ratio as the temperature is raised, with the transformation being complete at 90°C. Upon cooling of the sample back to room temperature, the three lines are restored to their original intensities and positions. This behavior was explained in terms of breakage of hydrogen bonding and a concomitant increase in site symmetry at higher temperatures. The explanation is consistent with the x-ray data of Mc-Cusker et al. [116], who argued that the structure refinement yields a lower residual when the lower symmetry space group of P6$_3$ is used instead of P6$_3$cm to describe the inequivalence of tetrahedral sites in VPI-5. Van Braam Houckgeest et al. [115] further proposed that the higher symmetry space group is consistent with the high-temperature spectrum depicting two phosphorus resonances in 2:1 intensity ratio, since such a symmetry would require a twofold higher number of one type of tetrahedral sites. On this basis, the resonances at −23 and −27 ppm could be assigned to phosphorus atoms in the six-rings and the one at −33 ppm

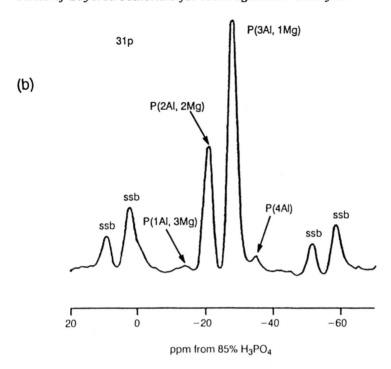

(b)

to that in the fused four-rings. These observations also fit well with the more recent findings by Perez et al. [117], who discussed the morphological changes caused by hydration levels and hydrolysis of Al-O-P linkages in large-pore 18-ring aluminophosphate systems. They pointed out that VPI-5 and H1 both convert to $AlPO_4$-8 structure upon such hydrolysis and that care should be taken in assigning the phosphorus resonances. The three resonances observed in VPI-5 and H1 transform into six resonances at low water content (see Fig. 21), and it is suggested that in a dry state the P-O-Al bond in every other four-ring is broken to produce a P = O site and a three-coordinated aluminum. Upon hydrating, the broken bond is hypothesized to heal with concomitant formation of six-coordinated aluminum (Fig. 23). The importance of water molecules in determining the structure was further illustrated by the deuterium NMR work of Goldfarb et al. [118]. In this study, two distinct types of water molecules were observed in $AlPO_4$-5 as well as in VPI-5. In the latter system, water was found to exhibit a higher degree of order, and the two-site jump between free and bound molecules (also found in $AlPO_4$-5) was associated with an additional threefold site jump.

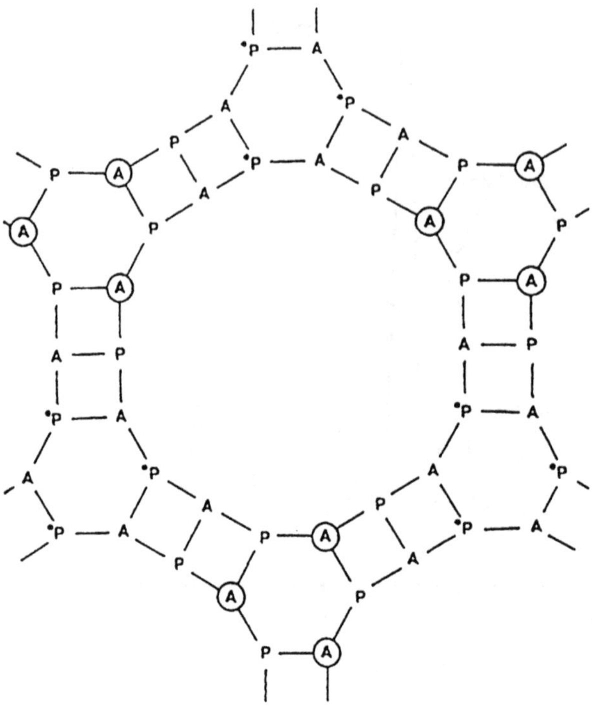

Figure 21   Distribution of T atoms in VPI-5: A, P, and encircled A represent $Al^{IV}O_4$, $P^{IV}O_4$, and $Al^{VI}O_4(H_2O)_2$, respectively. 'P stands for $P^{IV}O_4$ facing Al^{VI} in neighboring layers. (From Ref. 112.)

V. STUDIES OF LAYERED METAL SULFIDES

MoS_2 is a layered material that is the active ingredient in nearly all hydro-desulfurization catalysts of industrial importance today. Unfortunately this material is not amenable to NMR measurements because of the paramagnetism of the Mo^{5+} at the active defect site. Several papers concerning ESR of MoS_2 have been published, the most notable of which are due to Silbernagel [119–123]. It should be pointed out, however, that a significant amount of NMR has been done on precursors of such catalysts in an effort to better understand their high activity for particularly demanding reactions [123,124]. Other metal sulfides such as WS_2, TaS_2, and NbS_2, as well as layered materials that may be of catalytic importance, may also be studied by NMR. So far, though, there appear to be few published structural studies of metal sulfides pertain to adsorbates.

The work recently completed in Jonas's group is particularly interesting [123]. In these experiments deuterated pyridine is adsorbed in TaS_2, and the

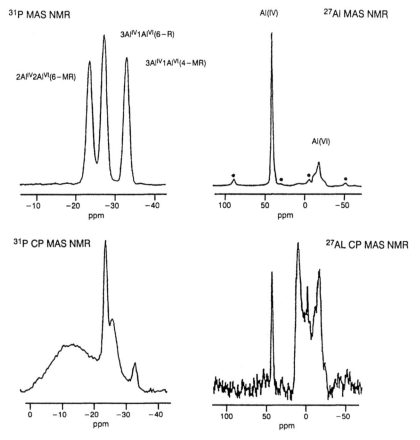

Figure 22 MAS NMR spectra of hydrated VPI-5. For the ^{31}P CP-MAS experiment, a ^{1}H pulse length of 6 μs, a contact time of 2 ms, and a recycle delay of 3 s was used. The [27]Al CP-MAS experiment was performed with a ^{1}H pulse length of 6 μs, a contact time of 0.5 ms, and a recycle delay of 1 s. Spinning side bands are indicated by dots. (From Ref. 112.)

dynamics of pyridine in the interlayer is studied as a function of temperature. By monitoring the observed doublets for pyridine-d_2 (deuterated at positions *ortho* to N) (Fig. 24) and comparing those to calculated averaged QCC's for particular types of anisotropic tumbling of the molecule in the interlayer, very specific details about the molecular motion can be discovered. The result found here is that the pyridine adsorbates undergo essentially one type of motion, namely, a rapid rotation about an axis perpendicular to the molecular C_2 axis in the molecular plane. It is argued that the orientation of the pyridine between layers is that of Fig. 24c.

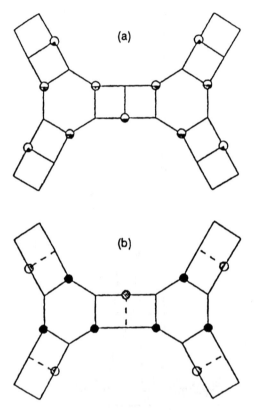

Figure 23 Schematic representation of the unique portion of the VPI-5 unit cell. The circles represent phosphorus atoms and their shadings the location and amount of this type of atom in the unit cell: (level a) ◐, ◔, P atoms in S2 at half and one-fourth occupancy, respectively; , P atoms in S1 site at half occupancy; (level b) ●, P at full occupancy in S1; ◉, ○, P at full and half occupancy, respectively, in S2. (From Ref. 117.)

A similar study has been carried out on a pyridine intercalated in $CdPS_3$ [123], but we will not cover the results since $CdPS_3$ is probably of little interest as a catalyst. The approach taken is similar to the TaS_2 study.

VI. STUDIES OF GRAPHITE AND ADSORBATES ON GRAPHITE

Graphite is certainly a layered material for which there is a large body of NMR literature, although the connection with catalysis is a bit more tenuous. In general, carbon supports (which may contain graphitic phases) for catalysts are rarely used at this point. There is, however, some interest in metal intercalated

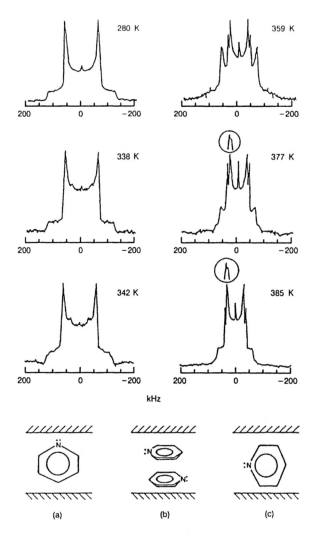

Figure 24 Variable-temperature ^{2}H spectra of perdeuterated pyridine adsorbed in TaS$_2$ (top). Proposed orientations of pyridine between TaS$_2$ layers (below): (a) C$_2$ axis perpendicular to layers; (b) bilayer in which C$_2$ axes and molecular planes are parallel to layers; (c) C$_2$ axis parallel to layers, molecular plane perpendicular to layers. (From Ref. 123.)

graphites for catalysis of specific organic reactions. For example, a 1976 patent describes the use of SbF$_5$ intercalated graphite as a cyclohexane isomerization catalyst [125].

Graphite consists of two-dimensional sheets of polynuclear aromatics stacked on top of each other. The one- and two-dimensional conductivity of graphite gives it a large anisotropic diamagnetic susceptibility, and this effect must be considered when interpreting spectra of materials adsorbed on graphite. Much of the literature concerning NMR of adsorbates on graphite, frequently in the form of relaxation studies, has been produced in the surface physics and surface thermodynamics community. Rather than try to review this body of literature comprehensively, we have selected some work that outlines techniques and problems in performing NMR measurements on graphitic systems.

Resing pioneered the area of NMR of adsorbates on graphites. Various papers concerning NMR of graphite, and particularly fluorinated intercalants in graphites, were published in the 1970s and 1980s. Most of this interest was due to the very high electrical conductivity of some intercalated graphites, and much of that work focused on studying charge transfer between the graphite layers and the intercalant [125–131]. Other studies were made of the graphite material itself. The ^{13}C homonuclear dipolar interaction was used in one investigation to determine the C-C bond distance in a highly oriented intercalated graphite [132]. A study of intercalant diffusion into the graphite was also carried out using a one-dimensional imaging technique to follow ^{19}F in an AsF_5 intercalant. The diffusion parallel to the graphite planes was found to be nearly that of the liquid [133].

More recently, work in the group of Boddenberg [134–136] has made use of ^2H NMR to examine adsorbate orientation and dynamics in graphitized carbon blacks. The interplay between the axially symmetric shielding tensor (due to diamagnetic susceptibility anisotropy) and the EFG tensor allows quite detailed information about orientation and fast rotational motion to be derived from the ^2H spectrum. In one of their latest reports [136], the surface diffusion coefficients of fluorobenzene and neopentane are estimated based on a model for diffusion of the molecule along the surface. The model assumes reorientation of the EFG tensor along a spherical trajectory, but clearly the diffusion is mainly two-dimensional if it is assumed that the molecule skates across the surface without rotation about its center of mass. The paper seems to imply that rotational diffusion of the molecule arises mainly from translational diffusion across a curved surface, and not from rotation of the molecule itself. Really what is measured is an angular diffusion coefficient, and it seems dangerous to translate this into a translational diffusion coefficient.

VII. CONCLUSIONS

There is clearly a large body of NMR literature from widely disparate fields pertaining to layered systems. What emerges from much of this work is an appreciation of the many and complex interactions in some of the layered materials. These interactions are due mainly to composition and often have little to do with the inherent anisotropy of the system. Some of the spin couplings are difficult to address using current solid-state NMR methods, although the development of averaging techniques for second-order quadrupolar interactions undoubtedly will help unravel quantitation issues in ^{27}Al NMR of clays and aluminophosphates. Other interactions remain more difficult to deal with, however. Broadening due to magnetic susceptibility effects can be addressed for some systems, but in other cases the situation is somewhat complicated. Also, paramagnetic effects can be both a blessing and a curse when studying clays high in Fe. Paramagnetism can bring about various effects depending on the strength of the hyperfine interaction, the electron relaxation time, and motional correlation times (e.g., broadening, shifting, splitting). Obviously, paramagnetism precludes the study by NMR of some metal sulfides. Graphite, too, has some particularly interesting properties that must be considered when making measurements. Its combination of structural anisotropy and high conductivity creates an especially rich environment for NMR. In summary, then, NMR is one of the most useful methods for studying layered systems, but some solid-state interactions remain to be dealt with adequately.

From a catalytic standpoint, we seem to be in a period of little industrial interest in layered catalysts such as clays, although there are undoubtedly instances where clays, and particularly pillared clays, may be useful for specific reactions. The initial excitement about using pillared clays as large-pore zeolites in cracking applications has given way to developments such as aluminophosphates, where much effort is now being focused. Some of the original interest in clays has also been channeled to materials such as synthetic zirconium phosphates, for which it may be possible to avoid some of the stability problems associated with pillared clays under hydrothermal conditions. Metal sulfides will probably be of interest catalytically as long as feeds containing sulfur are processed.

REFERENCES

1. R. Burch, ed., *Catalysis Today 2: 185* (1988).
2. R. M. Barrer and D. M. MacLeod, *Trans. Faraday Soc. 51*: 1290 (1955).
3. D. E. W. Vaughan, R. J. Lussier, and J. S. Magee, U.S. Patent, 4,176,090 (1979).
4. E. T. Lippmaa, M. A. Alla, T. J. Pehk, and G. Engelhardt, *J. Am. Chem. Soc. 100*: 1929 (1978).
5. G. Engelhardt and D. Michel, *High-Resolution Solid-State NMR of Silicates and Zeolites*, John Wiley and Sons, New York, 1987.
6. J. J. Fripiat, *Catalysis Today 2*: 281 (1988).
7. E. Lippmaa, M. Magi, A. Samoson, G. Engelhardt, and A. R. Grimmer, *J. Am. Chem. Soc. 102*: 4889 (1980).
8. T. Watanabe, H. Shimizu, A. Masuda, and H. Saito, *Chem. Lett.*: 1293 (1983).
9. P. F. Barron, M. A. Wilson, A. S. Campbell, and P. L. Frost, *Nature 299*: 616 (1982).

10. P. F. Barron, R. L. Frost, and J. O. Skjemstad, A. Koppi, *Nature 302*: 49 (1983).

11. J. Sanz and J. M. Serratosa, *J. Am. Chem. Soc. 106*: 4790 (1984).

12. J. Sanz and J. M. Serratosa, *Clay Min. 19*: 113 (1984).

13. J. G. Thompson, *Clay Min. 19*: 229 (1984).

14. B. A. Goodman and J. W. Stucki, *Clay Min. 19*: 663 (1984).

15. A. Samoson, E. Lippmaa, and A. Pines, *Mol. Physics 65*: 1013 (1988).

16. P. A. Diddams, J. M. Thomas, W. Jones, J. A. Ballantine, and J. H. Purnell, *J. Chem. Soc. Chem. Commun. C908*: 1340 (1984).

17. N. C. M. Alma, G. R. Hays, A. Samoson, and E. Lippmaa, *Anal. Chem. 56*: 729 (1984).

18. M. Lipsicas, R. H. Raythatha, T. J. Pinnavaia, I. D. Johnson, R. F. Giese, P. M. Constanzo, and J. L. Roberts, *Nature 309*: 604 (1984).

19. J. G. Thompson and P. F. Barron, *Clays Clay Min. 35*: 38 (1987).

20. S. Komarneni, C. A. Fyfe, and G. J. Kennedy, *Clay Min. 20*: 327 (1985).

21. R. A. Kinsey, R. J. Kirkpatrick, J. Hower, K. A. Smith, and E. Oldfield, *Am. Min. 70*: 537 (1985).

22. C. A. Weiss, Jr., S. P. Altaner, R. J. Kirkpatrick, *Am. Min. 72*: 935 (1987).

23. D. E. Woessner, *Am. Min. 74*: 203 (1989).

24. J. B. E. de la Caillerie and J. J. Fripiat, *Clays Clay Min. 39*: 270 (1991).

25. J. W. Akitt and A. Farthing, *J. Magn. Res. 32*: 345 (1978).

26. J. W. Akitt and A. Farthing, *J. Chem. Soc. Dalton Trans.*: 1617 (1981).

27. J. W. Akitt and A. Farthing, *J. Chem. Soc. Dalton Trans.*: 1624 (1981).

28. J. Y. Bottero, J. M. Cases, F. Fiessinger, J. E. Poirier, *J. Phys. Chem. 84*: 2933 (1980).

29. J. W. Akitt, *Prog. NMR Spectrom. 21*: 1 (1989).

30. D. Plee, F. Borg, L. Gatineau, and J. Fripiat, *J. Am. Chem. Soc. 107*: 2362 (1985).

31. T. J. Pinnavaia, S. D. Landau, M. Tzou, I. D. Johnson, and M. Lipsicas, *J. Am. Chem. Soc. 107*: 7222 (1985).

32. A. Schutz, W. E. E. Stone, G. Poncelot, and J. J. Fripiat, *Clays Clay Min. 35*: 251 (1987).

33. D. T. B. Tennakoon, W. Jones, and J. M. Thomas, *J. Chem. Soc. Faraday Trans. 82*: 3081 (1986).

34. F. Figueras, Z. Klapyta, P. Massiani, Z. Mountassir, D. Tichit, F. Fajula, C. Guegen, J. Bousquet, and A. Auroux, *Clays Clay Min. 38*: 257 (1990).

35. H. D. Morris, S. Bank, and P. D. Ellis, *J. Phys. Chem. 94*: 3121 (1990).

36. L. Zheng, Y. Hao, L. Tao, Y. Zhang, and Z. Xue, *Zeolites 12*: 374 (1992).

37. H. Lindgreen, H. Jacobsen, and H. J. Jakobsen, *Clays and Clay Min. 39*: 54 (1991).

38. H. J. Jakobsen, H. Jacobsen, and H. Lindgreen, *Fuel 67*: 727 (1988).

39. K. J. D. MacKenzie, M. E. Bowden, I. W. M. Brown, and R. H. Meinhold, *Clays Clay Min. 37*: 317 (1989).

40. B. A. Goodman, J. D. Russell, B. Montez, E. Oldfield, and R. J. Kirkpatrick, *Phys. Chem. Miner. 12*: 342 (1985).

41. M. A. Wilson, K. Wada, S-I. Wada, and Y. Kakuto, *Clay Miner. 23*: 175 (1988).

42. K. J. D. MacKenzie, M. E. Bowden, and R. H. Meinhold, *Clays Clay Min. 39*, 337 (1991).
43. J. Santaren, J. Sanz, and E. Ruiz-Hitzky, *Clays Clay Min. 38*: 63 (1990).
44. H. Pfeifer, *Colloids Surf. 45*: 1 (1990).
45. J. Conard, in *Magnetic Resonance in Colloid and Interface Science* (H. A. Resing and C. G. Wade, eds.), ACS Symp. Series 34, Washington, DC, 1976, p. 85.
46. S. Bank, J. F. Bank, and P. D. Ellis, *J. Phys. Chem. 93*: 4847 (1989).
47. V. Laperche, J. F. Lambert, R. Prost, and J. J. Fripiat, *J. Phys. Chem. 94*: 8821 (1990).
48. P. J. Chu, B. C. Gerstein, J. Nunan, and K. Klier, *J. Phys. Chem. 91*: 3588 (1987).
49. C. A. Weiss, Jr., R. J. Kirkpatrick, and S. P. Altaner, *Am. Min. 75*: 970 (1990).
50. C. A. Weiss, Jr., S. P. Altaner, and R. J. Kirkpatrick, *Am. Min. 72*: 982 (1990).
51. J. F. Lee, M. M. Mortland, S. A. Boyd, and C. T. Chiou, *J. Chem. Soc. Faraday Trans. 85*: 2953 (1989).
52. T. K. Pratum, *J. Phys. Chem. 96*: 4567 (1992).
53. D. E. Woessner and B. S. Snowden, *J. Colloid Interface Sci. 30*: 54 (1969).
54. D. E. Woessner and B. S. Snowden, *J. Chem. Phys. 50*: 1516 (1969).
55. A. M. Hecht and E. Geissler, *J. Colloid Interface Sci. 34*: 32 (1970).
56. D. E. Woessner, *J. Magn. Res. 16*: 483 (1974).
57. D. E. Woessner, *J. Magn. Res. 39*: 297 (1980).
58. W. E. E. Stone, in *Advanced Techniques for Clay Mineral Analysis* (J. J. Fripiat, ed.), Elsevier, Amsterdam, 1982, p. 77.
59. J. J. Fripiat, in *Magnetic Resonance in Colloid and Interface Science* (H. A. Resing and C. G. Wade, eds.), ACS Symp. Series 34, Washington, DC, 1976, p. 261.
60. H. A. Resing, D. Slotfeldt-Ellingsen, A. N. Garroway, T. J. Pinnavaia, and K. Unger, in *Magnetic Resonance in Colloids and Interface Science* (J. P. Fraissard and H. A. Resing, eds.), ACS Symp. Series, Washington, DC, 1980, p. 239.
61. C. A. Fyfe, J. M. Thomas, and J. R. Lyerla, *Angew. Chem. Int. Ed. Engl. 20*: 96 (1981).
62. N. J. Clayden and J. S. Waugh, *J. Chem. Soc. Chem. Commun.*: 292 (1983).
63. D. T. B. Tennakoon, R. Schlogl, T. Rayment, J. Klinowski, W. Jones, and J. M. Thomas, *Clay Min. 18*: 357 (1983).
64. D. T. B. Tennakoon, J. M. Thomas, W. Jones, T. A. Carpenter, and S. Ramdas, *J. Chem. Soc. Faraday Trans. 1 82*: 545 (1986).
65. K. A. Carrado, R. Hayatsu, R. E. Botto, and R. E. Winans, *Clays Clay Min. 38*: 250 (1990).
66. P. O'Brien, C. J. Williamson, and C. J. Groombridge, *Chem. Mater. 3*: 276 (1991).
67. J. G. Thompson, *Clays Clay Min. 33*: 173 (1985).
68. M. Raupach, P. F. Barron, and J. G. Thompson, *Clays Clay Min. 35*: 208 (1987).
69. Y. Sugahara, S. Satokawa, K. Kuroda, and C. Kato, *Clays Clay Min. 38*: 137 (1990).
70. P. Sidheswaran, A. N. Bhat, P. Ganguli, *Clays Clay Min. 38*: 29 (1990).

71a. G., Alberti and U. Costantino, in *Intercalation Chemistry* (M. S. Wittingham and J. A. Jacobson, eds.), Academic Press, New York, 1982, p. 147.

71b. A. Clearfield, *Comments Inorg. Chem. 10*: 89 (1990).

71c. A. Clearfield, ed., *Inorganic Ion Exchange Materials*, CRC Press, Boca Raton, FL, 1982.

72a. A. Clearfield and J. A. Stynes, *J. Inorg. Nucl. Chem. 26*: 117 (1964).

72b. A. Clearfield and G. D. Smith, *Inorg. Chem. 8*: 431 (1969).

73. A. Clearfield and S. P. Pack, *J. Inorg. Nucl. Chem. 37*: 1283 (1975).

74. A. Clearfield and J. R. Berman, *J. Inorg. Nucl. Chem. 43*: 2141 (1981).

75a. T. N. Frianeza and A. Clearfield, *J. Catal. 85*: 398 (1984).

75b. A. Clearfield, *J. Mol. Catal. 27*: 251 (1984).

76. A. Clearfield, R. H. Blessing, and J. A. Stynes, *J. Inorg. Nucl. Chem. 30*: 2249 (1968).

77. S. Yamanaka and M. Tanaka, *J. Inorg. Nucl. Chem. 41*: 45 (1979).

78. K. Segawa, Y. Kurusu, Y. Nakajima, and M. Kinoshita, *J. Catal. 94*: 491 (1985).

79. K. Segawa, Y. Nakajima, S. Nakata, S. Asaoka, and T. Takahashi, *J. Catal. 101*: 81 (1985).

80. D. E. C. Corbridge, *The Structure Chemistry of Phosphorus*, Elsevier Press, Amsterdam, 1974, p. 130.

81. S. Yamanaka and M. Hattori, *Chem. Lett.*: 1073 (1979).

82a. S. Yamanaka, K. Yamasaka, and M. Hattori, *J. Inorg. Nucl. Chem. 43*: 1659 (1981).

82b. S. Yamanaka and M. Hattori, *Inorg. Chem. 20*: 1929 (1981).

82c. S. Yamanaka, K. Sakamoto, and M. Hattori, *J. Phys. Chem. 88*: 2067 (1984).

83. M. B. Dines, R. E. Cooksey, P. C. Griffith, and R. H. Lane, *Inorg. Chem. 22*: 1003 (1983).

84. J. A. Barret, A. W. Dalziel, and M. K. Rahman, *Bull. Soc. Chim. Fr.*, special issue 1953 (1968).

85. N. J. Clayden, *J. Chem. Soc. Dalton Trans.*: 1877 (1987).

86a. R. J. B. Jakeman, A. K. Cheetham, N. J. Clayden, and C. M. Dobson, *J. Chem. Soc. Chem. Commun.*: 195 (1986).

86b. A. K. Cheetham, N. J. Clayden, C. M. Dobson, and R. J. B. Jakeman, *J. Am. Chem. Soc. 107*: 6249 (1985).

87. N. J. Clayden, *Solid State Ionics 24*: 117 (1987).

88. C. Y. Ortiz-Avila and A. Clearfield, *J. Chem. Soc. Dalton Trans.*: 1617 (1989).

89. D. J. MacLachlan and K. R. Morgan, *J. Phys. Chem. 94*: 7656 (1990).

90. D. J. MacLachlan and K. R. Morgan, *J. Phys. Chem. 96*: 3458 (1992).

91. L. Li, X. Liu, Y. Ge, L. Li, J. Klinowski, *J. Phys. Chem. 95*: 5910 (1991).

92. J. B. Moffat, *Catal. Rev. Sci. Eng. 18*: 199 (1978).

93a. S. T. Wilson, B. M. Lok, C. A. Messina, T. R. Cannan, and E. M. Flanigen, *J. Am. Chem. Soc. 104*: 1146 (1982).

93b. S. T. Wilson, B. M. Lok, C. A. Messina, T. R. Cannan, and E. M. Flanigen, *Intrazeolite Chemistry*, ACS Symp. 218, ACS, Washington DC, 1983, p 79.

94a. C. S. Blackwell and R. L. Patton, *J. Phys. Chem. 88*: 6135 (1984).

94b. C. S. Blackwell and R. L. Patton, *J. Phys. Chem. 92*: 3965 (1988).

95. D. Muller, E. Jahn, B. Fahlke, G. Ladwig, and U. Haubenreisser, *Zeolites 5*: 53 (1985).
96. M. A. Fedotov, I. L. Mudrakovskii, V. M. Mastikhin, V. P. Shmachkova, N. S. Kostarenko, *Izvestiya Akad.Nauk SSSR, Ser.Khim. 10*: 2340 (1987).
97. J. A. Martens, B. Verlinden, M. Mertens, P. J. Grobet, P. A. Jacobs, in *Zeolite Synthesis* (M. L. Occelli and H. E. Robson, eds.), ACS Symp. 398, ACS, Washington DC, 1989, p. 305.
98. R. H. Meinhold and N. J. Tapp, *Zeolites 11*: 401 (1991).
99a. E. M. Flanigen, B. M. Lok, R. L. Patton, and S. T. Wilson, in *New Developments in Zeolite Science and Technology* (S. Murakami, A. Iijima, J. W. Ward, eds.), Elsevier, Amsterdam, 1986, p. 103.
99b. D. R. Pyke, P. Wittney, and H. Houghton, *Appl. Catal. 18*: 173 (1985).
99c. O. V. Kikhtyanin, K. G. Ione, and V. M. Mastikhin, *Chem. Express. 1*: 721 (1986).
100. D. Freude, H. Ernst, M. Hunger, H. Pfeifer, and E. Jahn, *Chem. Phys. Lett. 143*: 477 (1988).
101. I. P. Appleyard, R. K. Harris, and F. R. Fitch, *Chem. Lett.*: 1747 (1985).
102. R. Wang, C. F. Lin, Y. S. Ho, L. J. Leu, and K. J. Chao, *Appl. Catal. 72*: 39 (1991).
103. D. Hasha, L. Sierra de Saldarriaga, C. Saldarriaga, P. E. Hathaway, D. F. Cox, and M. E. Davis, *J. Am. Chem. Soc. 110*: 2127 (1988).
104. M. E. Davis, C. Saldarriaga, C. Montes, and B. E. Hanson, *J. Phys. Chem. 92*: 2557 (1988).
105. N. Dumont, T. Ito, and E. G. Derouane, *Appl. Catal. 54*: L1 (1989).
106. M. W. Anderson, B. Sulikowski, P. J. Barrie, and J. Klinowski, *J. Phys. Chem. 94*: 2730 (1990).
107. G. H. Kuhl and K. D. Schmitt, *Zeolites 10*: 2 (1990).
108. C. Montes, M. E. Davis, B. D. Murray, and M. Narayana, *J. Phys. Chem. 94*: 6425, 6431 (1990).
109. P. J. Barrie and J. Klinowski, *J. Phys. Chem. 93*: 5972 (1989).
110a. M. E. Davis, C. Saldarriaga, C. Montes, J. Garces, and C. Crowder, *Nature 331*: 698 (1988).
110b. M. E. Davis, C. Montes, P. E. Hathaway, J. P. Arhancet, D. L. Hasha, and J. Garces, *J. Am. Chem. Soc. 111*: 3919 (1989).
110c. M. E. Davis, B. D. Murray, and M. Narayana, in *Novel Materials in Heterogeneous Catalysis* (R. T. K. Baker, L. L. Murrell, eds.), ACS Symp. 437, ACS, Washington DC, 1990, p. 48.
111. E. G. Derouane, L. Maistriau, Z. Gabelica, A. Tuel, J. B. Nagy, and R. Von Ballmoos, *Appl. Catal. 51*: L13 (1989).
112. P. J. Grobet, J. A. Martens, I. Balakrishnan, M. Mertens, and P. A. Jacobs, *Appl. Catal. 56*: L21 (1989).
113. Y. Wu, B. F. Chmelka, A. Pines, M. E. Davis, P. J. Grobet, and P. A. Jacobs, *Nature 346*: 550 (1990).
114. P. J. Grobet, A. Samoson, H. Geerts, J. A. Martens, and P. A. Jacobs, *J. Phys. Chem. 95*: 9620 (1991).

115. J. P. Van Braam Houckgeest, B. Kraushaar-Czarnetzki, R. J. Dogterom, and A. de Groot, *J. Chem. Soc. Chem. Commun.*: 666 (1991).

116. L. B. McCusker, C. Baerlocker, E. Jahn, and M. Bulow, *Zeolites 11*: 308 (1991).

117. J. O. Perez, P. J. Chu, and A. Clearfield, *J. Phys. Chem. 95*: 9994 (1991).

118. D. Goldfarb, H. X. Li, and M. E. Davis, *J. Am. Chem. Soc. 114*: 3690 (1992).

119. B. G. Silbernagel, *J. Catal. 56*: 315 (1979).

120. B. G. Silbernagel, T. A. Pecoraro, and R. R. Chianelli, *J. Catal. 78*: 380 (1972).

121. B. G. Silbernagel, in *Nuclear and Electron Resonance Spectroscopies Applied to Materials Science* (Kaufmann and Shenoy, eds.), Elsevier, Amsterdam 1981, p. 117.

122. D. C. Johnston, B. G. Silbernagel, M. Daage, and R. R. Chianelli, ACS Symposium on Nature of Active Sites in Heterogeneous Catalysts, Miami Beach, FL, 1985.

123. P. L. McDaniel, T. M. Barbara, and J. Jonas, *J. Phys. Chem. 92*: 626 (1988).

124. P. L. McDaniel, G. Liu, and J. Jonas, *J. Phys. Chem. 92*: 5055 (1988).

125. P. G. Rodewald, U.S. Patent No. 3,984, 352, Mobil Oil Corp., 1976.

126. G. R. Miller, H. A. Resing, F. L. Vogel, A. Pron, T. C. Wu, and D. Billaud, *J. Phys. Chem. 84*: 3333 (1980).

127. G. R. Miller, H. A. Resing, P. Brant, M. J. Moran, F. L. Vogel, T. C. Wu, D. Billaud, and A. Pron, *Synth. Metals 2*: 237 (1980).

128. L. G. Banks, H. A. Resing, D. C. Weber, C. Carosella, G. R. Miller, and P. Brant, *J. Phys. Chem. Solids 43*: 351 (1982).

129. H. A. Resing, M. J. Moran, and G. R. Miller, *J. Chem. Phys. 76*: 1706 (1982).

130. G. R. Miller, H. A. Resing, M. J. Moran, L. Banks, F. L. Vogel, A. Pron, and D. Billaud, *Synth. Metals 8*: 77 (1982).

131. M. J. Moran, G. R. Miller, R. A DeMarco, and H. A. Resing, *J. Phys. Chem. 88*: 1580 (1984).

132. G. R. Miller, C. F. Poranski, and H. A. Resing, *J. Chem. Phys. 80*: 1708 (1984).

133. G. C. Chingas, J. Milliken, H. A. Resing, and T. Tsang, *Synth. Metals 12*: 131 (1985).

134. R. Grosse and B. Boddenberg, *Z. Naturforsch, 41a*: 1361 (1986).

135. B. Boddenberg an G. Neue, *Z. Naturforsch, 42a*: 948 (1987).

136. B. Boddenberg, V. Grundke, and G. Auer, *Ber. Bunsenges Phys. Chem. 94*: 348 (1990).

7

New NMR Techniques for the Study of Catalysis

Waclaw Kolodziejski and Jacek Klinowski

University of Cambridge, Cambridge, England

I. INTRODUCTION

New applications of NMR to catalysis do not always accompany the development of new NMR techniques, perhaps because novel NMR ideas often originate from researchers who may be primarily interested in the technique itself rather than in any specific problem. On the other hand, the chemist who does have a practical problem often finds it hard to cope with the more refined methods and their physical background—particularly in solid-state NMR. We try to bridge this gap here by reviewing some modern NMR methods of particular interest for catalysis, stressing throughout the chemical aspects of the applications.

While the main recent advance in NMR has been the development of multidimensional spectroscopy, novel catalytic applications include in situ studies and two-dimensional (2D) solid-state techniques such as correlation spectroscopy, spin diffusion, and quadrupole nutation. Completely new techniques have appeared, such as multiple-quantum spin counting, and old ones have developed in quite unexpected directions. For example, cross-polarization, a 20-year-old experiment, has recently been applied to quadrupolar nuclei to yield important new information on heterogeneous catalysts. Magic-angle spinning (MAS) of quadrupolar nuclei has been extended to methods in which the sample is spun about two different angles either simultaneously or sequentially (DOR and DAS). These experiments have been made possible by the significant advances in NMR instrumentation in the last decade.

The division of the material discussed is somewhat arbitrary, and the ordering of the new NMR methods is not any indication of their relative importance or ease of application to catalysis. We hope with this selection to assist readers in finding fresh applications for these techniques, and we look forward to discussing them in future.

II. TECHNIQUES INVOLVING J-COUPLING

A. J-Resolved Spectroscopy

The application of 2D NMR to the separation of chemical shifts from scalar interactions is routine for liquid samples [1]. The technique is a powerful tool for signal assignment, and its performance has been examined in the solid state. However, the few published model studies mostly deal with "special" solids such as plastic crystals, where dipolar interactions are substantially reduced by molecular reorientation. Examples include adamantane [2,3], camphor [4,5], hexamethylbenzene [3], p-dimethoxybenzene and β-quinol methanol clathrate [6], camphene [7], and the organosilicon compound known as Q_8M_8 [8]. The first practically useful experiment of this kind [9] monitored the products of catalytic conversion of methanol to gasoline on zeolite H-ZSM-5, in which fast molecular reorientation mimics the liquid state. The ^{13}C spectrum (Fig. 1), recorded with the pulse sequence used for liquids, contains J-coupling information in the F_1 dimension and chemical shift information in the F_2 dimension. Although the experiment favors J-resolution of resonances from methyl groups because the dipolar interactions in these are reduced by fast group rotation [10], it is nonetheless very valuable for spectral assignment (Table 1).

With rigid solids, it is useful to employ simultaneous suppression of homonuclear ^1H dipolar interactions by multiple-pulse decoupling and of heteronuclear ^1H − ^{13}C dipolar interactions by MAS. Consider the homonuclear dipolar H − H ($\mathcal{H}_{II}$) and heteronuclear dipolar C − H ($\mathcal{H}_{IS}$) interactions. The combined $\mathcal{H}_{II} + \mathcal{H}_{IS}$ interaction is homogeneous, but homonuclear dipolar decoupling makes it inhomogeneous by suppressing $\mathcal{H}_{II}$. The broadening resulting from the residual inhomogeneous $\mathcal{H}_{IS}$ interaction can then be eliminated by MAS, even if the spinning frequency is lower than the dipolar width [12,13]. The heteronuclear scalar J-coupling survives, and its magnitude is scaled depending on the type of pulse sequence used for the homonuclear dipolar decoupling. Pulse sequences for 1D and 2D J-spectroscopy in solids are given in Fig. 2, and cross-sections of the 2D ^{13}C spectra of p-dimethoxybenzene [6] and dimethylsulfoxide (DMSO) intercalated into kaolinite [14] are shown in Fig. 3.

The technique does entail some practical difficulties. 2D ^{13}C J-resolved NMR spectra of solids can be obtained if a certain degree of local motion is present and if the transverse relaxation times are sufficiently long, but multiplets are often

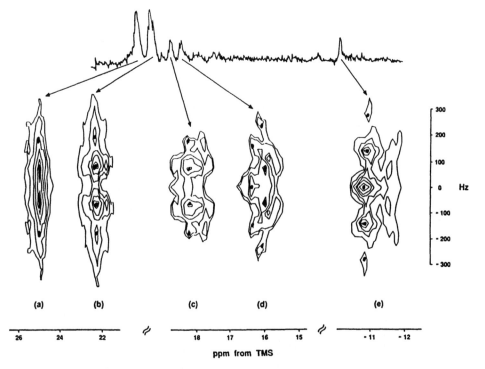

Figure 1 2D ^{13}C J-resolved MAS NMR spectrum of the products of a catalytic conversion of methanol to gasoline on zeolite H-ZSM-5 [9]. Methanol was enriched to 30% in the ^{13}C isotope.

Table 1 Parameters of the Two-Dimensional Spectrum Shown in Fig. 1

Signal	Chemical shift/ppm	Signal multiplicity	J-Coupling/ Hz	Tentative assignment [11]	2D J-Resolved assignment
a	24.7	4	135	Isobutane	Isobutane
b	22.2	4	135	*n*-Hexane Isopentane *n*-Heptane	Isopentane
c	18.7	4	135	Methyl- substituted benzenes	Methyl- substituted benzenes
d	16.7	3	130	Propane	Propane
	16.0	4	130		
e	−11.0	5	135	Methane	Methane

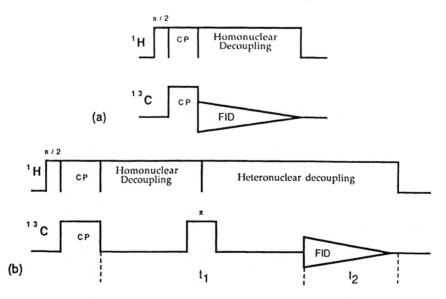

Figure 2 Pulse sequences for (a) 1D and (b) 2D J-spectroscopy in solids [3,5].

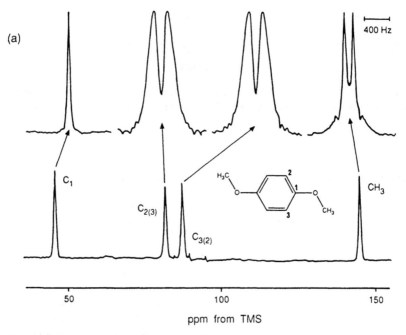

Figure 3 F_1 cross-sections of the 2D ^{13}C J-resolved spectra (top) and corresponding CP/MAS spectra (bottom) of: (a) p-dimethoxybenzene [6]; (b) kaolinite: DMSO intercalate [14]. The two methyl quartets in the intercalate spectrum come from inequivalent DMSO molecules between the kaolinite layers. Both J-resolved spectra were recorded using MREV-8 homonuclear decoupling at natural ^{13}C abundance.

(b)

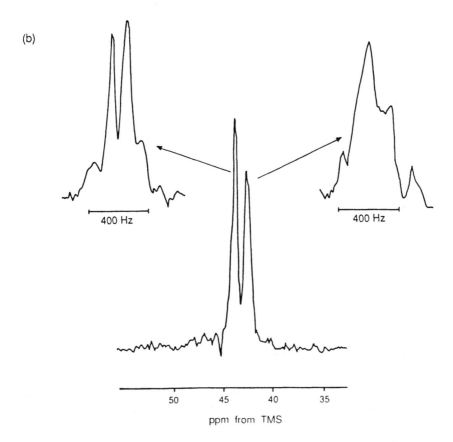

distorted and the intensities of the components are not in the ratios expected from Pascal's triangle. For example, the quartets in Fig. 3 do not have the ratio of intensities of 1:3:3:1. The following considerations may be helpful for setting up experiments:

1. The radiofrequency field strengths must be carefully calibrated, preferably using the 2D experiment described by Nielsen et al. [15]. Strong radiofrequency fields are most convenient because they increase the efficiency of the multiple-pulse homonuclear decoupling, while high stability of the power and phases of the radiofrequency pulses is crucial.

2. MREV-8 (or another homonuclear pulse sequence) must be set up very carefully. The ^1H frequency offset must be optimized using adamantane and the pulse sequence for 1D J-resolved spectroscopy in solids.

3. The F_1 scaling factor can be derived from a 2D ^{13}C J-resolved experiment on adamantane and the J_{C-H} data quoted in the literature [16].

4. Before the real sample is examined, the methylene and methine carbons (resonating in the range 43–44 ppm) in the spectrum of camphor must be clearly resolved [5].
5. The magic angle must be set very accurately on the real sample: an error of as little as 0.1° leads to line broadening of 50 Hz, which is comparable to the scaled coupling constants [6].
6. The revolution period of the rotor, t_r, must be an integer multiple of the duration of the decoupling cycle, t_c, and the F_1 dwell time must be an integer multiple of t_r. Also, t_c must be as short as possible, and the ratio t_r/t_c must be at least 3.

Several new assignment techniques for ^{13}C solid-state NMR spectra have recently been proposed: *S*pectral *E*diting by *Mu*ltiple-*Q*uantum *T*raps (SEMUT) [15], separated local field MAS [17–19], *W*indowless *I*sotropic *M*ixing for *S*pectral *E*diting (WIMSE) [20], *Off-R*esonance *P*roton *I*rradiation (ORPI) [21,22], and cross-polarization with fast MAS [23]. Their detailed performance has yet to be assessed, but it is already clear that the latter two methods are simple and eminently worthwhile.

B. Correlation Spectroscopy (COSY and INADEQUATE)

Two-dimensional homonuclear correlation spectroscopy (COSY) is well established for the study of liquids [24,25] and has recently been shown to be effective with solid samples [8,26–38], many of which are of catalytic interest [30,37]. The COSY spectrum contains diagonal peaks and off-diagonal cross-peaks. The cross-peaks arise because of coherence transfer between spins, and they indicate that the resonances at the relevant shift positions on the two axes are coupled. In the solid state, the necessary coherence transfer may occur through dipolar or scalar interactions. Experiments based on scalar couplings have been more popular, because in favorable circumstances they allow the spectroscopist to establish unambiguously the atomic connectivities within molecules in solid samples [8,27–38] including complex zeolitic frameworks [8,30–37].

However, several factors impose severe constraints on COSY experiments in the solid state. First, one is usually forced to use dilute nuclei embedded in a crystal lattice. The cross-peaks due to coupled spins are thus much less intense than diagonal peaks originating mostly from isolated spins, resulting in the strong diagonal peaks swamping the weaker cross-peaks. ^1H COSY is ineffective in the solid state mainly because of rapid spin diffusion [39] among protons. Solid-state COSY experiments on dilute nuclei at natural isotopic abundance have been successfully performed so far only with ^{29}Si in the organosilicon compound Q_8M_8 [8,28] and in highly siliceous zeolites [30–37], in both cases giving fairly sharp resonances. No ^{13}C COSY of organic solids at natural ^{13}C isotopic abundance has been reported. Nevertheless, this area of research will offer ample rewards to any spectroscopist able to overcome the difficulties. So far, the only COSY experiments reported on abundant nuclei are ^{31}P COSY in polycrystalline $\alpha - P_4S_7$ and in a solid phosphine complex of Hg(II) [29]. Isotopic enrichment in ^{13}C and [29]Si has been found useful in model and preliminary studies [26,27,30,31,38].

Most 2D NMR methods in solids have low sensitivity because of fast transverse relaxation. In particular, for the COSY experiment the system must evolve for sufficiently long time periods in both dimensions before satisfactory intensities of cross-peaks can be obtained. Accordingly, Fyfe et al. [30,31] introduced two extra delays into their solid-state COSY pulse sequence, a concept originally conceived for so-called "long range" or "delayed" COSY in liquids [40]. If transverse relaxation is too fast on the time scale of the required evolution and acquisition periods, there will be no cross-peak magnetization to detect. Furthermore, rapid transverse relaxation leads to wide lines, so that diaganol peaks can overlap with adjacent cross-peaks, which are already very weak because of the destructive interference of their broad antiphase components.

Rotation of powdered samples involves another difficulty [27,41,42]: magic-angle spinning makes the orientation of each crystallite time-dependent. In the presence of substantial chemical shift anisotropy, this modulation leads, after $\pi/2$ pulses, to unwanted rotor lines [41], which appear in both dimensions and form cross-peaks of their own. The situation becomes even more complicated when π pulses or a train of different pulses is involved, especially if the magic angle is set incorrectly [42]. In addition, the rotational sidebands can be enhanced and the observed peaks correspondingly broadened for certain MAS frequencies [27].

Attempts have been made to overcome the problem of the large diagonal peaks stemming mostly from isolated spins. These signals can be removed in the INADEQUATE experiment [43,44], which is, however, very insensitive even in liquids and requires a reasonably accurate prior estimation of the coupling constants [45]. Also, in the solid state the disadvantage of the rapid transverse relaxation is more acute in INADEQUATE, which uses a longer pulse sequence than conventional COSY. Nevertheless, there have been successful [29]Si 2D INADEQUATE experiments on highly siliceous zeolites [32–35,46,47], samples that are ideal for such work. It is also worth mentioning ^{13}C INADEQUATE experiments on solid camphor and adamantane [28,48,49] and a [119]Sn INADEQUATE experiment on solid $[Me_2SnS]_3$ [49], all of them at natural abundance. Surprisingly, the first solid-state INADEQUATE experiment was done *before* solid-state COSY by Menger et al. [50], who used a ^{13}C-enriched single crystal of glycine. This experiment is unique because it is based on dipolar rather than scalar couplings.

The excitation of a double-quantum coherence can also be used in double-quantum filtered COSY (DQF-COSY) [51]. The only useful experiment of this type was performed on a ^{29}Si-enriched sample of highly siliceous zeolite ZSM-39 [31]. In theory, DQF-COSY is only half as sensitive as ordinary COSY, but it filters out the signals of the uncoupled spins and does not require any knowledge of the coupling constants. We examined the performance of ^{13}C DQF-COSY while trying to find a method to trace carbon connectivities in mobile organic products of reactions on zeolites. Unfortunately, the method seems to work only with plastic crystals, such as adamantane, in which it is even possible to resolve $^{13}C-^{13}C$ J-splittings [52]. This situation is reminiscent of the results of Gay et al. [49], who succeeded with ^{13}C INADEQUATE on adamantane but failed with rigid solids. We conclude that excitation of the double-quantum coherence in solid-state ^{13}C NMR using J-couplings probably has no future because the linewidth usually exceeds the magnitude of the $^{13}C-^{13}C$ scalar interaction, leading to a cancellation of the antiphase doublets [49].

J-scaled COSY [53] has been used to overcome the cancellation problem (Fig. 4). The technique *scales up* the scalar splittings between the cross-peak components, thereby enhancing cross-peak intensities and consequently improving spectral resolution between adjacent diagonal and cross-peaks. The following example will show the advantages of ^{29}Si J-scaled COSY in particular and explain the application of COSY in general. The experiment was done at natural isotopic abundance on highly siliceous mordenite using a scaling factor of 5. The conventional ^{29}Si MAS NMR spectrum of highly siliceous mordenite (Fig. 5a) consists of three peaks in the intensity ratio of $2:1:3$ [54–56]. This pattern may be explained from the known structure of mordenite [57,58], which contains four distinct tetrahedral crystallographic sites in the intensity ratio $T1:T2:T3:T4 = 2:2:1:1$ (Fig. 6), with two of the peaks overlapping. Assignment of the signals has in the past relied on the correlation between ^{29}Si chemical shifts and the mean Si-O-Si bond angle, α. This dependence theoretically goes as $<\cos \alpha/(\cos \alpha - 1)>$, but follows an approximately linear relationship in the regime under investigation [59,60]. The published solutions of the structure for various forms

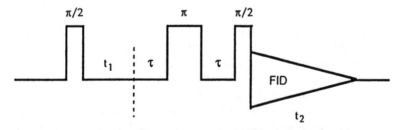

Figure 4 J-scaled COSY ^{29}Si MAS NMR pulse sequence. The relationship between τ and t_1 is chosen so that J-splittings are scaled up by a factor of 5.

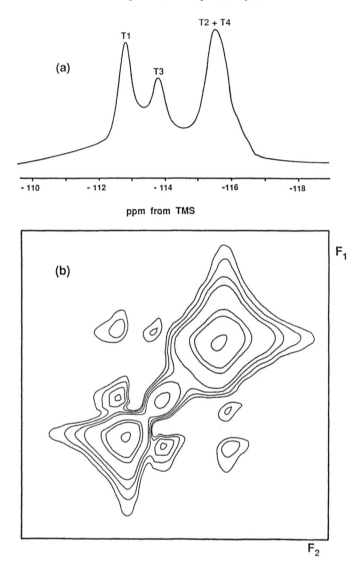

Figure 5 NMR spectra of highly siliceous mordenite. (a) ^{29}Si MAS NMR spectrum; (b) J-scaled COSY spectrum [37].

of mordenite [57,61,62] show that the mean T-O-T bond angles vary slightly with the degree of dealumination, cation type, and water content, while the relative values remain approximately constant. The values for a highly siliceous mordenite [61] given in Table 2 permit the immediate assignment of the down-field peak to the T1 site and show that the T2 site is a component of the strongest

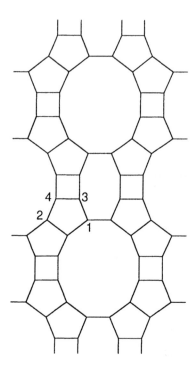

Figure 6 Structure of mordenite viewed along [001] plane [57,58]. The four kinds of crystallographic sites are indicated. Their relative populations (16:16:8:8 per unit cell) are not reflected in this projection.

peak. The <T3-O-T> and <T4-O-T> bond angles are similar, however, so it is not possible to assign the spectrum completely on the basis of bond angles alone. The two possible assignments of the three peaks in the spectrum are to T1:T4 : T2 + T3 (as in Refs. 54 and 56), or to T1:T3:T2 + T4 crystallographic sites.

The 2D J-scaled COSY spectrum of highly siliceous mordenite (Fig. 5b) reveals three cross-peaks that could not be resolved by COSY using two extra delays (as described by Fyfe et al. [30,31]). Attempts to do so yielded a spectrum with the cross-peaks obscured by the intense peaks on the main diagonal. Couplings between the tetrahedral sites are expected to be in the range of 10–15 Hz [33]—too small to give rise to prominent cross-peaks, unless they are scaled up by the particular pulse sequence used in J-scaled COSY. On the basis of the known connectivities of the mordenite structure (Table 2), only two cross-peaks are predicted for the T1:T4:T2 + T3 assignment, whereas the T1:T3:T2 + T4 assignment implies that three cross-peaks should be observed. Thus the detection of three cross-peaks in the 2D J-scaled COSY experiment shows that the correct

Table 2 Connectivities and Typical Mean T-O-T Bond Angles in the Mordenite Structure [61]

T-site	No. per unit cell	Neighboring sites	Mean T-O-T bond angle
T1	16	T1, T1, T2, T3	150.4°
T2	16	T1, T2, T2, T4	158.1°
T3	8	T1, T1, T3, T4	153.9°
T4	8	T2, T2, T3, T4	152.3°

interpretation is T1:T3:T2 + T4. This unambiguous assignment of the spectrum is not possible by 1D NMR or by conventional COSY.

III. TECHNIQUES INVOLVING DIPOLAR COUPLING

A. Spin Diffusion and Rotational Resonance

Spectral spin diffusion in the solid state involves simultaneous flipflop transitions of dipolar-coupled spins with *different* resonance frequencies [1,39,63–76], whereas spatial spin diffusion transports spin polarization between spatially separated *equivalent* spins. In this review we deal only with the first case. The interaction of spins undergoing spin diffusion with the proton reservoir provides compensation for the energy imbalance (extraneous spins mechanism) [68,70,73,74]. Spin diffusion results in an exchange of magnetization between the nuclei responsible for resolved NMR signals, which can be conveniently detected by observing the relevant cross-peaks in the 2D spin-diffusion spectrum [63–65]. This technique, formally analogous to the NOESY experiment in liquids, is already well established for solids and can also be applied to the study of catalysts.

Pulse sequences for 2D spin-diffusion experiments [63–65,76] are given in Fig. 7. Cross-polarization, or simply a $\pi/2$ pulse, creates transverse coherence, which precesses during the evolution period t_1. A $\pi/2$ pulse then converts the transverse coherence into longitudinal magnetization, which is allowed to migrate by spin diffusion for a time Δ without proton decoupling. The last $\pi/2$ pulse reconverts longitudinal magnetization into observable transverse coherence, the signal being detected during the acquisition period t_2. For ^{13}C in organic solids, proton decoupling during t_1 and t_2 is necessary, but for ^{31}P in aluminophosphates it is optional.

The few published ^{31}P solid-state spin-diffusion studies deal mostly with model cases of magnetization transfer between two sites [73,75,76]. We undertook a 2D ^{31}P NMR study of spin diffusion in aluminophosphate molecular

(a)

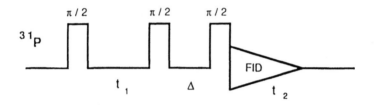

(b)

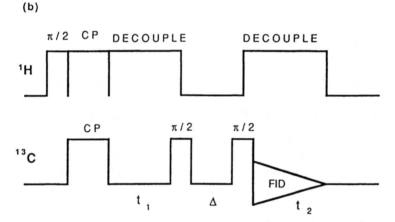

Figure 7 Pulse sequences for the spin-diffusion experiments described in the text: (a) ^{31}P spin diffusion [76]; (b) ^{13}C spin diffusion [63–65].

sieves and chose VPI-5 as a model sample [77]. VPI-5 is a crystalline aluminophosphate molecular sieve containing 18-membered rings of tetrahedral atoms [78,79]. The large channel diameter (ca. 12 Å) gives VPI-5 considerable potential for the separation of large molecules and for catalytic cracking of the heavy fractions of petroleum, which at present are discarded as bottom-of-the-barrel residue. On the basis of x-ray diffraction studies, the structure of VPI-5 was refined in the P6$_3$ space group [80] (see Fig. 8). The ^{13}P MAS NMR spectrum of hydrated VPI-5 shows three signals in an intensity ratio of 1:1:1. Variable-temperate studies [81] assigned the signals at -23 ppm (1) and -27 ppm (2) to phosphorus atoms in $6-4$ sites (P2 and P3) and the signal at -33 ppm (3) to phosphorus atoms in $4-4$ sites (P1) [80,81]. However, it is not known how signals 1 and 2 are to be assigned to *particular* P2 and P3 sites. The structural inequivalence of the P2 and P3 sites results from framework distortion caused by the coordination of two water molecules to Al1. This distortion is so small that

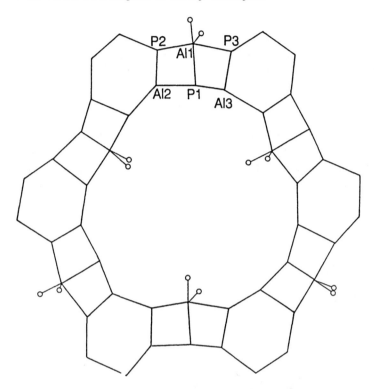

Figure 8 One layer of the framework structure of hydrated VPI-5 taken from the stereoscopic view along the [001] direction according to McCusker et al. [80], showing the deviation from P6₃cm symmetry. Aluminum and phosphorus atoms, linked via oxygen atoms (not shown for clarity), are located at the apices of the polygons. Sites located between two fused four-membered rings are known as 4–4 sites; those located between six-membered and four-membered rings are known as 6–4 sites. P2 and P3, and Al2 and Al3 sites are inequivalent as a result of the distortion. The Al1 site is six-coordinated as a result of bonding to four bridging oxygens and two "framework" water molecules. Other intracrystalline water is not shown.

the correlation between the ^{31}P chemical shift and mean Al-O-P angle [82] is not sufficiently sensitive to be useful for the assignment. A small difference between mean Si-O-Si angles in zeolites may lead to incorrect assignments of NMR signals on the basis of their [29]Si chemical shift [37], and therefore one must use more sophisticated methods, such as COSY, to arrive at the correct assignment [37].

2D NMR reveals that spin diffusion is present over a wide range of mixing times (from 0.1 s to 10 s) [77]. The spectra have a remarkably high signal-to-noise ratio (Fig. 9) and each pair of signals gives rise to cross-peaks. However, Fig. 9 shows that spin diffusion can be slowed down by fast MAS. We have proved that the proton reservoir is not relevant to the ^{31}P spin-diffusion process in VPI-5, and thus that this process is governed only by the $^{31}P - {}^{31}P$ dipolar interactions. Therefore, a reasonable assumption would be that only the strongest $^{31}P - {}^{31}P$ dipolar interactions survive under fast MAS, and this must be reflected in the dependence of the cross-peaks intensities on the rate of MAS. We compared the strongest $^{31}P - {}^{31}P$ dipolar interactions calculated from the XRD structure [80]: P2 – P3 (192 Hz) < P1 – P2 (281 Hz) < P1 – P3 (304 Hz) with the MAS rates (in brackets) corresponding to the inflection points of the curves in Fig. 10: 1 – 2 (5.91 kHz) < 1 – 3 (6.03 kHz) < 2 – 3 (8.07 kHz).

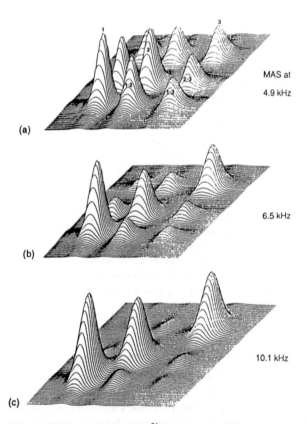

Figure 9 Experimental 2D ^{31}P NMR spin-diffusion spectra of hydrated VPI-5 recorded with 3-s mixing time at various MAS speeds [77]: (a) 4.9 kHz; (b) 6.5 kHz; and (c) 10.1 kHz. The three spectra are not on the same intensity scale, so only the relative intensities within each can be compared.

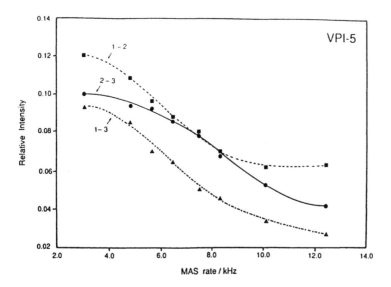

Figure 10 Relative intensity of the cross-peaks versus MAS rate for the ^{31}P NMR spin-diffusion experiments on hydrated VPI-5 with a 3-s mixing time [77]. The relative intensity is defined as a fraction of the total intensity of the 2D spectrum. Computer-fitted polynomial curves of the fourth order provide a guide for the eye and serve for the calculation of the inflection points (see text).

This analysis showed that signals 1, 2, and 3 should be assigned to sites P2, P3, and P1, respectively, and that spin diffusion could be useful for the assignment of spectral peaks in other phosphate molecular sieves.

Zeolite H-ZSM-5 is a powerful heterogeneous catalyst capable of converting methanol to gasoline. Three recent papers [9,11,83] report ^{13}C NMR spectra of the reaction products in situ using samples contained in specially designed capsules [84] spun at the magic angle. It was possible to identify a number of different organic species *in the adsorbed phase* and to monitor their fate during the course of the reaction [11,83]. In addition to conventional spectral assignment based on chemical shifts and signal intensities, 2D ^{13}C NMR was used to determine the number of protons coupled to each carbon in the various organics [9] (see Sec. II.A). A 2D solid-state spin-diffusion ^{13}C NMR experiment in situ [10] (Fig. 11) assists in the spectral assignment and, combined with model studies, is likely to provide new information concerning the distribution of hydrocarbon species in the intracrystalline space of the ZSM-5 catalyst.

Spin diffusion occurs between nuclei in adjacent functional groups within the same molecule (the intramolecular case) [67,69] or between nuclei in neighboring species mixed at the microscopic level (the intermolecular case) [65,69].

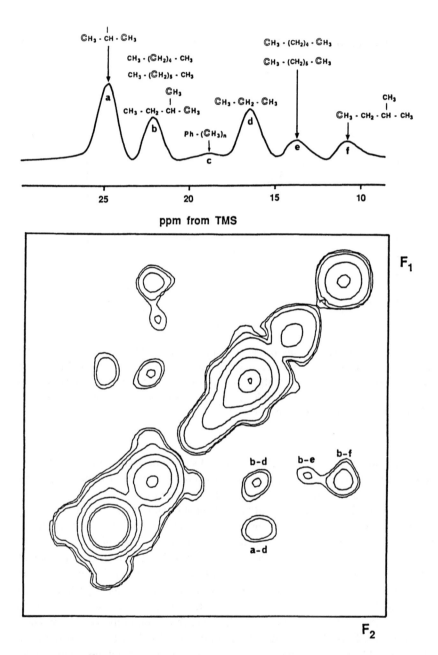

Figure 11 ¹³C NMR spin-diffusion spectrum of products of the conversion of methanol into gasoline over zeolite H-ZSM-5 with the projection at the top [10]. Carbon atoms to which individual resonances are assigned are highlighted. Methanol was enriched to 50% in the ¹³C isotope. The mixing time $\Delta = 1.5$ s was found to yield the best signal-to-noise ratio.

Both cases are observed in the ZSM-5 system under consideration (Table 3). However, the interpretation is not straightforward, since it requires some assumptions based on the system features and the nature of the spin-diffusion phenomenon. Thus zeolite ZSM-5 contains no cages, and its channel diameter only allows the hydrocarbon species in the channels to be lined up sequentially. For any two molecules to exchange positions, access to an unoccupied channel crossing is required, a condition difficult to satisfy at high adsorbate loadings (30% w/w in this case). Hydrocarbon molecules are capable of limited motion along the channels, which disfavors *intermolecular* spin diffusion. Free isotropic molecular rotation cannot occur, so that *intramolecular* dipolar interactions are present even for quite mobile functional groups and thereby make intramolecular spin diffusion possible. Intramolecular spin diffusion is preferred here to intermolecular spin diffusion, which is exactly opposite to the case of adamantane reported by Bronnimann et al. [64]. We note that the optimal mixing time (producing the most intense cross-peaks) is generally different for each pair of

Table 3 Assignment of the 2D ^{13}C NMR Spin-Diffusion Spectrum Shown in Fig. 11 [10]

	Diagonal peaks			
Signal	Chemical shift (ppm)	Group	Assignment	Ref.
a	24.7	CH_3	Isobutane	9, 11, 83
b	22.3	$CH_3CH_2CH(\underline{C}H_3)_2$	Isopentane	9, 11, 83
		CH_2	*n*-Hexane + *n*-heptane	11, 83
c	18.7	CH_3	Methyl-substituted benzenes	9, 11, 83
d	16.7	$CH_3 + CH_2$	Propane	9, 11, 83
e	14.3	CH_3	*n*-Hexane + *n*-heptane	11, 83
f	11.2	$\underline{C}H_3CH_2CH(CH_3)_2$	Isopentane	11, 83

	Cross-peaks	
Signals	Assignment	Type of spin diffusion
a-d	Isobutane-propane	Intermolecular
b-d	Isopentane-propane	Intermolecular
b-e	*n*-Hexane	Intramolecular
	n-Heptane	Intramolecular
b-f	Isopentane	Intramolecular

the participating diagonal peaks, and in unfavorable cases short longitudinal relaxation times can hinder or even prevent the detection of cross-peaks. Therefore, no conclusions can be drawn from the absence of some cross-peaks and the presence of others, unless the dipolar interactions underlying the corresponding spin-diffusion cases are closely similar (involving the same functional groups in similar molecules), and unless the relevant longitudinal relaxation times T_1 are alike.

Signal a (see Table 3) produces only one cross-peak (with signal d), and its assignment is obvious since a and d each belong to a single compound. The b − d cross-peak must be classified as intermolecular and may be assigned variously to: isopentane-propane, n-hexane (CH_2)-propane, or n-heptane (CH_2)-propane spin diffusion. Clearly, the assignment poses severe problems. Spin diffusion between propane and n-hexane or n-heptane would have to involve, first, their CH_3 groups, but the d − e cross-peak is missing. This indication, while not conclusive (see above), inclines us to suggest that the b − d cross-peak be assigned to isopentane − propane. The b − e and b − f cross-peaks can come from either intermolecular or intramolecular spin diffusion, the latter being more likely.

Further work, especially model studies, is needed if spin diffusion is to be a useful tool for the investigation of molecular catalysis on zeolites. By adsorbing various compounds and their mixtures, typical mixing times for the communication between various functional groups in various molecules under intermolecular and intramolecular spin diffusion could be established. This determination would provide insight into the redistribution of the intermediates and reaction products on the catalyst and allow us to address the unresolved questions. For example, we note that prominent intermolecular a − d and b − d cross-peaks are present, but there is no intermolecular a − b cross-peak. It would be interesting to learn whether branched hydrocarbons such as isobutane and isopentane occupy channel crossings, and whether their consequent remoteness prevents the spin diffusion. Further, one could enquire whether propane molecules mostly occupy zeolite channels, so that efficient spin diffusion between propane and adjacent isobutane or isopentane may take place. Such studies are currently in progress.

We have so far been concerned with 2D spin-diffusion spectroscopy. There are, however, two 1D experiments that are likely to be applied to catalytic problems: selective excitation [72,85,86] and rotational resonance [87–93]. Selective excitation of selected resonances using the DANTE pulse trains [94] can be used to measure specific $^{13}C − ^{13}C$ connectivity in complex, multiple ^{13}C-labeled solids [86]. Rotational resonance can be achieved by adjusting the MAS rate ω_r to an integer fraction of the chemical shift difference ω_Δ^{iso} between two selected carbon resonances: $\omega_\Delta^{iso} = n\omega_r$, n being a small integer. Under this condition the energy mismatch of the spin flipflop process is balanced by the mechanical rotation energy, leading to rotor-driven spin diffusion. Dramatic broadening and splitting of lines may also be observed. The current theory of spin diffusion and lineshapes in rotational resonance [89,93] applies to magnetically dilute spin pairs, which can be created by selective ^{13}C labeling. In particular, the magnetic dilution condition is fulfilled for catalytic surfaces at low loadings. Protons need to be decoupled to eliminate their influence on the Zeeman magnetization transfer. Theory enables us to simulate the lineshapes and the magnetization transfer dynamics in rotational resonance using the internuclear distance, the isotropic shifts, the principal values of the shift tensors and their mutual orientation, and the decay time constant of the zero-quantum coherence of the two spins involved. The principal values of the shift tensors are known for many functional groups, and the zero-quantum relaxation rate can be estimated as the sum of relevant single-quantum transverse relaxation rates. Therefore, by matching experimental results with numerical simulations, it is possible to determine the internuclear distance accurately and to deduce the relative orientation of the molecular groups from the relative orientation of their shift tensors. Rotational resonance NMR is capable of measuring the distance between two homonuclear sites separated by as much as 0.505 nm [92] (see Fig. 12).

B. Multiple-Quantum NMR

Multiple-quantum NMR [95–98] is a fascinating "spin counting" tool for characterizing homonuclear spin clusters in the solid state [99–106]. The method is particularly well suited for studies of species adsorbed at surfaces [99,102] or in the voids of molecular sieves [104,106]. There have been not many applications to catalysis so far, but we believe the method will enjoy wider popularity in future.

The technique employs a specially designed 2D NMR pulse sequence [100], which forces nuclear spins to act collectively via their dipolar couplings, thereby creating unobservable multiple-quantum (MQ) coherences. The MQ coherences then evolve in the t_1 time domain and, after conversion into observable single-quantum coherences, are indirectly detected over the acquisition time t_2. The multiple-quantum information is contained in the F_1 dimension of the 2D spectrum (Fig. 13). The modified phase-incremented experiment [105] proceeds with the evolution held fixed and offers ease of operation and more accurate intensity distributions in return for loss of information contained in the fine structure and shapes of the MQ peaks.

The collective behavior of spins can be roughly understood as a simultaneous flip of two (double-quantum coherence), three (triple-quantum coherence), or more spins (multiple-quantum coherence). In other words, the spin system coherently exchanges two, three, or ΔM photons, respectively, with the resonant

(a)

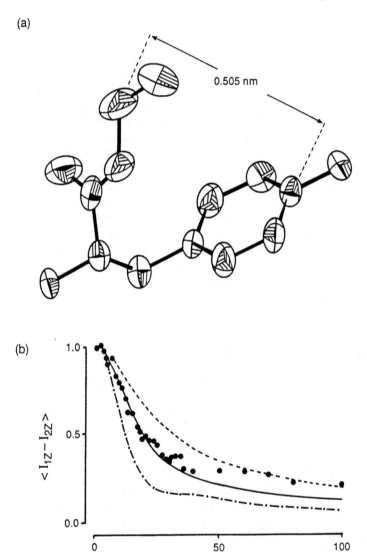

0.505 nm

(b)

Figure 12   (a) Structure of tyrosine ethyl ester showing the separation between the labeled carbons. (b) Calculated and experimental evolution of the difference polarization for the $n = 1$ rotational resonance in tyrosine ethyl ester at 9.5 T and $\omega_r/2\pi = 9.400$ kHz [92]. The solid curve is calculated for 0.505 nm, and the upper and lower dotted curves for 0.555 and 0.455 nm, respectively.

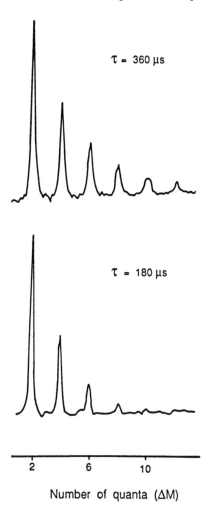

Figure 13 F_1 cross-sections of the 2D ^{1}H MQ NMR spectra of hydrogenated amorphous silicon containing 50 at. % of hydrogen (adapted from Ref. 102). The spectra were recorded at two different preparation times τ given.

radiation (M being the magnetic quantum number). This action can occur only if the preparation time τ is sufficient to establish communication, via pairwise dipolar interactions, within the group of spins intended to be excited [95,99,101]. Dipolar interaction renders two spins correlated after a time period roughly proportional to the cube of their separation. It follows that multiple-quantum coherences develop at a rate determined by the spatial distribution of spins in the sample. For a uniform distribution, the MQ transition order ΔM grows monotonically with increasing τ. However, if the system is a collection of diluted, isolated clusters, the MQ count will reach a plateau corresponding to ΔM equal to the number of spins in the cluster (Fig. 14). For MQ transition orders of more than six, it is recommended [101,102,106] to fit the MQ intensity envelope to a Gaussian function and to derive the spin network size N from the standard deviation $\sqrt{\dfrac{N}{2}}$. If a sufficiently long propagation time τ is used, the system can be driven into even higher modes of coherence, and this behavior occurs when the clusters begin to communicate with each other. For the technique to be successful one must be able to identify the plateau, which means that the intercluster dipolar interactions should be suitably weaker than the intracluster dipolar interactions.

C. Spin-Echo Double Resonance (SEDOR)

Spin-Echo Double Resonance (SEDOR) [107–109] is an NMR technique capable of measuring local heteronuclear dipolar interactions between selected (by the spectroscopist) nuclei of different species I and S. An ordinary spin-echo sequence (Fig. 15a) observed for spins I refocuses their chemical shift, while also undoing the dephasing effects of the magnetic field inhomogeneity. The echo is phase-modulated by the homonuclear I − I dipolar coupling, but no modulation by the heteronuclear I − S dipolar coupling is observed if spins S are not irradiated. However, if they are affected by an additional π pulse in the double-resonance experiment (Fig. 15b and 15c), the local magnetic field seen by nearby spins I is modified, and this action changes their precessional frequencies and scrambles their precessional phases. As a result, the spin echo is attenuated. The magnitude of the effect is related to the magnitude of the heteronuclear I − S dipolar interactions, from which the relative spatial disposition of spins I and S can be deduced. It is usually necessary to carry out, in parallel, a normal spin-echo experiment with the S pulse absent to account for the echo decay due to homonuclear I − I spin interactions.

Two main SEDOR pulse sequences are in use at present (Fig. 15b and 15c). In the conventional experiment (Fig. 15b) [107,108,110,111], the π pulses in both channels are simultaneous, and the reduction of the echo amplitude by the S pulse (called the SEDOR decay [111]) is monitored as a function of the time interval τ. In the latest pulse sequence (Fig. 15c) [99,112–114], the time period τ between the I pulses is fixed, and the S pulse is applied between the two I pulses. The amplitude of the spin-echo signal of spins I is measured as a function of the time delay τ' between the S pulse and the first I pulse. The reduction in the spin-echo amplitude divided by the original amplitude (that is, the amplitude obtained without the S spins π pulse) is called the SEDOR fraction. For nuclei I not

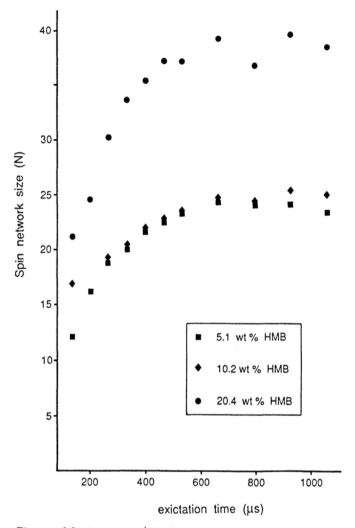

exictation time (μs)

Figure 14 Results of ^{1}H MQ NMR experiments for hexamethylbenzene (HMB) adsorbed at 573 K on dehydrated zeolite NaY [106]. The HMB molecule contains 18 hydrogens, so the spin cluster size is 18 per one molecule. At lower loadings (5.1 and 10.2 wt %) the plateau corresponds to about one HMB molecule per supercage and at higher loading (20.4 wt %) the plateau reflects an occupancy of two HMB guests per supercage.

(a)

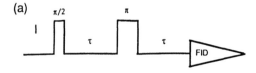

(b)

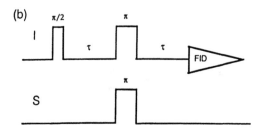

(c)

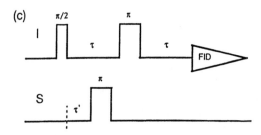

Figure 15 Spin-echo pulse sequences: (a) ordinary single-resonance spin-echo, (b) conventional SEDOR pulse sequence [107,108,110,111], (c) the latest SEDOR pulse sequence [99,112–114].

coupled to nuclei S, the extra S pulse makes no difference to the I signal and the SEDOR fraction is therefore zero. Both SEDOR functions, namely the SEDOR decay and the SEDOR fraction, which depend on τ and τ', respectively, contain information on heteronuclear dipolar interactions. The typical protocol of data refinement requires one to consider various possible models of the $I - S$ disposition and their relative motion, followed by calculation of the corresponding dipolar interactions and theoretical SEDOR functions. Comparison with the experimental result then allows one to choose the correct model.

A useful feature of the SEDOR experiment is that by measuring the dipolar coupling between selected nuclei, it probes only their immediate vicinity. This localization is achieved because the dipolar interaction is inversely proportional to the cube of the internuclear distance, so that it falls off very quickly with internuclear separation. To illustrate the point, let us consider what is, to our knowledge, the first application of SEDOR to heterogeneous catalysis. Makowka et al. [110] addressed the problem of CO chemisorption on small Pt particles supported on η-alumina. They studied the SEDOR effect on the ^{195}Pt signal during the irradiation of ^{13}C nuclei in the adsorbed monolayer of CO enriched to 90% in ^{13}C (see the pulse sequence in Fig. 15b). The spectrum in Fig. 16 was recorded at a fixed frequency of 73.1 MHz and with a fixed echo delay, while varying the static magnetic field. The SEDOR signal is due only to those ^{195}Pt nuclei that lie at the surface (being close to the CO molecules), whereas the ordinary spin echo detects all the ^{195}Pt nuclei in the sample. The relative areas of the signals provide information on Pt dispersion. For the particular cases shown in Fig. 16, the fraction of Pt atoms at the surface of the particles is about 40%, which is in fair agreement with the results of hydrogen chemisorption studies.

SEDOR is a very accurate technique for measuring internuclear distances in the heteronuclear case. It is indispensable if the dipolar couplings cannot be

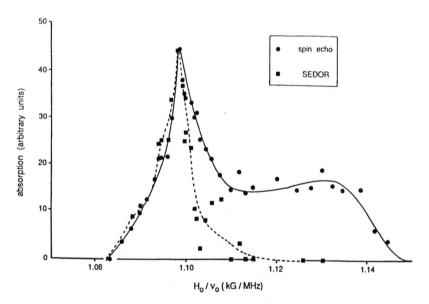

Figure 16 ^{195}Pt – ^{13}C SEDOR and ^{195}Pt spin-echo signals for Pt particles covered with a chemisorbed monolayer of ^{13}CO molecules [110]. The asymmetrical shape results from variations in the chemical and Knight shifts across the surface, from anisotropies in the chemical and Knight shifts, or from a superposition of lines with different widths with slightly different positions due to inequivalent Pt bonding.

determined from lineshape analysis because the signal is obscured either by field inhomogeneity due to catalyst magnetic susceptibility or by chemical shift anisotropy [99]. For surface-adsorbed species, SEDOR data on internuclear distances are of great importance because they allow one to examine the surface-adsorbate bonding. For instance, a Pt − C separation of about 1.5 Å was found for CO chemisorbed on highly dispersed platinum [110]. Since this distance is shorter than the bond length in any Pt carbonyl, the CO molecules must be bonded to the surface Pt atoms via the C atoms. The SEDOR-measured C − O bond length of 1.20 ± 0.03 Å for CO chemisorbed on Pd particles [113] clearly favors the bridge type over the linear bonding configuration, since it is typical of that in bridging carbonyls. Recently, $^{27}Al − ^{31}P$ SEDOR has been used to measure Al − P distances in aluminophosphate molecular sieves [114]. Such information can provide valuable insight into the local structure of these catalysts. For example, the technique seems capable of distinguishing between framework and extraframework Al in zeolites.

SEDOR is highly suitable for identifying intermediate species at catalytic surfaces. Sometimes it is even possible to draw conclusions simply by inspecting the behavior of the SEDOR fraction for long τ' intervals. Note that as τ' increases, the SEDOR fraction approaches an asymptotic value, which is roughly the fraction of I atoms having neighbor S atoms. How the SEDOR fraction approaches its asymptotic value depends on I − S dipolar couplings characteristic for the I − S grouping under study. Consider, for example, the results of Wang et al. [112], who investigated ^{13}C-ethylene adsorption on small Pt clusters supported on η-alumina. The following species formed from ethylene adsorbed at room temperature were considered: $CH − CH_2$, $CH − CH_3$, $CH_2 − CH_2$, and $C − CH_3$. $^{13}C − ^1H$ SEDOR proves (Fig. 17) that only about half of the ^{13}C nuclei are in the close vicinity of the ^1H nuclei, thus ruling out immediately the first three models. The data agree well with the model of the ethylidene species $C − CH_3$, with the methyl group rotating at 77 K about C − C axis at a rate much faster than the NMR linewidth of 10 kHz. For sensitivity, the SEDOR measurements were made at low temperature. To give the reader an idea of the wealth of information contained in SEDOR results, we mention the other conclusions from the same publication [112]. The authors determined the C − C bond length of C − CH_3 species to be 1.49 ± 0.02 Å. Up to 390 K, the C − C bond length remains the same, suggesting the same surface species. C − C bond scission takes place above 390 K and is complete at about 480 ·K, where predominantly single carbon atoms with no hydrogen attached and a small amount of CH_3 species are formed.

SEDOR has certain limitations. The method is less reliable when the I − I interactions are stronger than the I − S interactions, or when strong S − S interactions are present. Therefore, the technique seems to be well suited for the study of catalysis on surfaces where the above limitations are not severe, at least

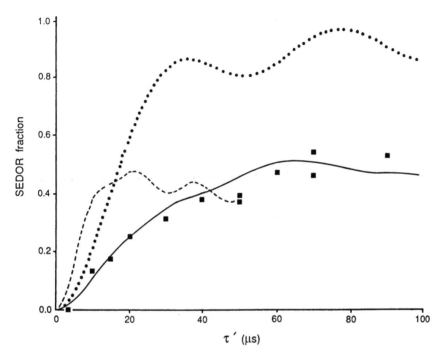

Figure 17 ^{13}C – ^{1}H SEDOR fraction versus τ' at 77K for ^{13}C$_2$H$_4$ on Pt clusters [112]. The solid line is the theoretical prediction for the ethylidene species C-CH$_3$ with the methyl group rotating about the C-C axis, the dashed line is that of frozen C-CH$_3$, and the dotted line is that of CH-CH$_3$ with rotating methyl group.

at low loadings. The surface studies are insensitive, however, and in most cases isotopic enrichment and a decrease in temperature to 77 K or to 4 K was necessary. [99,110,112,113] For the CO chemisorbed on Pd particles it would have taken 21 days to collect each data point at 77 K on the SEDOR-fraction-vs.-τ' curve, using the pulse sequence in Fig. 15c [113]. Measuring time was successfully reduced to 3 h by performing the experiment at 4 K and by replacing the normal $\pi/2 - \tau - \pi - \tau -$ acquisition spin-echo train with the Carr-Purcell pulse sequence [113]. The sensitivity problem does not arise with abundant nuclei, such as ^{27}Al and ^{31}P [114]. One should also bear in mind the heteronuclear J-couplings, which also contribute to SEDOR decay. This problem is particularly serious for heavier nuclei.

Finally, we mention double-resonance techniques tailored to study the de-focusing process on rotational echoes generated by MAS rather than on static spin echoes. *R*otational Spin-*E*cho *D*ouble *R*esonance (REDOR) [115], *R*otary *R*esonance *R*ecoupling (RRR) [116], and *T*ransferred-*E*cho *D*ouble *R*esonance

(TEDOR) [117] are capable of selectively measuring individual heteronuclear dipolar couplings in crystalline solids, giving rise to multiple-peak MAS NMR spectra. All these methods have potential catalytic applications. Fyfe et al. [118] used dipolar-dephasing MAS NMR for the study of mixed pairs of quadrupolar and spin-1/2 nuclei in VPI-5. Dipolar connectivities were examined in both directions between ^{31}P and [27]Al, and REDOR and TEDOR experiments were performed (including a 2D extension of TEDOR). More recently, van Eck and Veeman [119] developed a novel 2D-heteronuclear correlation pulse sequence for solids under MAS, which enables the correlation of spectra of two different nuclei through their weak heteronuclear dipolar coupling. The sequence provides an alternative to other 2D techniques for establishing through-space correlations and employs a coherence transfer pulse sequence somewhat similar to the TEDOR experiment. Their 2D heteronuclear correlation spectrum of VPI-5 is shown in Fig. 18. Such experiments, which provide information about connectivities and distances between coupled nuclei, clearly have considerable potential for structural elucidation.

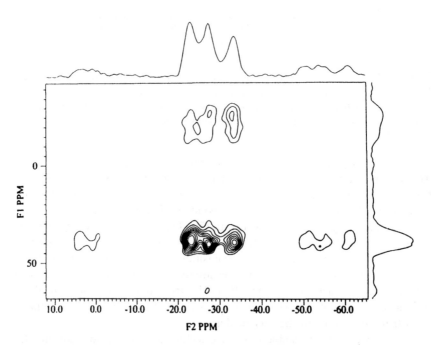

Figure 18   2D heteronuclear correlation spectrum of VPI-5 [119]. The projection of the [27]Al spectrum is displayed on the right of the 2D spectrum, the projection of the ^{31}P spectrum at the top.

D. Cross-Polarization to Quadrupolar Nuclei

Cross-polarization [120] (CP) has proved to be useful for the enhancement of signals from rare spin-1/2 nuclei, especially ^{13}C in organic solids. In addition, variable-contact-time CP has been used for signal assignment and for the study of slow molecular motions, which can be monitored by the proton relaxation time $T_{1\rho}$ derived from the experiment. Until recently, CP experiments on quadrupolar nuclei were restricted almost exclusively to model, single-crystal samples [121–123] because strong quadrupolar interactions interfere with the polarization transfer, particularly in polycrystalline materials [124]. The problem is not so severe, however, for noninteger spin nuclei, if only the central transition is involved. Thus CP transfer from ^1H to ^{27}Al, ^{17}O, and ^{11}B has recently been demonstrated on powder samples of catalytic interest [125–131]. This possibility opens new perspectives for studies of framework and extraframework sites in the proximity of protons in catalytic supports such as aluminas [126], natural silicates [129], or glasses [130], in pillared layered materials [131], clays [127], and molecular sieves [125,128]. The technique permits easy assignment of "protonated" versus "nonprotonated" heteroatoms, the quadrupolar nuclei of which are observed (Fig. 19). Also, CP observation of Ca^{2+} and Na^+ ions in voids of catalysts seems possible, since the appropriate ^{43}Ca and ^{23}Na experiments on standards have been reported [132,133].

For the selective pulse experiment involving the central transition of quadrupolar nuclei with noninteger spin S, the Hartmann-Hahn condition [134] becomes [129]

$$\left(S + \frac{1}{2}\right) \gamma_x B_{1x} = \gamma_H B_{1H}$$

To assess the experimental details, we refer to Morris et al. [127], who considered the reduction of the ^{27}Al CP/MAS signal in comparison with the MAS signal from kaolinite. The problems involved are: (1) the thermodynamic loss associated with the CP match between two spin populations, (2) short $T_{1\rho} = 458$ μs of ^{27}Al in the material, and (3) the disturbing effect of sample spinning on the ^{27}Al spin-lock field. Experimental results [127] showed that the ^{27}Al CP signal of kaolinite relative to that from Bloch decay was reduced to 48% without MAS and to 7.8% with MAS. This destructive effect of sample spinning (Fig. 20) was explained by a combination of factors, namely ω_{1Al} not being large relative to resonance offset and the subsequent modulation of the ω_{eff} by rapid MAS. Note that, as a consequence of the (S + 1/2) factor in the Hartmann-Hahn condition for this experiment, ω_{1Al} is three times smaller than ω_{1H}. Spin locking of half-integer quadrupolar nuclei under MAS has been treated theoretically, and the results verified experimentally for ^{23}Na in $NaClO_3$ [135]. In our experience, the experiment is not difficult and $^1H - ^{27}Al$ CP/MAS can be matched in one

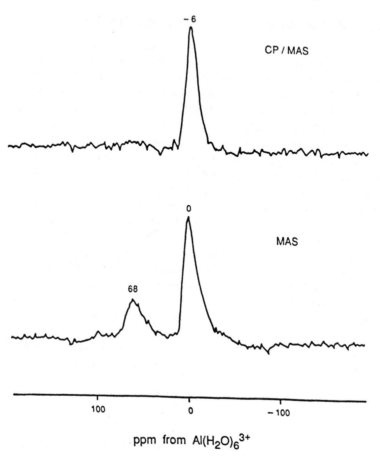

Figure 19 ^{27}Al MAS and CP/MAS spectra of γ-alumina [126]. In the MAS spectrum the resonances come from O_h (68 ppm) and T_d (0 ppm) sites in the bulk material. In the CP/MAS spectrum, the peak at -6 ppm comes from the surface O_h sites.

scan on the kaolinite sample using 500 ms contact time and an MAS rate of 7 kHz [128,131]. Still, one has to be cautious about two more problems. First, it has been reported that some factor associated with the ^{27}Al quadrupole in a T_d geometry may affect the $T_{1\rho}$ for this site, making it too short for CP to be effective [126]. A similar problem arises if paramagnetic impurities are present [127]. Second, rigorous CYCLOPS phase cycling is compulsory to avoid significant amounts of spin-locked non-CP magnetization in the observed signal [126,127]. Otherwise, such signals can arise because CP is no longer being used for ^{27}Al NMR as a signal enhancement technique, but rather as a signal filter.

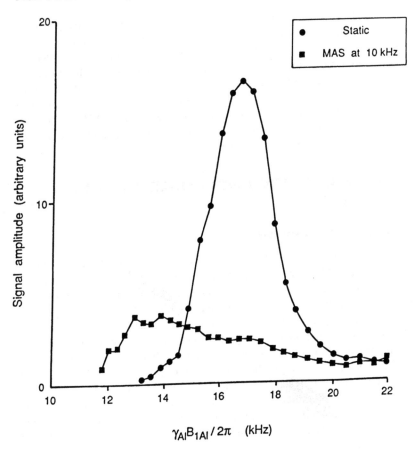

Figure 20 $^1H - {}^{27}Al$ Hartmann-Hahn match for kaolinite observed with $\gamma_H B_{1H}/2\pi = 50$ kHz [128].

Signal enhancements were obtained in $^1H - {}^{17}O$ cross-polarization experiments without spinning, and reliable second-order quadrupolar powder patterns were observed [129]. Relaxation parameters involved in cross-polarization transfer were shown to be characteristic of the various sites, so that they can be used for signal assignment. In addition, in some cases the differences in cross-polarization rates were used to "edit" spectra by a selective enhancement of protonated oxygen resonances, such as those from surface hydroxyl groups in amorphous silica. The latter method can be applied to complicated systems, provided dipolar $^1H - {}^{17}O$ interactions for the various sites are different. We illustrate this procedure by using a static $^1H - {}^{17}O$ CP spectrum [126] of talc,

$Mg_3Si_4O_{10}(OH)_2$ (Fig. 21). Talc contains three chemically distinct oxygen sites: Si-O-Mg, Si-O-Si, and Mg-O-H, in the population ratio $2:3:1$. In the spectrum there are three overlapping second-order quadrupolar powder patterns, which can be closely simulated assuming the same intensity ratio as the site population ratio. The quadrupole coupling constants and isotropic chemical shifts found for Si-O-Mg and Si-O-Si sites are similar to those in other silicates, allowing spectral assignment to be made. The assignment of the Mg-O-H signal is obvious, since it has spectral parameters similar to those in $Mg(OH)_2$ and is the only one observed under cross-polarization.

IV. TECHNIQUES INVOLVING QUADRUPOLAR COUPLING

A. General Comments

Quadrupolar nuclei, those with $I > 1/2$, have a nonspherical distribution of nuclear charge and therefore interact with the electric field gradient in the solid. Since 74% of all nuclei with spin are quadrupolar, their study is of considerable interest. The quadrupole interaction broadens and shifts the NMR signals and may also affect their relative intensities. Since many chemists working in catalysis are unfamiliar with quadrupolar effects, we give a brief summary of the most important considerations to be borne in mind. In practice, two distinct cases must be considered: nuclei with noninteger spin (of which ^{27}Al is perhaps the most important to catalysis) and nuclei with integer spin, such as ^2H.

The electric field gradient is described by the traceless and symmetrical tensor V. V, in turn, is quantified in terms of its largest component, $V_{zz} = eQ$, and the asymmetry parameter, η. In the principal axis system, the quadrupolar Hamiltonian is

$$\mathcal{H}_Q = \frac{e^2 qQ}{4I(2I-1)} \left[3I_z^2 - I^2 + \eta \left(I_x^2 - I_y^2 \right) \right].$$

In high magnetic fields, $\mathcal{H}_z \gg \mathcal{H}_Q$ and the quadrupole shift can be calculated using perturbation theory. Taking $\eta = 0$ for simplicity, the first-order shift is

$$v_m^{(1)} = \frac{1}{2} v_Q \left(m - \frac{1}{2} \right) (3\cos^2\theta - 1)$$

where

$$v_Q = \frac{3e^2 qQ}{2h \, I(2I-1)}$$

Thus for a *half-integer spin*, the central ($1/2 \leftrightarrow -1/2$) transition is not affected by the first-order quadrupolar interaction. The spectrum contains contributions from the other transitions, but these are usually too broad to be detected.

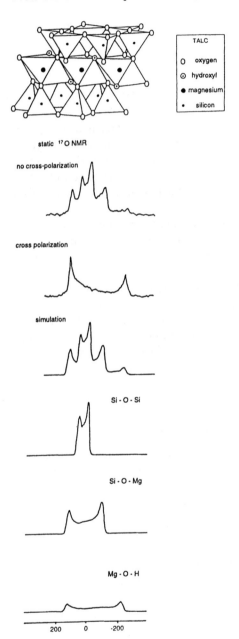

Figure 21 Static ^{17}O NMR spectra of polycrystalline talc recorded with a ^{17}O-enriched sample at 67.8 MHz with the high-power proton decoupling [129].

For *integer spins*, there is no central transition, and the first-order shift may be reduced by MAS.

The second-order shift for the central transition is

$$\nu_{1/2}^{(2)} = \frac{-\nu_Q^2}{16\nu_L} \left[I(I + 1) - \frac{3}{4} \right] (1 - \cos^2 \theta)(9\cos^2 \theta - 1)$$

where ν_L is the Larmor frequency. The second-order interaction is thus not averaged to zero by MAS and consequently effects the positions of NMR signals, including that of the central transition. The effect is reduced by working at a high magnetic field, which increases the Larmor frequency term. The linewidth of the central transition is only reduced by a factor of approximately 4 in comparison with a static spectrum.

The dependence of the second-order quadrupolar shift on the angle θ means that the peak shape for each resonance position can consist of more than one maximum if quadrupolar effects are significant. This is a major difficulty in the interpretation and quantification of spectra. An additional problem is that peak intensities for sites with different quadrupolar coupling constants are not necessarily quantitative unless very short pulses are used [136]. For example, aluminum in zeolites can be quantitatively monitored by NMR [137–139] provided certain experimental conditions are met: when the radiofrequency pulse is short enough, signal intensity is independent of the strength of the quadrupolar interaction. Computation of line intensities for various spin systems has been described in detail [140–146].

B. ^{2}H NMR Study of Motional Processes

The very small quadrupole interactions of ^{2}H make it an important integer spin nucleus. The quadrupole coupling constant of ^{2}H in organic compounds is of the order of 200 kHz, compared to approximately 5 MHz for ^{14}N and 80 MHz for ^{35}Cl. This situation makes ^{2}H NMR particularly useful in chemical studies. The spectrum of a single deuteron in a crystal is a doublet, the peak separation of which is related to the orientation of the C $-$ ^{2}H bond in the external magnetic field. Dipolar interactions of ^{2}H are up to 10 kHz and manifest themselves as spectral broadening. Since the quadrupolar interactions of ^{2}H are sensitive to molecular motion, the nucleus is useful for chemical studies of molecular motion at a wide range of frequencies. ^{2}H NMR is a convenient tool for the study of molecular ordering and molecular motion of adsorbed probe molecules and may become an important complementary technique for the study of interface systems in general and liquid crystals in particular. ^{2}H NMR experiments are done with *static* samples, and dynamic information is extracted by comparing spectra measured at different temperatures with model computer simulations. An enormous amount of ^{2}H work has been reported but, surprisingly, there are not many recent papers that are directly relevant to catalytic studies [147–153].

Gottlieb and Luz [147] recorded ^{2}H spectra of perdeuterated benzene, p-xylene, pyridine, acetone, acetonitrile, cyclohexane, methanol, and water adsorbed on active alumina. They were able to interpret the results using a model in which each crystallite surface supports an ordered layer of molecules that are in fast dynamic equilibrium with disordered "bulk" molecules associated with the same surface. In the case of acetone, the spectrum appears to be a superposition of spectra from two distinct sites. Meirovitch and Belsky [148] used the technique to study guest conformation, dynamics, and interchannel orientation in several cyclophosphazene inclusion compounds. Eckman and Vega [149] investigated the dynamics of small organic molecules, such as methanol, benzene, toluene, and p-xylene, adsorbed on zeolites ZSM-5, mordenite, Y, and erionite over a range of temperatures. They obtained information on the dynamics of the filling of the intracrystalline space, the motion of the adsorbed species, and site-selective adsorption. Kustanovich et al. [150] interpreted the ^{2}H spectra of two deuterated species of p-xylene ($CH_3C_6D_4CH_3$ and $CD_3C_6H_4CD_3$) on zeolite Na-ZSM-5 in terms of possible dynamic states and sorption sites of the guest molecules. Five dynamic states were identified, with relative populations varying according to the level of loading and the temperature: (1) static, rigidly sorbed molecules; (2) and (3) molecules whose phenyl rings undergo, respectively, discrete 180° flips and free continuous rotation about their para axis; and (4) and (5) molecules whose molecular axes exhibit twofold jumps through 90° and 112°, respectively. The latter two states correspond to molecules located close to channel intersections and undergoing two-site jumps between the two kinds of channels present in ZSM-5. Deuterium NMR results on p-xylene-d$_6$, toluene-d$_3$, and benzene-d$_6$ sorbed on H-ZSM-5, [151] as well as on mono-, di-, and trimethylamine adsorbed on zeolites ZK-5 and Y[152], have been reported by the same group, in all cases yielding a wealth of dynamic information.

Intercalation of layered compounds, including clay minerals, with polar guest molecules is of interest to catalysis. Duer et al. [153] studied the structure and motion of the guest in the interlayer space of the kaolinite: DMSO intercalate by ^{2}H NMR of static samples. The guest DMSO molecules are hydrogen-bonded via the oxygen atom to the kaolinite hydroxyls that face the interlayer space, and there is a degree of interaction between the sulfur atom of the DMSO molecule and the siliceous matrix of kaolinite. There are two independent sites for the DMSO methyl groups with effective ^{2}H quadrupole coupling constants of 59 and 67 kHz, and the population of at least one site is in motion. The difference in these couplings probably reflects different geometries of the methyl groups in the two sites, caused by their partial "keying" into the siliceous matrix of kaolinite. Three motional models produce simulations in agreement with the experimental spectra. One fit is consistent with the keying of one methyl group of a DMSO molecule into the kaolinite layer. Above 339 K, further exchange processes must be taken into account in lineshape simulations, and they may trigger structural transformations of the intercalate. We believe that ^2H NMR has considerable potential for the study of catalytic reactions in solids.

C. Quadrupole Nutation

Quadrupole nutation NMR of nuclei with half-integer spin in powdered samples [139–144] provides a useful means for selective determination of quadrupolar parameters in correlation with chemical shift information. This technique can distinguish between nuclei subjected to different quadrupole interactions (the signals from which would overlap in conventional NMR spectra), but has so far been used mainly to determine the local environment of Al zeolitic catalysts.

A quadrupole nutation experiment consists of three periods: preparation, evolution, and detection. During the preparation period the spin system reaches equilibrium, and then during evolution the sample is irradiated with a radiofrequency field with pulse length t_1. The detection period, t_2, corresponds to the acquisition of the FID. By keeping the sampling time t_2 constant and increasing t_1 by equal increments at regular intervals, a series of FIDs is acquired. A double Fourier transformation in t_2 and t_1 gives a two-dimensional NMR spectrum, with the axes F_2 containing the chemical shift and the second-order quadrupolar shift while F_1 contains only quadrupolar information. The projection on the F_2 axis gives a normal powder lineshape attributable to the combined effect of the chemical shift and quadrupolar interactions. The projection onto F_1 gives the precession frequencies in the rotating frame around the static radiofrequency field, which depend on the ratio of quadrupolar parameters, $\omega_Q = e^2qQ/h$ and η, and on the spin quantum number I and radiofrequency field amplitude ω_{rf}.

For quadrupolar nuclei, the amplitude and length of the radiofrequency pulse affect the lineshape and the relative intensities [139,142,146]. It is therefore important to specify the irradiation conditions of the spin system. The strength of the quadrupolar parameter, ω_Q, can be up to several MHz depending on the nucleus and the structure of the compound, whereas the amplitude of the pulse is not usually large enough to allow the quadrupole interaction during irradiation of the spin system to be neglected.

Early quadrupole nutation experiments concerned the chemical status of nonframework Al in dealuminated zeolite Y of static samples or with MAS at moderate rates [137–139,154]. However, it has been shown that ^{27}Al quadrupole nutation with fast (12.5 kHz) MAS and radiofrequency excitation fields up to 220 kHz provides significantly better spectral resolution [155], enabling, for example, the resonances at 0, 30, 56, and 62 ppm observed in ultrastable zeolite Y to be assigned to different aluminous species. The various quadrupolar parameters can also be determined. Thus the quadrupole coupling constant of the site responsible for the 30-ppm signal changes significantly with the degree of dealumination, from 4.5–5 MHz (at Si/Al = 4.0) to greater than 6 MHz (at Si/Al = 9). The corresponding isotropic chemical shifts are in the range 41–57 ppm. The conclusion was that the nonframework matter contains four-, five- and six-coordinated aluminum.

Hydrated aluminophosphates have also been studied by ^{27}Al quadrupole nutation with MAS [128,156]. AlPO$_4$-8 was found to contain four and possibly five crystallographically inequivalent sites for aluminum. Above 353 K, the variable-temperature quadrupole nutation spectrum of the molecular sieve VPI-5 (Fig. 22) contains only one signal from four-coordinated Al (apart from the signal from six-coordinated Al), indicating that the material undergoes a high-temperature-phase transformation to a higher framework symmetry. In the range 190–220 K, the quadrupole interaction parameters and isotropic chemical shift of the resonance from six-coordinated Al (obtained by computer line fitting) go through extremes, indicating that hydrated VPI-5 also undergoes a low-temperature structural transformation. Both transformations are fully reversible.

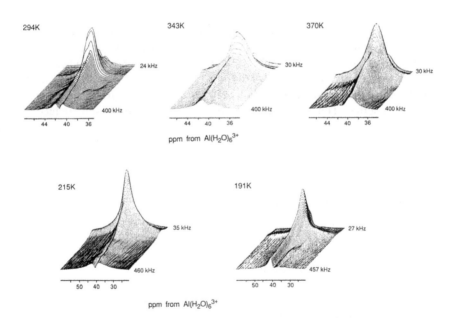

Figure 22 Variable-temperature ^{27}Al quadrupole nutation spectra in the region of four-coordinated Al of hydrated VPI-5 measured with $\omega_{rf}/2\pi = 115 \pm 5$ kHz radiofrequency field [156].

Quadrupole nutation spectra of several aluminophosphate molecular sieves recorded with fast MAS have been compared with the DOR spectra (see below) [157]. Enhanced resolution was obtained in the quadrupole nutation experiment at certain radiofrequency pulse strengths. This extra resolution can be comparable to that attainable using DOR and does not introduce spinning sidebands.

D. Double Rotation (DOR) and Dynamic Angle Spinning (DAS)

A major recent advance in the study of quadrupolar nuclei by solid-state NMR is the development of double-rotation (DOR) and dynamic-angle spinning (DAS) [158,159,163]. These techniques are capable of removing not only the first-order effects such as chemical shift anisotropy, but also the second-order quadrupolar interaction, which, as we have seen above, affects *all* quadrupolar nuclei. The second-order interaction can be expanded as a function of Wigner rotation matrices $D_l^{mm'}$, so that the shift of the central transition is

$$\nu_{1/2}^{(2)} = \frac{\nu_Q^2}{\nu_L} \left[I(I + 1) - \frac{3}{4} \right] \sum_{k=0}^{2} \sum_{n=-k}^{k} A_{2k}^{2n} D_{2k}^{2n,o} (\phi, \theta, 0)$$

where the constants A_{2k}^{2n} depend on the value of the asymmetry parameter η.

Rotation of the sample about an angle θ with an angular frequency ω_R makes the Wigner rotation matrices time dependent. Over an integer number, N, of periods, the average of $D_l^{mo} (\omega_R t, \theta, 0)$ vanishes for $m \neq 0$, irrespective of the orientation of the rotation axis. This leaves the contributions from the D_2^{00}, D_4^{00}, and higher terms. Thus the average second-order quadrupolar shift reduces to

$$\nu_{1/2}^{(2)} = \frac{\nu_Q^2}{\nu_L} \left[I(I + 1) - \frac{3}{4} \right] \left[A_o^o + B_2^o(\alpha, \beta) P_2 (\cos \theta) \right.$$

$$\left. + B_4^o (\alpha, \beta) P_4 (\cos \theta) \right]$$

where the $P_n (\cos \theta)$ are the Legendre polynomials

$$P_2 (\cos \theta) = \frac{1}{2} (3 \cos^2 \theta - 1)$$

$$P_4 (\cos \theta) = \frac{1}{8} (35 \cos^4 \theta - 30 \cos^2 \theta + 3)$$

A_o^o, B_2^o, and B_4^o depend on the asymmetry parameter η.

There is clearly no value of θ for which both the $P_2 (\cos \theta)$ and the $P_4 (\cos \theta)$ terms are zero. However, two ways can be envisaged to cancel these terms together. If the sample is spun *simultaneously* about two different axes θ_1 and θ_2, we have

$P_2 (\cos \theta_1) = 0$

$P_4 (\cos \theta_2) = 0$

with solutions $\theta_1 = 54.74°$ (the conventional "magic angle") and $\theta_2 = 30.56°$ or $70.12°$. The resulting complicated trajectory averages the anisotropic broadening to zero. This scheme is the basis of the "double-rotation" (DOR) experiment [159]. The design of a DOR probehead poses considerable engineering problems, but these have been successfully overcome [160,161]. Another problem involves spinning sidebands from the outer rotor, which spins at rates below about 1 kHz. However, Samoson and Lippmaa [162] demonstrated that when data acquisition in DOR is synchronized with spinner revolution, every second spinning sideband is suppressed.

Second-order quadrupole interaction may also be removed by dynamic-angle spinning (DAS) [163], in which the sample is rotated *sequentially* about two different axes, θ_1' and θ_2', chosen so that

$P_2 (\cos \theta_1') = - P_2 (\cos \theta_2')$

$P_4 (\cos \theta_1') = - P_4 (\cos \theta_2')$

with the solutions $\theta_1 = 37.38°$ and $\theta_2 = 79.19°$. By suitable choice of angles, higher orders of interaction may be eliminated as well. Since switching the rotation axis cannot be performed instantaneously, an echo pulse sequence must be used to preserve the transverse magnetization during the angle-switching period. This procedure does, however, require that the sample have a sufficiently long spin-lattice relaxation time, or else the magnetization will be lost during the angle switching period. Thus it is theoretically possible to remove the second-order interaction completely. The enhanced resolution afforded by DAS has been demonstrated using ^{23}Na spectra of polycrystalline sodium oxalate and ^{17}O spectra of isotopically enriched cristobalite [163].

The DOR experiment is somewhat easier to implement than DAS, because it does not involve the rapid switching of rotation axes. It is also less affected by the T_1 relaxation time. Note that DAS and DOR only remove the angular dependence, so there will still be the second-order quadrupolar shift in resonance position arising from the A_0^0 term.

DOR and DAS open up whole new areas of research into the solid state, but there have so far been very few published applications of these techniques [163–167]. Chmelka et al. [164] used ^{17}O DOR and DAS to distinguish the three inequivalent oxygen sites known to be present in the mineral diopside, $CaMgSi_2O_6$. The remaining applications all involve ^{27}Al in aluminophosphate molecular sieves. Wu et al. [165] showed that ^{27}Al DOR is capable of resolving discrete framework aluminum sites in VPI-5, permitting quantitative investigation of site-specific adsorbate interactions (typically H_2O interactions) with the

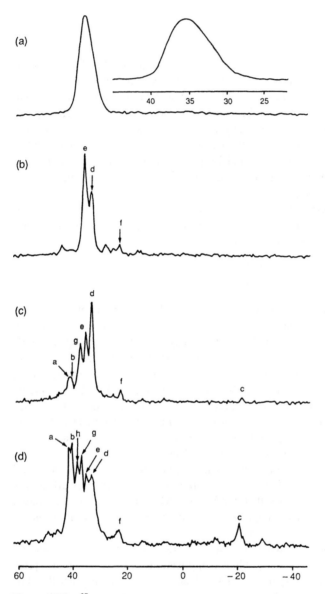

Figure 23 ^{27}Al DOR spectra of dehydrated and partially rehydrated VPI-5. (a) MAS spectrum of dehydrated VPI-5; (b) DOR spectrum of dehydrated VPI-5; (c) DOR spectrum after 2 days of rehydration; (d) DOR spectrum after 23 days of rehydration [165].

framework. Figure 23a and 23b show the ^{27}Al MAS spectrum of dehydrated VPI-5 and the DOR spectrum of the same sample. Two peaks unresolved in the MAS spectrum, d at 33.3 ppm and e at 35.9 ppm, are observed in the DOR spectrum. From the 1:2 intensity ratio of these two peaks, peak d is assigned to **Al1** sites and peak e to **Al2** sites. During dehydration, six-coordinated Al sites are converted to four-coordinated sites, consistent with the disappearance of peak c at -18.4 ppm. Furthermore, in hydrated VPI-5, ^{27}Al tetrahedral sites (peaks a and b) are altered by dehydration to yield different ^{27}Al tetrahedral environments (peaks d and e). Two sharp lines that emerge upon slow rehydration, g at 37.3 ppm and h at 38.8 ppm, arise from **Al2** species influenced by adsorbed water molecules.

Jelinek et al. [166] demonstrated that ^{27}Al DOR in a magnetic field of 11.7 T distinguishes the extremely distorted five-coordinated aluminum sites in the molecular sieve precursor AlPO$_4$-21. Upon calcination, AlPO$_4$-21 transforms to AlPO$_4$-25, where there are two four-coordinated aluminum sites with similar isotropic chemical shifts, which cannot be resolved in an 11.7-T field. The two tetrahedral environments have different quadrupole coupling constants, however, and are distinguished by DOR at 4.2 T. Barrie et al. [167] used DOR to monitor the nature of the phase transition of AlPO$_4$-11 observed when water is adsorbed. In the hydrated material, water is strongly attached to approximately 1/5 of the framework aluminum, producing six-coordinated Al. All five suggested crystallographic sites are hydrated to an equal extent, and consequently the ^{27}Al NMR signal due to the six-coordinated species does not completely narrow upon DOR.

V. NMR STUDIES OF SURFACES

Although NMR is widely used in materials research for the study of structure and dynamics, its inherently low sensitivity makes it useful primarily for the study of bulk phenomena. Only a few studies o films have been reported, and they were performed on high surface-area materials such as graphitized carbon black with a surface area of approximately 70 m^2/g or exfoliated graphite. Raftery et al. [168] observed NMR of ^{129}Xe in thin films of xenon frozen onto the surfaces of glass sample cells with various geometries. The ^{129}Xe polarization was enhanced by optical pumping, and the xenon was then transferred to a high-field pulsed NMR spectrometer allowing the observation of strong signals from xenon films about 1 μm thick. The NMR lineshape was found to depend on the film geometry, because of the bulk diamagnetic susceptibility of solid xenon. The spectral lineshape and resonance frequency also depend on temperature. Given that most heterogeneous catalytic reactions take place on the gas/solid interface, the method of increasing the sensitivity of the NMR experiment reported by Raftery et al. is an exciting development.

VI. THE PASADENA EFFECT

Bowers and Weitekamp predicted [169] and then experimentally demonstrated [170] a method, which they termed the PASADENA effect, of obtaining very large nuclear spin polarizations on molecules formed by addition of para-hydrogen such that the dihydrogen protons become magnetically inequivalent. The principle of the method is as follows. Para- and ortho- forms of molecular hydrogen differ from one another in their nuclear spin states. A consequence of the symmetrization postulate of quantum mechanics is that these two forms are associated only with specific states of molecular rotation. The nuclear spin order of parahydrogen can be converted by chemical reaction into large nonequlibrium NMR signals. The PASADENA effect was first demonstrated using the reaction of hydrogenation of acrylonitrile, CH_2CHCN, to proprionitrile, CH_3CH_2CN, catalyzed by Wilkinson's catalyst at ambient temperature and pressure [170]. Large transient ^1H NMR signals were observed both in proprionitrile transitions and in the hydride region of the hydrogenated catalyst. The amplitude of these lines was at least two orders of magnitude greater than would result from the equilibrium magnetization of the product formed. The advantages of the technique, which is of obvious interest to heterogeneous catalysis, where the concentrations of the product are often small, are its great sensitivity and the fact that the products can be monitored in a time-resolved fashion. The PASADENA effect has been used to study the adsorption of H_2 on the surface of zinc oxide [171].

VII. MECHANICAL DETECTION OF MAGNETIC RESONANCE

Conventional techniques for measuring magnetic resonance involve the detection of electromagnetic signals induced in a coil or microwave cavity by the collective precession of magnetic moments (from nuclei or electrons) excited by an alternating magnetic field. In a different approach [172], isolated electron spins have been detected by scanning tunneling microscopy, with the spin precession inducing a radiofrequency modulation in the tunneling current. Rugar et al. [173] recently described a new and extremely sensitive method of detection, the principles of which derive from magnetic force microscopy. They measured the small, oscillatory magnetic force (10^{-14} N) acting on a paramagnetic sample (a few grains of diphenylpicrylhydrazil weighing <30 ng), which was excited into magnetic resonance in the presence of an inhomogeneous magnetic field. This force was detected by optically sensing the angstrom-scale vibration of a micro-mechanical cantilever on which the sample is mounted (see Fig. 24). The

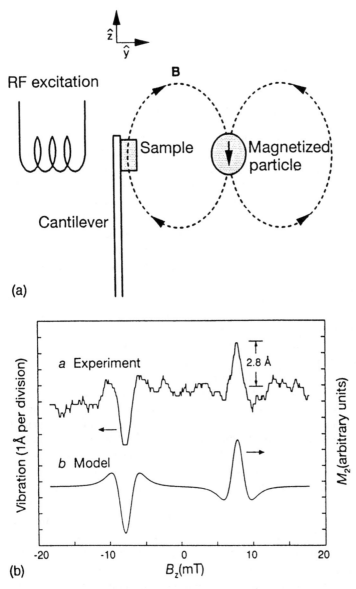

Figure 24 Basic configuration used to detect magnetic resonance through magnetic force interactions [173]. (a) The sample experiences a force due to the inhomogeneous magnetic field emanating from a nearby magnetic particle. Spins in the sample are excited by an applied radiofrequency field, and a magnetic resonance modulation technique is used to modulate the magnetization of the sample. This causes the force on the sample to oscillate and the cantilever to vibrate. (b) Amplitude of cantilever vibration signal plotted as a function of the polarizing field for an applied modulation of 1 mT and radiofrequency excitation of 220 MHz. Vibration peaks with amplitude ± 2.8 Å occur at field values corresponding to electron spin resonance (± 7.9 mT).

sensitivity of the technique to the spatial distribution of the spins suggests that mechanical detection of magnetic resonance has the potential for imaging microscopic samples in three dimensions. So far, the authors achieved a spatial resolution of 19 μm in one dimension.

VIII. NMR IMAGING IN CATALYSIS

Following the pioneering experiments of Lauterbur [174], NMR imaging (NMRI) has become well established in clinical medicine. Early applications used ^{1}H only and relied on the varying water content and relaxation properties of healthy and diseased tissue [175]. Recent work has concentrated on fast imaging techniques enabling image acquisition within less than a second [176,177] and on imaging of other nuclei. In addition, the scope of the technique has moved toward chemical-shift selective imaging, which allows imaging of a particular molecular species [178], and toward spatially localized spectroscopy [179].

The potential of the technique for applications in materials science has only recently begun to be realized. Among these are the study of the displacement of oil in rock cores [180], diffusion of solvents into polymeric materials [181], and imaging of liquid-containing voids in porous solids [182]. These applications have all involved imaging the liquid phase in the porous material. Imaging of the solid itself places more stringent requirements on the hardware, but advances in this area are also being made [173–186]. Günther et al., for example, describe ^{13}C NMRI spectra of solid polyethylene [186].

NMRI has proved to be useful in studying problems concerning commercial oxide-supported metal catalysts. When the reaction is not subject to strong diffusion limitations, optimum catalytic performance is expected from a catalyst support of uniform porosity such that the diffusion of reactants and products into and out of the catalyst, respectively, is the same throughout the support. Another problem associated with catalysis is the distribution of the catalytically active metal during wet impregnation techniques. Both the impregnation and drying of the catalyst are important stages in determining the final distribution of catalytically active metal throughout the support [187]. Experimental [188] and theoretical [189] studies have shown the nonuniform activity profiles can be important in determining catalytic performance; such profiles are thought to arise from varying distribution of metal in a uniformly porous support.

The voids in commercial porous silica and alumina supports in the form of spheres or extrudates are often far from uniform [190]. Large regions of low porosity are observed, and in the case of extruded pellets, the exterior surface phase often has voidage characteristics very different from the bulk. Figure 25 shows a ^{1}H image (128 × 128 pixels) of a 0.7-mm slice through a silica sphere 3 mm in diameter impregnated with water. The image was acquired at 200 MHz

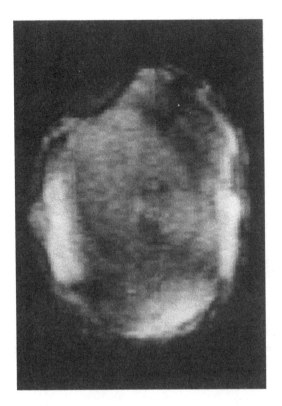

Figure 25 Image of a 0.7-mm slice through a silica sphere 3 mm in diameter acquired with 16 scans [190].

using gradient reversal [191] with an acquisition time of 4 ms and repetition time of 200 ms. Real-time observation of the drying of water-saturated supports allows the transport of water through the support to be monitored during the drying process. Figure 26 clearly shows that the removal of water from the support is inconsistent with existing models of the drying stage in catalyst synthesis, where typically the solvent is modeled as evaporating from a uniformly porous pellet with the drying front moving inward from the pellet surface.

Imaging of flows of water and other hydrogen-containing liquids through pipes, orifices, and porous media is of considerable interest. In medicine, the distribution of blood velocity in the aorta can be charted in a noninvasive manner. An example relevant to chemical engineering is illustrated in Fig. 27 [192].

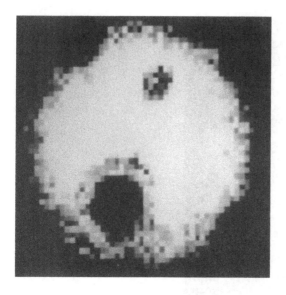

(a)

Figure 26 64 × 64 images of a 0.7-mm slice through a 3.0-mm-diameter silica sphere drying in air [190]. Times from commencement of drying are: (a) 0 min, (b) 26 min, and (c) 48 min.

Finally, we note an exciting recent imaging experiment. Barrall et al. [193] have shown that NMR spatial imaging data may be acquired, processed, and interpreted in ways that supply information directly analogous to diffraction experiments, with length scales determined by gradient strengths rather than radiation wavelengths. This approach provides access to autocorrelations of sample density that statistically characterize small-scale density variations. These NMR "Patterson functions" can be acquired orders of magnitude more rapidly than comparably resolved NMR images and are suitable for spatial characterization of small features in bulk samples, such as morphology of materials (see Fig. 28). Unlike hindered diffusion approaches, moreover, neither mobility, penetrants, nor transport time is required for examining granularity and porosity.

IX. CLOSING COMMENTS

The future of NMR in the study of catalysis looks bright. The structure of the solid catalyst itself can now be probed using various 2D NMR techniques, and

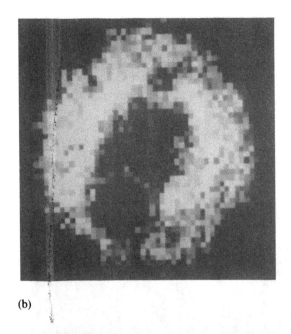

(b)

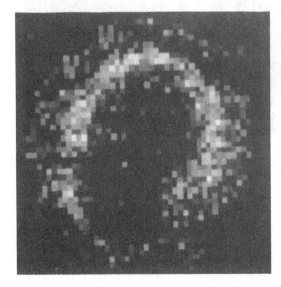

(c)

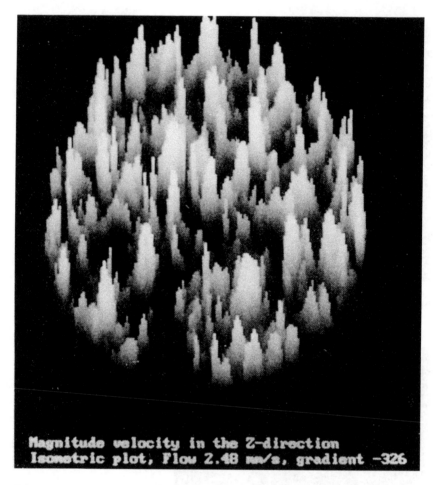

Figure 27   Flow-sensitive image of water flowing (2.24 mm s^{-1}) through the central 200-μm area of the sintered silica medium [192].

further developments are underway. Inspection of the results from the recent experiment of Kumar and Opella [194], who measured a *three-dimensional* ^{1}H/^{15}N heteronuclear correlation NMR spectrum of a single crystal of glycylglycine (Fig. 29), are sufficient to convince anyone that multidimensional NMR of solids has considerable prospects. Quadrupole nutation NMR probes the environment of nuclei in rigid solid catalysts, rotational resonance measures accurately the distance between specific sites, and surface-sensitive techniques, such as SE-DOR with its modifications and ^{2}H NMR, overcome the traditional limitations of

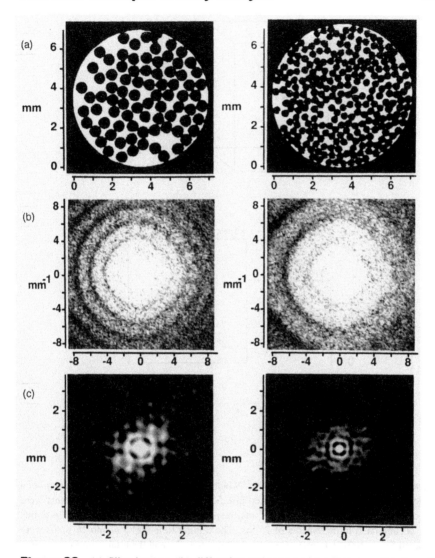

Figure 28 (a) Slice images, (b) diffraction patterns, and (c) Patterson functions for fibers immersed in water in a 7-mm tube [193]. The image shown in (a) is derived by Fourier transformation of NMR data, whose square, the NMR "diffraction pattern," shown in (b) reflects packing statistics. Fourier transformation of the diffraction pattern yields the Patterson function shown in (c), whose central disk reflects the fiber size and whose nearest-neighbor ring indicates packing density. (Left) 0.56-mm fibers, (right) 0.33-mm fibers. The images and Patterson functions of the smaller fibers on the left, shown on the same scale, are smaller than those on the right, but the diffraction pattern on the right is expanded since it scales according to spatial frequency.

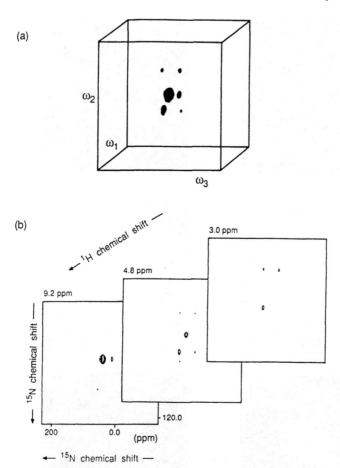

Figure 29   (a) Three-dimensional $^1H/^{15}N$ heteronuclear correlation NMR spectrum of a single crystal of glycylglycine measured at 8.5 T. A total of 16 t_1 and 20 t_2 points were collected in increments of 48 and 175 ms, respectively. In the t_3 dimension, 128 points were accumulated in increments of 75 ms. The mixing time was 3 s, and 16 scans were averaged for each (t_1, t_2) point. The recycle delay was 3 s and the total experimental time was 34 hours. (b) Two-dimensional slices through the three-dimensional spectrum shown in (a). (From Ref. 194.)

nuclear magnetic resonance. In situ techniques can identify the intermediates and products of catalytic reactions and elucidate their mechanism. Multiple-quantum NMR can count molecules present in porous solids. Finally, DOR and DAS effectively average second-order quadrupolar interactions. Applications of most of these techniques in catalysis have hardly begun. At the same time, high magnetic fields and constant advances in NMR hardware have made it possible to monitor virtually all chemical elements in solids. We believe that the next few years will witness many exciting new experiments.

REFERENCES

1. R. R. Ernst, G. Bodenhausen, and A. Wokaun, *Principles of Nuclear Magnetic Resonance in One and Two Dimensions*, Oxford University Press, London, 1987.
2. T. Terao, H. Miura, and A. Saika, *J. Magn. Reson. 49*:365 (1982).
3. W. K. Zilm and D. M. Grant, *J. Magn. Reson. 48*:524 (1982).
4. T. Terao, H. Miura, and A. Saika, *J. Am. Chem. Soc. 104*:5228 (1982).
5. C. L. Mayne, R. J. Pugmire, and D. M. Grant, *J. Magn. Reson. 56*:151 (1984).
6. H. Miura, T. Terao, and A. Saika, *J. Magn. Reson. 68*:593 (1986).
7. R. K. Harris, P. Jackson, L. H. Merwin, and B. J. Say, *J. Chem. Soc. Faraday Trans. 1 84*:3649 (1988).
8. W. Kolodziejski and J. Klinowski, *Solid State NMR 1*:41 (1992).
9. M. W. Anderson and J. Klinowski, *Chem. Phys. Lett. 172*:275 (1990).
10. W. Kolodziejski and J. Klinowski, *Appl. Catal. A81*:133 (1992).
11. M. W. Anderson and J. Klinowski, *J. Am. Chem. Soc. 112*:10 (1990).
12. M. Maricq and J. S. Waugh, *Chem. Phys. Lett. 47*:327 (1977).
13. M. Maricq and J. S. Waugh, *J. Chem. Phys. 70*:3300 (1979).
14. J. Rocha, W. Kolodziejski, and J. Klinowski, *Chem. Phys. Lett. 176*:395 (1991).
15. N. C. Nielsen, H. Bildsøe, H. J. Jakobsen, and O. W. Sørensen, *J. Magn. Reson. 79*:554 (1988).
16. T. Terao, H. Miura, and A. Saika, *J. Chem. Phys. 75*:1573 (1981).
17. K. W. Zilm and G. G. Webb, *Fuel 67*:707 (1988).
18. G. G. Webb and K. W. Zilm, *J. Am. Chem. Soc. 111*:2455 (1989).
19. N. K. Sethi, *J. Magn. Reson. 94*:352 (1991).
20. D. P. Burum and A. Bielecki, *J. Magn. Reson. 95*:184 (1991).
21. D. L. VanderHart, Poster M32, 30th Conference on Experimental Nuclear Magnetic Resonance Spectroscopy, Asilomar, California, April 2–6, 1989.
22. D. L. VanderHart and G. C. Campbell, Poster MP89, 31st Conference on Experimental Nuclear Magnetic Resonance Spectroscopy, Asilomar, California, April 1–5, 1990.
23. W. Kolodziejski and J. Klinowski, *J. Magn. Reson. 99*:611 (1992).
24. A. Bax, *Two-Dimensional NMR in Liquids*, Delft University Press, Delft, 1982.
25. G. E. Martin and A. S. Zektzer, *Two-Dimensional NMR Methods for Establishing Molecular Connectivity*, VCH Publishers, Weinham, 1988.
26. E. M. Menger, S. Vega, and R. G. Griffin, *J. Am. Chem. Soc. 108*:2215 (1986).
27. D. P. Raleigh, G. S. Harbison, T. G. Neiss, J. E. Roberts, and R. G. Griffin, *Chem. Phys. Lett. 138*:285 (1987).
28. R. Benn, H. Grondey, C. Brévard, and A. Pagelot, *J. Chem. Soc. Chem. Commun.* 102 (1988).
29. T. Bjornholm, *Chem. Phys. Lett. 143*:259 (1988); T. Allman, *J. Magn. Reson. 83*:637 (1989).

30. C. A. Fyfe, H. Gies, and Y. Feng, *J. Chem. Soc. Chem. Commun.* 1240 (1989).
31. C.A. Fyfe, H. Gies, and Y. Feng, *J. Am. Chem. Soc. 111*:7702 (1989).
32. C. A. Fyfe, Y. Feng, and G. T. Kokotailo, *Nature 341*:223 (1989).
33. C. A. Fyfe, Y. Feng, H. Gies, H. Grondey, and G. T. Kokotailo, *J. Am. Chem. Soc. 112*:3264 (1990).
34. C. A. Fyfe, H. Grondey, Y. Feng, and G. T. Kokotailo, *J. Am. Chem. Soc. 112*:8812 (1990).
35. C. A. Fyfe, H. Gies, Y. Feng, and H. Grondey, *Zeolites 10*: 278 (1990).
36. C. A. Fyfe, Y. Feng, H. Grondey, G. T. Kokotailo, and H. Gies, *Chem. Rev. 91*:1525 (1991).
37. W. Kolodziejski, P. J. Barrie, H. He, and J. Klinowski, *J. Chem. Soc. Chem. Commun.* 961 (1991).
38. C. T. G. Knight, R. J. Kirkpatrick, and E. Oldfield, *J. Non-Cryst. Solids 116*:140 (1990).
39. A. Abragam, *The Principles of Nuclear Magnetism*, Oxford University Press, London, 1961.
40. A. Bax and R. Freeman, *J. Magn. Reson. 44*:542 (1981).
41. E. T. Olejniczak, S. Vega, and R. G. Griffin, *J. Chem. Phys. 81*:4804 (1984).
42. H. Miura, T. Terao, and A. Saika, *J. Chem. Phys. 85*:2458 (1986).
43. A. Bax, R. Freeman, and S. P. Kempsell, *J. Am. Chem. Soc. 102*:4849 (1980).
44. A. Bax, S. P. Kempsell, and R. Freeman, *J. Magn. Reson. 41*:349 (1980).
45. J. Buddrus and H. Bauer, *Angew. Chem. Int. Ed. Engl. 26*:625 (1987).
46. C. A. Fyfe, H. Grondey, Y. Feng, and G. T. Kokotailo, *Chem. Phys. Lett. 173*:211 (1990).
47. C. A. Fyfe, H. Grondey, Y. Feng, G. T. Kokotailo, S. Ernst, and J. Weitkamp, *Zeolites 12*:50 (1992).
48. T. A. Early, B. K. John, and L. F. Johnson, *J. Magn. Reson. 75*:134 (1987).
49. I. D. Gay, C. H. W. Jones, and R. D. Sharma, *J. Magn. Reson. 91*:186 (1991).
50. E. M. Menger, S. Vega, and R. G. Griffin, *J. Magn. Reson. 56*:338 (1984).
51. U. Piantini, O. W. Sørensen, and R. R. Ernst, *J. Am. Chem. Soc. 104*:6800 (1982).
52. W. Kolodziejski and J. Klinowski, unpublished results.
53. R. V. Hosur, K. V. R. Chary, and M. R. Kumar, *Chem. Phys. Lett. 116*:105 (1985).
54. J. M. Thomas, J. Klinowski, S. Ramdas, B. K. Hunter, and D. T. B. Tennakoon, *Chem. Phys. Lett. 102*:158 (1983).
55. C. A. Fyfe, G. C. Gobbi, W. J. Murphy, R. S. Ozubko, and D. A. Slack, *J. Am. Chem. Soc. 106*:4435 (1984).
56. P. Bodart, J. B. Nagy, G. Debras, Z. Gabelica, and P. A. Jacobs, *J. Phys. Chem. 90*:5183 (1986).
57. W. M. Meier, *Z. Kristallogr. 115*:439 (1961).
58. W. M. Meier and D. H. Olson, *Atlas of Zeolite Structure Types*, Butterworth-Heinemann, London, 1992.
59. G. Engelhardt and D. Michel, *High-Resolution Solid-State NMR of Silicates and Zeolites*, Wiley, New York, 1987.

60. G. Engelhardt in Recent advances in zeolite science (J. Klinowski and P. J. Barrie, eds.), *Stud. Surf. Sci. Catal.* 52:151 (1989).
61. J. L. Schlenker, J. J. Pluth, and J. V. Smith, *Mat. Res. Bull.* 14:849 (1979) and references therein.
62. W. J. Mortier, J. J. Pluth, and J.V. Smith, in *Natural Zeolites, Occurrence, Properties, Use* (L. B. Sand and F. A. Mumpton, eds.), Pergamon Press, Oxford, 1978, p. 53.
63. N. M. Szeverenyi, M. J. Sullivan, and G. E. Maciel, *J. Magn. Reson.* 47:462 (1982).
64. C. E. Bronnimann, N. M. Szeverenyi, and G. E. Maciel, *J. Chem. Phys.* 79:3694 (1983).
65. P. Caravatti, J. A. Deli, G. Bodenhausen, and R. R. Ernst, *J. Am. Chem. Soc.* 104:5506 (1982).
66. P. Caravatti, G. Bodenhausen, and R. R. Ernst, *J. Magn. Reson.* 55:88 (1983).
67. M. H. Frey and S. J. Opella, *J. Am. Chem. Soc.* 106:4942 (1984).
68. D. Suter and R. R. Ernst, *Phys. Rev. B* 32:5608 (1985).
69. M. Linder, P. M. Henrichs, J. M. Hewitt, and D. J. Massa, *J. Chem. Phys.* 82:1585 (1985).
70. P. M. Henrichs, M. Linder, and J. M. Hewitt, *J. Chem. Phys.* 85:7077 (1986).
71. K. Takegoshi and C. A. McDowell, *J. Chem. Phys.* 84:2084 (1986).
72. D. L. VanderHart, *J. Magn. Reson.* 72:13 (1987).
73. A. Kubo and C. A. McDowell, *J. Chem. Phys.* 89:63 (1988).
74. A. Kubo and C. A. McDowell, *J. Chem. Soc. Faraday Trans. I* 84:3713 (1988).
75. C. Connor, A. Naito, K. Takegoshi, and C. A. McDowell, *Chem. Phys. Lett.* 113:123 (1985).
76. N. J. Clayden, *J. Magn. Reson.* 68:360 (1986).
77. W. Kolodziejski, H. He, and J. Klinowski, *Chem. Phys. Lett.* 191:117 (1992).
78. M. E. Davis, C. Saldarriaga, C. Montes, J. Garces, and C. Crowder, *Nature* 331:698 (1988).
79. M. E. Davis, C. Saldarriaga, C. Montes, J. Garces, and C. Crowder, *Zeolites* 8:362 (1988).
80. L. B. McCusker, Ch. Baerlocher, E. Jahn, and M. Bülow, *Zeolites 11:* 308 (1991).
81. J. P. van Braam Houckgeest, B. Kraushaar-Czarnetzki, R. J. Dogterom, and A. de Groot, *J. Chem. Soc. Chem. Commun.* 666 (1991).
82. D. Müller, E. Jahn, G. Ladwig, and U. Haubenreisser, *Chem. Phys. Lett.* 109:332 (1984).
83. M. W. Anderson and J. Klinowski, *Nature 339:*200 (1989).
84. T. A. Carpenter, J. Klinowski, D. T. B. Tennakoon, C. J. Smith, and D. C. Edwards, *J. Magn. Reson.* 68:561 (1986).
85. P. Caravatti, G. Bodenhausen, and R. R. Ernst, *J. Magn. Reson.* 55:88 (1983).
86. V. Bork and J. Schaefer, *J. Magn. Reson.* 78:348 (1988).
87. E. R. Andrew, A. Bradbury, R. G. Eades, and T. Wynn, *Phys. Lett.* 4:99 (1963).
88. E. R. Andrew, L. Farnell, L. F. Gledhill, and T. D. Roberts, *Phys. Lett.* 21:505 (1966).

89. D. P. Raleigh, M. H. Levitt, and R. G. Griffin, *Chem. Phys. Lett. 146*:71 (1988).

90. M. G. Colombo, B. H. Meier, and R. R. Ernst, *Chem. Phys. Lett. 146*:189 (1988).

91. W. E. J. R. Maas and W.S. Veeman, *Chem. Phys. Lett. 149*:170 (1988).

92. D. P. Raleigh, F. Creuzet, S. K. Das Gupta, M. H. Levitt, and R. G. Griffin, *J. Am. Chem. Soc. 111*:4502 (1989).

93. M. H. Levitt, D. P. Raleigh, F. Creuzet, and R. G. Griffin, *J. Chem. Phys. 92*:6347 (1990).

94. G. A. Morris and R. Freeman, *J. Magn. Reson. 29*:433 (1978).

95. M. Munowitz and A. Pines, *Science 233*: 525 (1986).

96. M. Munowitz and A. Pines, *Adv. Chem. Phys. 66*:1 (1987).

97. A. Pines, in *Proceedings of the 100th Fermi School of Physics* (B. Maraviglia, ed.), North Holland, Amsterdam, 1988.

98. M. Munowitz, *Coherence and NMR*, Wiley, New York, 1988.

99. P-K. Wang, C. P. Slichter, and J. H. Sinfelt, *Phys. Rev. Lett. 53*:82 (1984).

100. J. Baum, M. Munowitz, A. N. Garroway, and A. Pines, *J. Chem. Phys. 83*:2015 (1985).

101. J. Baum and A. Pines, *J. Am. Chem. Soc. 108*:7447 (1986).

102. J. Baum, K. K. Gleason, A. Pines, A. N. Garroway, and J. A. Reimer, *Phys. Rev. Lett. 56*:1377 (1986).

103. M. Munowitz, A. Pines, and M. Mehring, *J. Chem. Phys. 86*:3172 (1987).

104. R. Ryoo, S.-B. Liu, L. C. de Menorval, K. Takegoshi, B. Chmelka, M. Trecoske, and A. Pines, *J. Phys. Chem. 91*:6575 (1987).

105. D. N. Shykind, J. Baum, S-B. Liu, and A. Pines, *J. Magn. Reson. 76*:149 (1988).

106. B. F. Chmelka, J. G. Pearson, S-B. Liu, R. Ryoo, L. C. de Menorval, and A. Pines, *J. Phys. Chem. 95*:303 (1991).

107. D. E. Kaplan and E. L. Hahn, *J. Phys. Radium 19*:821 (1958).

108. M. Emshwiller, E. L. Hahn, and D. Kaplan, *Phys. Rev. 118*:414 (1960).

109. C. P. Slichter, *Principles of Nuclear Magnetic Resonance*, Springer-Verlag, New York, 1978, p. 233.

110. C. D. Makowka, C. P. Slichter, and J. H. Sinfelt, *Phys. Rev. Lett. 49*:379 (1982).

111. J. B. Boyce and S. E. Ready, *Phys. Rev. B 38*:11008 (1988).

112. P. K. Wang, C. P. Slichter, and J. H. Sinfelt, *J. Phys. Chem. 89*:3606 (1985).

113. S. E. Shore, J. P. Ansermet, C. P. Slichter, and J. H. Sinfelt, *Phys. Rev. Lett. 58*:953 (1987).

114. E. R. H. van Eck and W. S. Veeman, *Solid State NMR 1*:1 (1992).

115. T. Gullion and J. Schaefer, *J. Magn. Reson. 81*:196 (1989).

116. T. G. Oas, R. G. Griffin, and M. H. Levitt, *J. Chem. Phys. 89*:692 (1988).

117. A. W. Hing, S. Vega, and J. Schaefer, *J. Magn. Reson. 96*:205 (1992).

118. C. A. Fyfe, K. T. Mueller, H. Grondey, and K. C. Wong-Moon, *Chem. Phys. Lett. 199*:198 (1992).

119. E. R. H. van Eck and W. S. Veeman, *J. Am. Chem. Soc. 115*:1168 (1993).

120. A. Pines, M. G. Gibby, and J. S. Waugh, *J. Chem. Phys. 59*:569 (1973).

121. S. Vega, T. W. Shattuck, and A. Pines, *Phys. Rev. A 22*:638 (1980).

122. P. Brunner, M. Reinhold, and R. R. Ernst, *J. Chem. Phys. 73*:1086 (1980).

123. S. Vega, *Phys. Rev. A 23*:3152 (1981).

124. T. K. Pratum and M. P. Klein, *J. Magn. Reson. 55*:421 (1983).

125. C. S. Blackwell and R. L. Patton, *J. Phys. Chem.* 88:6135 (1984).

126. H. D. Morris and P. D. Ellis, *J. Am. Chem. Soc.* 111:6045 (1989).

127. H. D. Morris, S. Bank, and P. D. Ellis, *J. Phys. Chem.* 94:3121 (1990).

128. J. Rocha, Z. Liu, and J. Klinowski, *Chem. Phys. Lett.* 182:531 (1991).

129. T. H. Walter, G. L. Turner, and E. Oldfield, *J. Magn Reson.* 76:106 (1988).

130. D. E. Woessner, *Z. Phys. Chem. (Munich)* 152:51 (1987).

131. A. Lerf, E. Lalik, W. Kolodziejski, and J. Klinowski, *J. Phys. Chem.* 96:7389 (1992); W. Kolodziejski, E. Lalik, A. Lerf, and J. Klinowski, *Chem. Phys. Lett.* 194:429 (1992).

132. R. G. Bryant, S. Ganapathy, and S. Kennedy, *J. Magn. Reson.* 72:376 (1987).

133. R. K. Harris and G. J. Nesbitt, *J. Magn. Reson.* 78:245 (1988).

134. S. R. Hartmann and E. L. Hahn, *Phys. Rev.* 128:2042 (1962).

135. A. J. Vega, *J. Magn. Reson.* 96:50 (1992).

136. D. Fenzke, D. Freude, T. Fröhlich, and J. Haase, *Chem. Phys. Lett.* 111:171 (1984).

137. P. P. Man and J. Klinowski, *J. Chem. Soc. Chem. Commun.* 1291 (1988).

138. P. P. Man and J. Klinowski, *Chem. Phys. Lett.* 147:581 (1988).

139. P. P. Man, J. Klinowski, A. Trokiner, H. Zanni, and P. Papon, *Chem. Phys. Lett.* 151:143 (1988).

140. A. Samoson and E. Lippmaa, *Phys. Rev. B* 28:6567 (1983).

141. A. Samoson and E. Lippmaa, *Chem. Phys. Lett.* 102:205 (1983).

142. A. P. M. Kentgens, J. J. M. Lemmens, F. M. M. Geurts, and W. S. Veeman, *J. Magn. Reson.* 71:62 (1987).

143. A. Samoson and E. Lippmaa, *J. Magn. Reson.* 79:255–68 (1988).

144. F. M. M. Geurts, A. P. M. Kentgens, and W. S. Veeman, *Chem. Phys. Lett.* 120:206 (1985).

145. D. Fenzke, D. Freude, T. Fröhlich, and J. Haase, *Chem. Phys. Lett.* 111:171 (1984).

146. P. P. Man, *J. Magn. Reson.* 67:78 (1986).

147. H. E. Gottlieb and Z. Luz, *J. Magn. Reson.* 54:257 (1983).

148. E. Meirovitch and I. Belsky, *J. Phys. Chem.* 88:4308 (1984).

149. R. R. Eckman and A. J. Vega, *J. Phys. Chem.* 90:4679 (1986).

150. I. Kustanovich, D. Fraenkel, Z. Luz, S. Vega, and H. Zimmermann, *J. Phys. Chem.* 92:4134 (1988).

151. I. Kustanovich, H. M. Vieth, Z. Luz, and S. Vega, *J. Phys. Chem.* 93:7427 (1989).

152. I. Kustanovich, Z. Luz, S. Vega, and A. L. Vega, *J. Phys. Chem.* 94:3138 (1990).

153. M. J. Duer, J. Rocha, and J. Klinowski, *J. Am. Chem. Soc.* 114:6867 (1992).

154. A. Samoson, E. Lippmaa, G. Engelhardt, U. Lohse, and H-G. Jerschkewitz, *Chem. Phys. Lett.* 134:589 (1987).

155. J. Rocha, S. W. Carr, and J. Klinowski, *Chem. Phys. Lett.* 187:401 (1991).

156. J. Rocha, W. Kolodziejski, H. He, and J. Klinowski, *J. Am. Chem. Soc.* 114:4884 (1992).

157. J. Rocha, J. Klinowski, P. J. Barrie, R. Jelinek, and A. Pines, *Solid State NMR* 1:217 (1992).

158. A. Llor and J. Virlet, *Chem. Phys. Lett.* 152:248 (1988).

159. A. Samoson, E. Lippmaa, and A. Pines, *Mol. Phys. 65*:1013 (1988).

160. A. Samoson and A. Pines, *Rev. Sci. Instrum. 60*:3239 (1989).

161. Y. Wu, B. Q. Sun, A. Pines, A. Samoson, and E. Lippmaa, *J. Magn. Reson. 89*:297 (1990).

162. A. Samoson and E. Lippmaa, *J. Magn. Reson. 84*:410 (1989).

163. K. T. Mueller, B. Q. Sun, G. C. Chingas, J. W. Zwanziger, T. Terao, and A. Pines, *J. Magn. Reson. 86*:470 (1990).

164. B. F. Chmelka, K. T. Mueller, A. Pines, J. Stebbins, Y. Wu, and J. W. Zwanziger, *Nature 339*:42 (1989).

165. Y. Wu, B. F. Chmelka, A. Pines, M. E. Davis, P. J. Grobet, and P. A. Jacobs, *Nature 346*:550 (1990).

166. R. Jelinek, B. F. Chmelka, Y. Wu, P. J. Grandinetti, A. Pines, P. J. Barrie, and J. Klinowski, *J. Am. Chem. Soc. 113*:4097 (1991).

167. P. J. Barrie, M. E. Smith, and J. Klinowski, *Chem. Phys. Lett. 180*:6 (1991).

168. D. Raftery, H. Long, L. Reven, P. Tang, and A. Pines, *Chem. Phys. Lett. 191*:385 (1992).

169. C. R. Bowers and D. P. Weitekamp, *Phys. Rev. Lett. 57*:2645 (1986).

170. C. R. Bowers and D. P. Weitekamp, *J. Am. Chem. Soc. 109*:5541 (1987).

171. D. P. Weitekamp (in press).

172. Y. Manassen, R. J. Hamers, J. E. Demuth, and A. J. Castellano, Jr., *Phys. Rev. Lett. 62*:2531 (1989).

173. D. Rugar, C. S. Yannoni, and J. A. Sidles, *Nature 360*:563 (1992).

174. P. C. Lauterbur, *Nature 242*:190 (1973).

175. P. Mansfield and P. G. Morris, *Adv. Magn. Reson.* (Suppl. 2) (1982).

176. P. Mansfield, *J. Phys. C 10*:155 (1977).

177. A. Haase, *Magn. Reson. Med. 13*:77 (1990).

178. L. D. Hall and V. Rajanayagam, *J. Magn. Reson. 74*:139 (1986).

179. J. Frahm, K. D. Merboldt, and W. Hanicke, *J. Magn. Reson. 72*:502 (1987).

180. W. P. Rothwell and H. J. Vinegar, *Appl. Opt. 24*:3969 (1985).

181. W. P. Rothwell, D. R. Holecek, and J. A. Kershaw, *J. Polym. Sci. 22*:241 (1984).

182. S. N. Sarkar, E. W. Wooten, and R. A. Komoroski, *Appl. Spectrosc. 45*:619 (1991).

183. G. C. Chingas, J. B. Miller, and A. N. Garroway, *J. Magn. Reson. 66*:530 (1986).

184. J. B. Miller, D. G. Cory, and A. N. Garroway, *Phil. Trans. Roy. Soc. Lond. A333*:413 (1990).

185. P. Jezzard, J. J. Attard, T. A. Carpenter, and L. D. Hall, *Progr. NMR Spectrosc. 23*:1 (1991).

186. E. Günther, H. Raich, and B. Blümich, *Bruker Rep.* 15 (1991/92).

187. M. Komiyama, *Catal. Rev.-Sci. Eng. 27*:341 (1985).

188. J. C. Summers and L. L. Hegedus, *J. Catal. 51*:185 (1978).

189. E. R. Becker and J. Wei, *J. Catal. 46*:372 (1977).

190. M. P. Hollewand and L. F. Gladden, *Trans. IChemE. 70A*:183 (1992).

191. W. A. Edelstein, J. M. S. Hutchinson, G. Johnson, and T. W. Redpath, *Phys. Med. Biol. 25*:751 (1980).

192. B. Blümich and W. Kuhn, eds., *Magnetic Resonance Microscopy Methods and Applications in Materials Science, Agriculture and Biomedicine*, VCH, Weinheim, 1992, p. 304.
193. G. A. Barrall, L. Frydman, and G. C. Chingas, *Science* 255:714 (1992).
194. B. S. A. Kumar and S. J. Opella, *J. Magn. Reson.* 95:417 (1991).

Index

Milton Keynes UK
Ingram Content Group UK Ltd.
UKHW021845071024
449327UK00021B/1540